# Methods in Bioengineering

## Microdevices in Biology and Medicine

## The Artech House Methods in Bioengineering Series

Series Editors-in-Chief

Martin L. Yarmush, M.D., Ph.D.
Robert S. Langer, Sc.D.

*Methods in Bioengineering: Biomicrofabrication and Biomicrofluidics*, Jeffrey D. Zahn and Luke P. Lee, editors

*Methods in Bioengineering: Microdevices in Biology and Medicine*, Yaakov Nahmias and Sangeeta N. Bhatia, editors

*Methods in Bioengineering: Nanoscale Bioengineering and Nanomedicine*, Kaushal Rege and Igor Medintz, editors

*Methods in Bioengineering: Stem Cell Bioengineering*, Biju Parekkadan and Martin L. Yarmush, editors

*Methods in Bioengineering: Systems Analysis of Biological Networks*, Arul Jayaraman and Juergen Hahn, editors

# Methods in Bioengineering

## Microdevices in Biology and Medicine

Yaakov Nahmias
Massachusetts General Hospital, Harvard Medical School
Bioengineering Program, Hebrew University of Jerusalem

Sangeeta N. Bhatia
Department of Electrical Engineering, Massachusetts Institute of Technology
Howard Hughes Medical Institute

Editors

ARTECH HOUSE
BOSTON | LONDON
artechhouse.com

**Library of Congress Cataloging-in-Publication Data**
A catalog record for this book is available from the U.S. Library of Congress.

**British Library Cataloguing in Publication Data**
A catalogue record for this book is available from the British Library.

ISBN-13: 978-1-59693-404-7

**Cover design by Yekaterina Ratner**


**685 Canton Street**
**Norwood, MA 02760**



10 9 8 7 6 5 4 3 2 1

# Contents

# Preface

Microfabrication technology has already changed the world around us. Hiding under the shiny coat of our cars, iPods, cellular phones, laptops, and televisions, the integrated circuit and silicon microchip have changed the way we live forever. Features a thousand times smaller than a single millimeter enable an unparallel control over electrical signals resulting in nearly magical computational, communication, and memory powers. At the dawn of the twenty-first century, a similar revolution is changing the study of biology and the practice of medicine. Microscale patterns, three-dimensional features, and the physics of small places offer to radically change our ability to screen thousands of conditions, control the cellular microenvironment, and provide innovative tools for the diagnosis and treatment of disease. Notably, microdevices that have already reached the market are gaining increasing popularity.

Perhaps the most celebrated application of microtechnology is the Affymetrix GeneChip, a DNA microarray capable of screening the relative transcription of tens of thousands of genes, essentially the entire genome, in a single experiment. First published in 1995 by the Patrick O. Brown group at Stanford University, the microarray spotting approach has spawned many variants such as chromatin immuneprecipitation on chip (ChIP-on-chip) and SNP profiling. The GeneChip microarray has become a standard tool for the screening of complex genetic information. Another commercially available system that is rapidly growing in popularity is the Agilent Bioanalyzer, a microfluidics-based microchip that uses electrophoresis for the separation of RNA, DNA, and proteins. The newest models allow for on-chip staining and flow cytometry analysis of a small number of cells by replacing electrophoresis with a pressure-driven flow. Finally, PillCam is a commercially available microdevice that conjures up visions of the 1966 film *The Fantastic Voyage*. Developed by Given Imaging, PillCam is a capsule measuring 11 by 26 mm and weighing less than 4 grams. It contains a miniaturized imaging device that takes up to 14 images per second as it passes down the gastrointestinal tract. Currently approved by the FDA for the detection of esophageal and small intestine disorders, such as Crohn's disease or tumors, it is hoped to ultimately replace the much dreaded colonoscopy.

The success of these early microdevices has brought us to realize the need for a methods-based book that will provide timely insight into the technology of newly developed bio-MEMS devices. *Methods in Bioengineering: Microdevices in Biology and Medicine* is intended for students and scientists who wish to apply these tools for basic science or clinical diagnostics and for clinicians who wish to familiarize themselves with the science of this emerging technology. As part of the Artech House Methods in Bioengineering Series, this book presents the science behind microscale device design as well as the engineering of its fabrication. Each chapter includes a detailed, step-by-step methodology as well as a troubleshooting table designed to enable the rapid dissemination of microfabrication technology. Supported by dozens of full-color illustrations, this book covers the microfabrication technology involved in developing microdevices for biological applications and from bench to bedside. Readers will gain a unique perspective on the challenges and emerging opportunities in developing microdevices for cell capture from whole blood, study of transcriptional dynamics in living cells, temporal control of cell-cell interactions, nanoscale measurements of cellular forces, immobilization of living organisms, optical and electrical on-chip cell sorting, human-on-chip models of drug metabolism, microreactors for tissue engineering, and 3-D control of the cellular microenvironment.

We hope you enjoy this book as much as we enjoyed putting it together.

*Yaakov Nahmias, Ph.D.*
*Massachusetts General Hospital*
*Boston, Massachusetts*

*Sangeeta N. Bhatia, M.D., Ph.D.*
*Massachusetts Institute of Technology*
*Boston, Massachusetts*
*Editors*
*July 2009*

CHAPTER

# 1

# Immunoaffinity Capture of Cells from Whole Blood

Kenneth T. Kotz,[1,2,3] Daniel Irimia,[1,2,3] Ronald G. Tompkins,[1,2,3] and Mehmet Toner[1,2,3]

[1]BioMEMS Resource Center, Surgical Services, and Center for Engineering in Medicine, Massachusetts General Hospital, Boston, MA
[2]Department of Surgery, Shriners Hospital for Children, Boston, MA
[3]Harvard Medical School, Boston, MA

## Abstract

Cellular-based diagnostics are of increasing importance in health and disease monitoring as well as basic science research. Isolating purified, homogeneous cells from complex biological samples, however, is a lengthy process suited for specialized, well-equipped research laboratories and difficult to implement in clinical medicine.

Here we outline a rapid and easy process for utilizing state-of-the-art microfluidic technology to isolate leukocyte subpopulations directly from whole blood using neutrophils as an example.

Key terms
microfluidics
PDMS
immunoaffinity capture
whole blood
fractionation

## 1.1 Introduction

Information at the cellular level is critical for many clinical diagnostics and for basic biological research. Current cellular diagnostics range from the complete blood count (CBC), one of the most commonly ordered screening test in an emergency room setting [1], to more advanced diagnostics such as the CD4+ T lymphocyte count, which is a direct surrogate marker for HIV status [2]. Cellular phenotype is also of interest in immunology, where researchers seek to understand the immune system through its individual cellular and molecular components. Standard techniques exist for cellular enumeration and cellular fractionation from complex samples, including density gradient centrifugation, negative selection techniques such as RosetteSep, and positive selection techniques such as fluorescence-activated cell sorting (FACS) and magnetic-activated cell sorting (MACS). These different methods typically require highly trained technical staff processing samples over a period of hours. In the case of FACS and centrifugation techniques, they require large, specialized equipment as well.

Our lab is interested in building tools that enable clinicians to rapidly and easily study the immune system to help doctors predict clinical outcomes in patients. We have developed a set of tools that rapidly and efficiently captures cells on the walls of microfluidic devices using antibody-affinity isolation [3]. The well-defined fluid flow in microfluidic channels translates into precise shear forces seen by cells near the surfaces of the microfluidic device. This, combined with the specificity of monoclonal antibodies for cell-surface antigens, leads to highly specific capture of cells in these devices [4, 5]. A given cell type with a specific cell-surface antigen can thus be isolated by designing a microfluidic device, coating it with a specific antibody, and flowing the biological sample through the device at a specific flow rate.

This protocol outlines the overall process for the design, manufacture, and testing of devices for rapid, specific isolation of granulocytes directly from whole blood. Granulocytes are a particularly challenging cell to process using standard density gradient techniques because they are easily activated by sample processing and because they are short-lived [6, 7]. The design shown in Figure 1.1 consists of a set of microfluidic channels in a parallel plate geometry. For this geometry, the shear stress at the surface of the channels is given by $\tau = {}^{6Q\mu}\!/_{wh^2}$, where $\tau$ is the shear stress at the surface, $\mu$ is the dynamic viscosity, $Q$ is the volumetric flow rate, and $w$ and $h$ are the width and height, respectively. The design maximizes the total device width across the long dimension of a standard microscope slide in order to maximize the flow rate for a given shear stress and a given height, thus minimizing processing time. Device operation consists of flowing the blood through the device for 5 minutes, washing the device with physiological saline buffer for 5 minutes, and then processing the captured cells for downstream analysis.

The devices are made through standard PDMS rapid-prototyping methods [8]. PDMS is a flexible elastomer that can be chemically modified with biomolecules. This protocol outlines the process by which PDMS devices are fabricated, coated with antibodies, and used to capture and process cells. The design used here has been used by engineers, scientists, and technicians with equal success. It is capable of capturing extremely pure (>98%) cellular populations and processing them for enumeration, genomics, and high-throughput proteomics. While the protocol here describes capture of granulocytes, we have adapted it for capture of T and B lymphocytes, monocytes, and other rarer populations found in circulating blood.

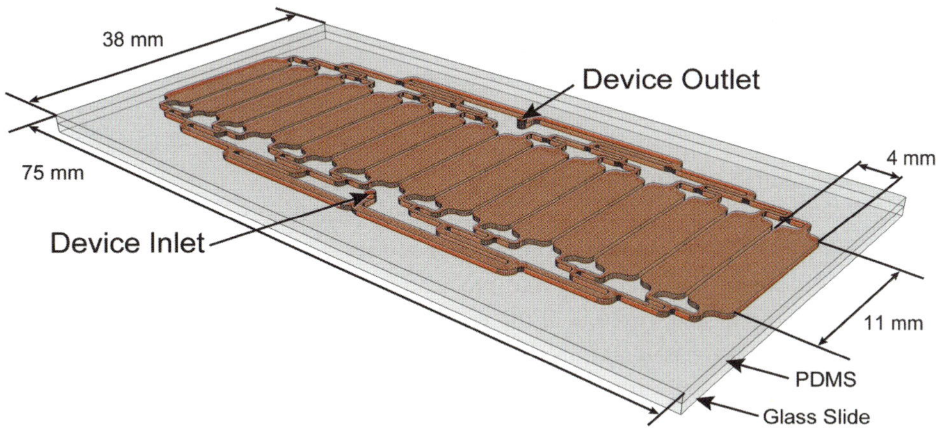


**Figure 1.1** Schematic overview of microfluidic neutrophil capture cassette. Microfluidic channel height is 50 $\mu$m.

## 1.2 Experimental Design

This protocol outlines the positive selection of cells by antibodies coated on the surface of microfluidic channels at a particular shear stress. The main design considerations therefore are the particular antibody-antigen pair that will be used to capture cells and the device geometry and fluid flow conditions that give the proper shear forces at the surface of the interaction. The optimal antibody-antigen interaction is typically determined by consulting standard resources that list cell-surface antigens as well as their distribution on the cell surface [9]. Once a set of suitable antibodies has been identified, typically one can validate the presence and uniqueness of the interaction on a flow cytometer. A discussion of flow cytometry is outside the scope of this protocol, and the reader is referred to many excellent reviews on the subject [10].

The next design parameter is the optimal shear stress for cell capture for a given cell and antibody. While specific microfluidic devices have been designed to determine optimal flow conditions for capture [3, 11, 12], generally it is straightforward to run multiple devices with a design as given in Figure 1.1 at multiple shear rates spanning a range of 0.2 to 5 dynes/cm$^2$. Capture is assessed for purity, total number, and efficiency at each condition and design changes are made to meet target requirements. The design in Figure 1.1 was optimized for throughput for a target efficiency of approximately 50%. Efficiency generally can be increased by decreasing the distance between the parallel plates (data not shown), increasing the length of the channel [11], or adding obstacles to increase surface area and break up flow streamlines along the length of the device [13]. Purity is mainly determined by the uniqueness of a specific antibody-antigen pair.

Any experiment involving cells or human samples adds additional variables to experimental design. Samples that contain cells can be heterogeneous in surface marker expression, and cell expression of surface antigens can change over time. Antibody capture of cells can cause crosslinking of proteins, which can lead to activation of cell-signaling pathways that affects cellular phenotype. Clinical samples can be extremely variable in cell numbers, activation states, and antigen expression. For small proof-of-concept studies, 6 to 10 subjects are usually sampled, with multiple devices used for different downstream analysis. Despite complications in cellular materials, the microfluidic capture described in this protocol is extremely robust and used by many researchers without formal training in microfluidics.

The last set of design parameters is determined by the downstream application of captured cells. For genomic studies, it is of utmost importance to maintain nuclease-free conditions while preparing and processing samples. For proteomics, it is necessary to assess the design in terms of material compatibility with downstream processes. Mass spectrometry–based methods are particularly prone to chemical contaminants and protein background. Furthermore, for either method, it is helpful to develop a specification for total protein or nucleic acid that is needed for downstream analysis. The device is then scaled to capture sufficient numbers of cells to meet the processing specification.

## 1.3 Materials

The microfluidic cell isolation devices used in this protocol are rapid-prototyped with a PDMS elastomer molded off a silicon wafer master. These devices are then chemically coated with antibodies to capture cells from biological fluids. The reagents, materials and common supplies, and equipment necessary for these processes are outlined below in Tables 1.1 to 1.3, respectively. Many laboratories have devised unique solutions to interfacing microfluidic devices with macroscopic fluid-handling devices (pumps, valves, and so forth). Figure 1.2 is a brief sample of tools that our lab uses on a routine basis for interfacing biological fluids.

**Table 1.1** Common Reagents Required for Microfluidic Cell Isolation

| Reagent | Supplier | Protocol Section |
|---|---|---|
| 2-propanol | Sigma-Aldrich | Device fabrication |
| Sylgard 184 polydimethyl siloxane | Dow Corning | Device fabrication |
| Ethanol, anhydrous, denatured | Sigma-Aldrich | Surface modification |
| 3-mercaptopropyl silane | Gelest | Surface modification |
| GMBS | Pierce | Surface modification |
| Nuclease-free water | Ambion | Surface modification |
| Nuclease-free PBS | Ambion | Surface modification and cell capture |
| Neutravidin | Pierce | Surface modification |
| CD66b | Serotec | Surface modification |
| Bovine serum albumin | Sigma-Aldrich | Surface modification and cell capture |
| Cytofix | Becton Dickenson | Postprocessing |
| Methanol | Sigma-Aldrich | Postprocessing |
| Giemsa Stain | Sigma-Aldrich | Postprocessing |
| CD66-PE | Becton Dickenson | Postprocessing |
| CD14-APC | Becton Dickenson | Postprocessing |
| CD3-AF488 | Invitrogen | Postprocessing |
| DAPI | Invitrogen | Postprocessing |
| RLT buffer | Qiagen | Postprocessing |

**Table 1.2** Common Supplies Required for Microfluidic Cell Isolation

| | Supplier | Use in Protocol |
|---|---|---|
| Syringes | | |
| 1 mL | Becton Dickenson | General fluid handling |
| 3 mL | Becton Dickenson | General fluid handling |
| Needles | | |
| 20G SS blunt tip | Small Parts | Cutting holes in PDMS |
| 22G SS blunt tip | Small Parts | Directly injecting liquids into devices during surface modification and RLT lysis |
| 23G SS blunt tip | Small Parts | Fits into 0.02″ Tygon tubing and 24G Teflon tubing |
| 23G SS tubing cut 0.5″ | Small Parts | Fits into 0.02″ Tygon and 24G Teflon tubing and into holes in PDMS for blood and buffer injection into devices |
| Tubing | | |
| 0.02″ ID (fits 23G needles) | Small Parts | ID fits 23G needles and 23G tubing |
| Teflon 24G (fits 23G needles) | Small Parts | ID fits 23G needles for RLT lysis output |
| Other miscellaneous supplies | | |
| Vacutainer blood-collection system | Becton Dickenson | Drawing blood |
| QIAshredder column | Qiagen | Cell lysate homogenization |
| 1.5″ × 3″ glass slides | Fisher | PDMS bonding |
| Weigh boats | Fisher | Weighing out PDMS |
| Plastic forks | Many | Mixing PDMS |
| Surgical knife with No. 11 blade | Fisher | Cutting PDMS |
| Lab coats, gloves, hair nets, face masks | Many | Personal protective equipment |

**Table 1.3** Equipment for Microfluidic Cell Isolation

| | Supplier | Use in Protocol |
|---|---|---|
| Balance | Fisher | Weighing PDMS |
| Vacuum jar | Fisher | Degassing PDMS with house vacuum |
| Cutting surface | | PDMS cutting |
| UV-ozone source | Novascan | Surface oxidation and PDMS bonding |
| Dry bag (AtmosBag) | Sigma-Aldrich | Surface functionalization |
| Hot plate | Fisher | Annealing silane to surface |
| Syringe pump | Harvard Apparatus | Pumping fluids |
| Microscope | Many | Visualizing microscopic device features; visualizing cells |
| Syringe pump stand | Custom | Holding syringe pump vertical [Figure 1.6(b)] |

## 1.4 Methods

### 1.4.1 Device fabrication

The following protocol describes microfluidic device fabrication using standard PDMS rapid-prototyping techniques [8]. Device fabrication can be divided into three main parts: (1) generation of a master of SU-8 on a silicon wafer; (2) replicating the part with a flexible elastomer, including demolding, sectioning, and cutting fluidic interconnect ports; and (3) device bonding to a PDMS or glass substrate.

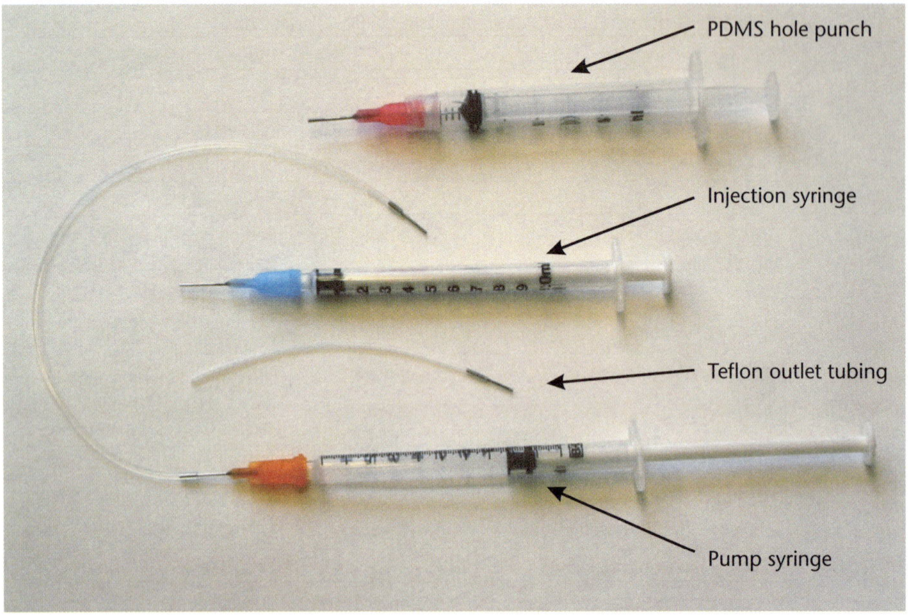


**Figure 1.2** Tools used for interfacing external fluids to PDMS devices. PDMS hole punch is made from a 3 mL syringe body, 20G SS blunt-tip needle (pink) sharpened with a twist drill, and a plug ejector made from 22G SS wire. The injection syringe is a standard 1 mL syringe with a 22G SS blunt-tip needle (blue). The Teflon outlet tubing is made from a 0.5″ 23G SS tubing fed into a 3″ section of 24G Teflon tubing. The pump syringe is a standard 1 mL syringe body connected to a 23G blunt-tip needle (orange) attached to a 6″ length of 0.02″ ID Tygon tubing capped with a 0.5″ 23G SS tube.

#### 1.4.1.1 SU-8 master fabrication

The master generation proceeds through a series of photolithographic steps by which layers of SU-8 photoresist are deposited onto a silicon wafer. Device features are photocrosslinked onto the silicon wafer with a UV lamp and a transparency mask containing the device. The generation of a master with SU-8 on a silicon wafer typically requires special facilities (Class 1000 clean room) or contracting from outside vendors. The process by which a master is generated is reviewed elsewhere in this book (see Chapter 2 or Chapter 5) and will not be repeated here.

Once a completed master is obtained, it is ready to replicate with PDMS. SU-8-on-silicon masters, if handled carefully, typically last for more than 100 molding cycles, as described below. The following molding procedures in our lab are carried out in a large, class 100,000 clean room environment in order to minimize defects caused by environmental particles. In a smaller setting, however, PDMS molding can be done on a benchtop, preferably equipped with a HEPA-filtered laminar flow hood. Figure 1.3 depicts an overview of the process.

#### 1.4.1.2 Replicating with PDMS

PDMS is a soft, flexible, two-part elastomer. It is mixed in a container, poured over the master, degassed, and cured overnight. Once cured, it can be peeled off the master, creating a negative cast of the master. The process is outlined next.

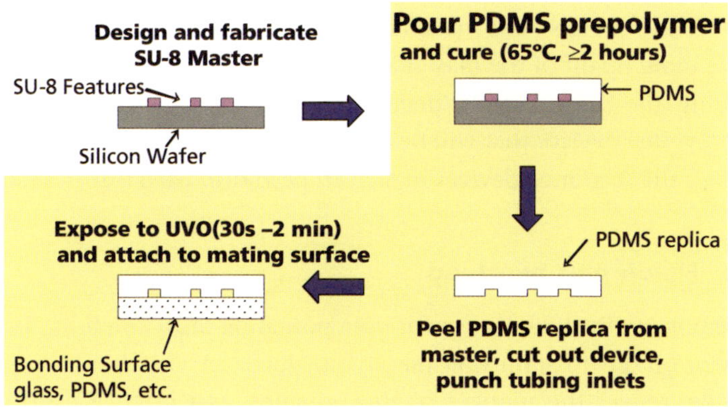


**Figure 1.3** Schematic overview of PDMS device fabrication. Sections highlighted in yellow are covered in this section.

#### 1.4.1.3 Pouring PDMS

1. Put on a clean pair of gloves, lab coat, and face mask.
2. Remove the silicon master from its protective case and place it in a petri dish secured by tape, with SU-8 features facing upwards. For a 4″ wafer, we use standard 150 mm petri dishes. Blow the dish with a nitrogen gun to remove any dust that may be in the dish.
3. On a top-loading balance, weigh out and mix 55g total of PDMS elastomer with a 1:10 ratio of hardener to resin. Do this by first weighing out 5g of curing agent, then 50g of polymer base. This amount of elastomer is sufficient to produce a mold 3 to 4 mm thick. This thickness is ideal for punching clean holes that will seal against 23G stainless steel (SS) tubing used in the cell-capture experiments.
4. Mix the precured PDMS with a mixing fork. Be sure to both swirl and fold the mixture to ensure that the curing agent is evenly distributed.
5. Pour the PDMS into the SU-8 master mold placed in a petri dish.
6. Degas the PDMS by placing the mixed precured PDMS in the vacuum desiccator and evacuating the chamber. Bubbles will appear, rise to the surface of the mixture, and pop. Degas the mixture for a minimum of 30 minutes. Degassing is complete when bubbles are no longer visible in the mixture. Once all bubbles have been removed, cover the petri dish and place in an oven at 65°C to 80°C for 3 to 6 hours or overnight to cure the PDMS.

#### 1.4.1.4 PDMS demolding

1. Remove the PDMS casting from the oven and place on a clean benchtop.
2. Using an X-ACTO knife with a new No. 11 blade, make a clean vertical cut along the edge of the silicon master. To make the cut, sink the point of the knife vertically into the PDMS until it reaches the silicon substrate and drag it in the direction of the cut. Make sure to maintain pressure on the knife such that the tip is always in contact with the silicon substrate, and be careful not to cut through the tape holding the master to the petri dish.

3. Once the cut has been made around the outside of the master, use tweezers to peel the mold up off of the master and place upside down onto a clean cutting surface.
4. Using the same cutting device, cut the PDMS mold into sections containing individual devices that will be bonded as described below.
5. Place the sectioned devices in a clean petri dish with features facing up.

### 1.4.2 Fluidic port punching

1. Remove individual devices for hole punching onto a well-lit, flat cutting surface.
2. Wipe off the tip of the hole puncher (Figure 1.2) with the alcohol-soaked clean room wipe, retract the plunger of the puncher, and bring the tip of the needle into alignment with the first port you will punch.
3. Adjust the plunger of the puncher so that it is as vertical as possible. Push the puncher through the PDMS until you hit the bottom. Do not rotate or rock the puncher as this will release microscopic PDMS particles onto the surface of the device.
4. With tweezers, lift the PDMS device off the cutting/punching surface and push the plunger into punched hole to drive out the cored section of PDMS.
5. Retrieve the cored section from the under side of the device using a pair of forceps and discard.
6. Retract the plunger, place the device back onto the cutting surface, and pull the needle out of the PDMS device in one straight motion again to minimize the release of loose particles of PDMS onto the device surface.
7. Repeat steps 2 to 6 for each port of each device.
8. Place the punched PDMS device onto a petri dish with feature side up.

Once a device has been poured, cut, and punched, it can be held in a petri dish to await bonding and surface-chemical modification. In our labs, PDMS replicas and glass slides are prepared for bonding with an oxygen plasma (100 mW, 2% oxygen, 35s) in a PX-250 plasma chamber (March Instruments, Concord, Massachusetts), then immediately placed in contact to bond the surfaces irreversibly. Chambers are then baked at 70°C for 5 minutes following bonding. An alternative method using a commercial UV-ozone (UVO) source can be used with equal effect as outlined below.

#### 1.4.2.1 Device bonding (with UVO surface treatment)

1. Lift cover off of UVO machine (Novascan PSD-UV or Jelight UVO 42), wipe metal platform with a cloth wetted with 2-propanol (isopropyl alcohol, or IPA), and blow dry with clean, dry nitrogen gas.
2. Using tweezers, place the PDMS device with the feature side facing upwards on the metal platform.
3. Using tweezers, place clean glass slides next to the device to be bonded.
4. If there are any visible dust particles on the device or slide, wipe with a clean, lint-free cloth soaked in IPA.
5. Place cover on the UV-ozone source making sure that the device is approximately 3 to 5 mm from the UV lamp, which is housed in the cover.

6. Expose the device to UV for 3 to 5 minutes. The optimal time will be determined by the distance from the UV lamp, environmental factors (humidity, temperature, and so forth.), and lamp power.
7. Remove cover and, using tweezers, grasp PDMS slab from its side and flip device over onto the glass side so that the features are bonded against the glass.
8. Place the device on a hot plate at 70°C for 5 to 10 minutes to facilitate irreversible chemical bonding between PDMS and glass surface.

The process of UVO surface treatment causes surface-localized oxidation. On PDMS, the reactive silanol bonds that form at the surface will slowly diffuse back into the bulk of the PDMS elastomer, especially at elevated temperatures. Therefore, chemical modification of the PDMS surface should immediately follow the oxygen plasma/ozone bonding as outlined below.

### 1.4.3 Surface modification

The surface functionalization protocol, described below, is for the covalent attachment of Neutravidin to the microfluidic channel surfaces [3]. Covalent linking of the protein to the surface creates a very stable method for attaching any general biotinylated antibody to the surface of the device. In order to create a stable, long-lasting surface, all aqueous solutions used should be filter-sterilized. When isolating cells for downstream nucleic acid assays, it is beneficial to use nuclease-free solutions as mentioned in the protocol below. If the cassettes will not be used for nucleic acid work, any general reagents can be used. The protocol described in Figure 1.4 has been adapted from earlier work and involves: (1) coating the device surfaces with a mercaptosilane, (2) using the thiol groups on the surface to covalently attach Neutravidin, and (3) attaching a biotinylated antibody to the surface of the device for immunoaffinity capture of cells.

#### 1.4.3.1 Silanization of surface—anhydrous method for GMBS attachment of proteins

1. Wipe down all working surfaces of chemical fume hood with ethanol or bleach to minimize contamination on the working surfaces with dust, bacteria, or mold.

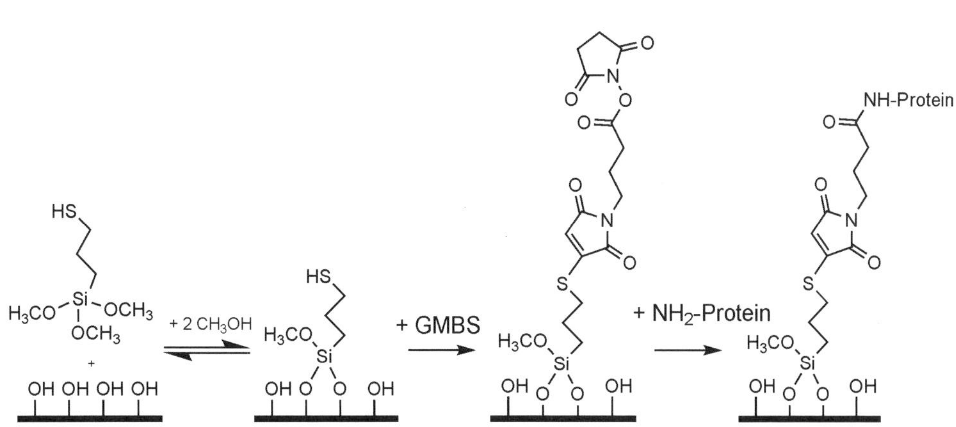


**Figure 1.4** Overview of surface functionalization.

2. Wipe down all working surfaces of chemical fume hood with Kimwipe wetted with RNase Away (Molecular Bio-Products) to remove any environmental contamination by nucleases.
3. Place the following reagents and supplies into a dry glove bag (Sigma AtmosBag with nitrogen atmosphere or equivalent) in chemical fume hood: 3-mercaptopropyl trimethoxysilane (3-MPS, Gelest), two 50 mL conical vials (Corning RNase, DNase Free) in a vial rack, bottle of denatured anhydrous ethanol (Sigma, 277649-1L or equivalent), two thin strips (1 cm wide) of Parafilm, a 1 mL pipetteman with barrier tip.
4. Fill the glove bag with dry nitrogen and prepare a 5% v/v solution of 3-MPS in ethanol. The ethanol and silane are resealed in the glove bag using Parafilm, and all reagents are brought out of the glove bag for device functionalization in the fume hood. For approximately 60 devices, prepare 30 mL of total 5% 3-MPS solution. Once prepared, the 3-MPS solution can be taken out of the glove bag, and the devices are functionalized on the benchtop of the chemical fume hood.
5. Each device is flushed with the 5% silane solution (four to five times the device dead volume) from step 4 using a 1 mL injection syringe (see Figure 1.2), and the silane solution reacts with the device at room temperature for 15 to 30 minutes.
6. The device is flushed with excess *anhydrous* ethanol (~1 mL/device) with a fresh injection syringe.
7. The device is placed on a hot plate at 80°C to 100°C for 15 to 60 minutes until the ethanol has evaporated to anneal the silane onto the surface. At this point the device can be stored in a desiccator at room temperature for more than 4 weeks.

#### 1.4.3.2 Nuclease-free covalent attachment of NeutrAvidin biotin-binding protein with 4-maleimidobutyric acid N-hydroxysuccinimide ester (GMBS)

1. If necessary, prepare a GMBS stock solution by adding 0.5 mL of anhydrous DMSO (Sigma-Aldrich 276855 or equivalent) to a 50 mg bottle of GMBS (Pierce) for a final stock concentration of 100 mg/mL. This should be prepared in a nitrogen-filled glove bag as in steps 3 and 4 of Section 1.4.3.1. The stock solution can then be stored in a 4°C flammable refrigerator for months. Before use, the stock solution should be warmed to room temperature.
2. Again, in a nitrogen-filled glove bag, add 28 $\mu$L of GMBS stock solution to 10 mL of anhydrous ethanol; this is usually sufficient for 20 devices. Reseal the GMBS stock solution with Parafilm. Remove reagents from the glove bag, and functionalize devices on the chemical fume hood bench surface.
3. Immediately flush devices (four to five times the device dead volume) with GMBS solution using an injection syringe (Figure 1.2).
4. While GMBS solution is reacting with the surface, prepare 50 $\mu$g/mL (1:20 dilution) of Neutravidin protein (Pierce) in PBS (nuclease-free, pH 7.4, Ambion). Plan to add six device volumes of the protein solution.
5. Let GMBS react for 30 minutes at room temperature. After this time, flush devices with nuclease-free water (Ambion), making sure that the ethanol has been removed from the device. *Be certain that there are no air bubbles at this stage.* Air bubbles must be cleared from the device using an ethanol solution. If you inject the device with air

during the addition of water, then replace the water with fresh ethanol, remove the bubbles, and repeat the water rinse.

6. Add three times the device dead volume of solution from step 4 to each device through the device inlet. After 15 minutes, add an additional three device volumes to the device outlet. This ensures even protein coverage onto the surfaces of the cell-capture cassettes.
7. Let react for more than 1 hour at room temperature or preferably overnight at 4°C.
8. Before adding antibody solution, flush devices with ice-cold BSA wash buffer (see below).

At this point, if solutions have been filter-sterilized (0.22 $\mu$m), devices can stored in a 4°C refrigerator for up to 3 months. For longer storage, the devices should be flushed with saline buffer containing a small amount (0.05%) of sodium azide (Sigma 71289 or equivalent) as a preservative. For any storage, it is helpful to place one or more devices in a petri dish containing 1 to 2 mL of saline buffer and seal the dish with Parafilm to reduce evaporation of the fluid-primed channels.

#### 1.4.3.3 Immobilization of antibody

1. Prepare 1:20 to 1:100 dilution of biotinylated monoclonal antibody in BSA wash buffer for a final concentration of biotinylated IgG of 1 to 30 $\mu$g/mL.
   The cell-capture protocol described below specifically isolates CD66b-positive granulocytes directly from whole blood. For this procedure, use CD66b (Serotec) diluted to 20 $\mu$g/mL.
2. Flow two to four times the device dead volume with solution from step 1 to each device using a syringe pump equipped with pump syringe (Figure 1.2), manually with an injection syringe (Figure 1.2), or manually using a pipetteman with a gel-loading tip.
3. Let react for 30 minutes and then repeat step 2, injecting into the opposite port of the device.
4. Let react for more than 30 minutes at room temperature or overnight at 4°C.

At this point, the device is ready to capture cells. It can also be stored at this stage. The length of storage depends upon the stability of the antibody at 4°C. The devices with antibody should be sealed to prevent evaporation in the channels by connecting the inlet and the outlet of the device with a length of Tygon tubing primed with saline buffer. The devices should also be sealed in a petri dish containing 1 to 2 mL of sterile saline buffer. To troubleshoot this procedure, see the Troubleshooting Table.

### 1.4.4 Cell capture

Cell capture is a simple two-step process. First, a suspension of cells is flowed through the device at a prescribed flow rate, and then fresh buffer is passed through the channel to wash away any cells that were not captured by antibodies. The specificity of capture is determined by the specificity of the antibody-antigen interaction and by the distribution of the antigen on the different cells' surfaces. The cell suspension can be derived from multiple sources: cultured cells in buffer, cultured cells spiked into whole blood, whole blood itself, preprocessed fractions of cells from whole blood, cells in urine, or cells contained in lavage samples (bronchoalveolar, peritoneal, ductal, and so forth). Figure 1.5 outlines specific cell capture of CD66b-positive granulocytes directly from whole blood.

WARNING! Exposure to human blood products and human tissue samples poses a potentially significant health risk to laboratory and research personnel, including the transmission of communicable disease. It is the responsibility of the participating investigator and his or her research institution to ensure that all individuals who may come into contact with human blood and tissue products have been fully informed of the associated risks and provided the appropriate training and personal protection to minimize those risks. It is strongly recommended that these procedures be performed, where possible, in a biosafety hood (BSL-2 or greater) to reduce the risk of microbial contamination of the samples and exposure of the technician or research nurse performing the procedure.

#### 1.4.4.1 Preparing equipment and reagents

1. Remove the Parafilm from the petri dish and open up the dish. The cassette should be left in the bottom half of the petri dish during the cell capture to act as a secondary containment in case of spills.
2. Check the device for damage. Check for cracks or any other major physical damage to the device. If device is damaged, do not use it for cell isolation.
3. As outlined in the protocol above, the device is stored preprimed with 1% BSA or 1% BSA-antibody solution. Make certain that there is a droplet of liquid (1× PBS buffer) at both the inlet and the outlet of the device.
4. Using tweezers, carefully unplug one end of the tubing from *one* of the device ports. It does not matter from which port the tubing is unplugged.
5. At least 2 hours prior to the capture experiment, flush the devices with BSA wash buffer using an injection syringe (Figure 1.2). All syringes, needles, and tubing that will be used to flow blood or cells into the microfluidic device are also filled with BSA solution. This minimizes nonspecific depletion of cells onto the walls of the external connections to the device.
6. The capture experiment can be performed at room temperature. If devices need to be run throughout the day, they should be held in a cold room at 4°C until needed.
7. The open end of the tubing that was just unplugged is the waste outflow. Open a 1.5 mL microcentrifuge tube, invert it, and guide the waste outflow tube to the bottom (pointed end) of the tube. Next, lay the tube on its side to collect the outflow.
8. Prepare three pump syringes outlined in Figure 1.2. These will be used for cell loading, device washing, and cell fixing.
9. Load a syringe pump so that the pump syringe will be pointing downwards. This ensures that if any settling occurs, the denser cellular fraction tends to flow through the device. An image of the pump holder, made out of polycarbonate and aluminum blocks, is shown in Figure 1.6(b).

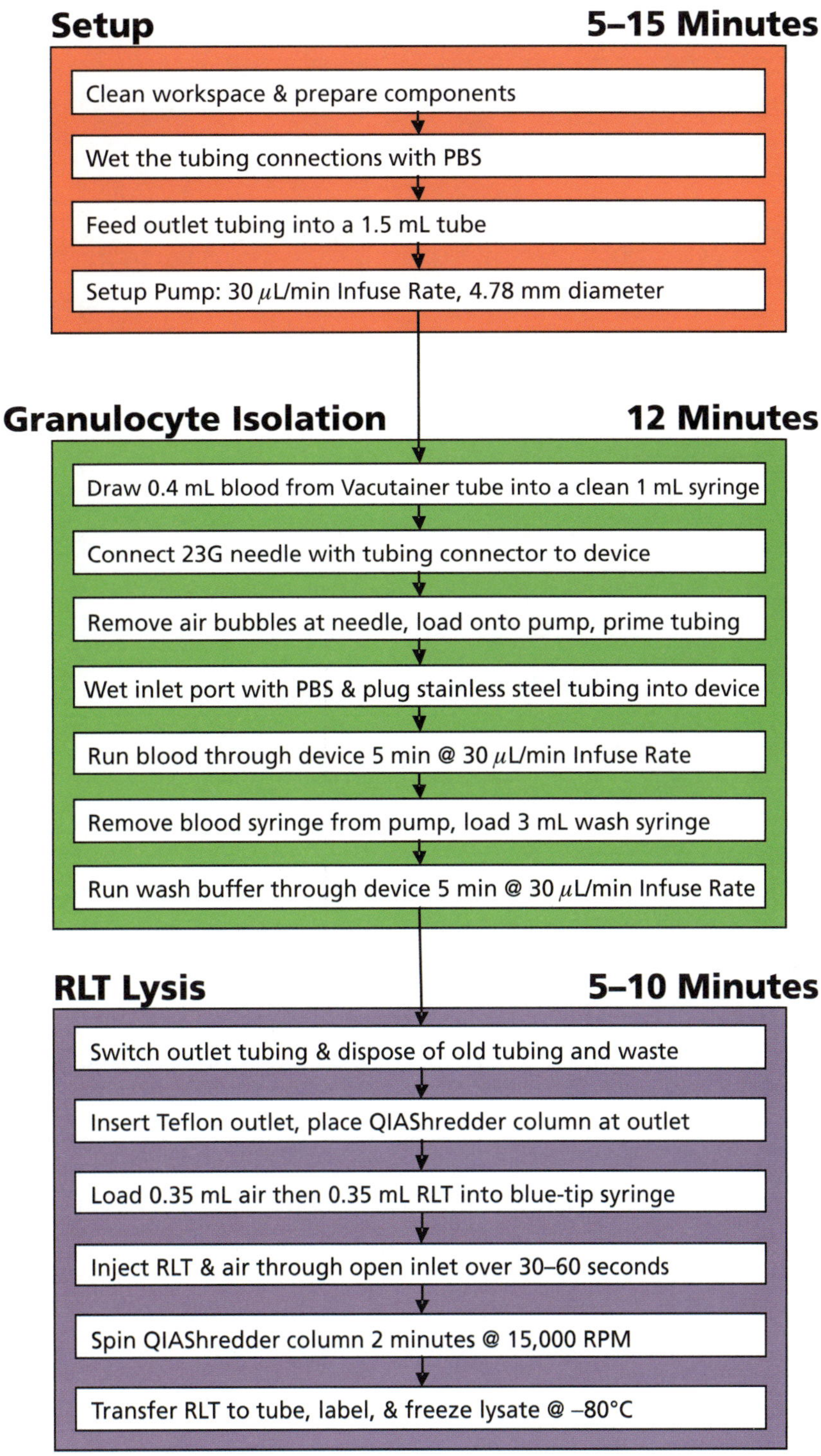


**Figure 1.5** Flowchart of granulocyte capture with approximate process times.

### 1.4.5 Injecting blood into cassette

1. Blood is collected by venipuncture by an experienced phlebotomist using a Vacutainer (Becton-Dickinson) blood-collection system. Venous whole blood is

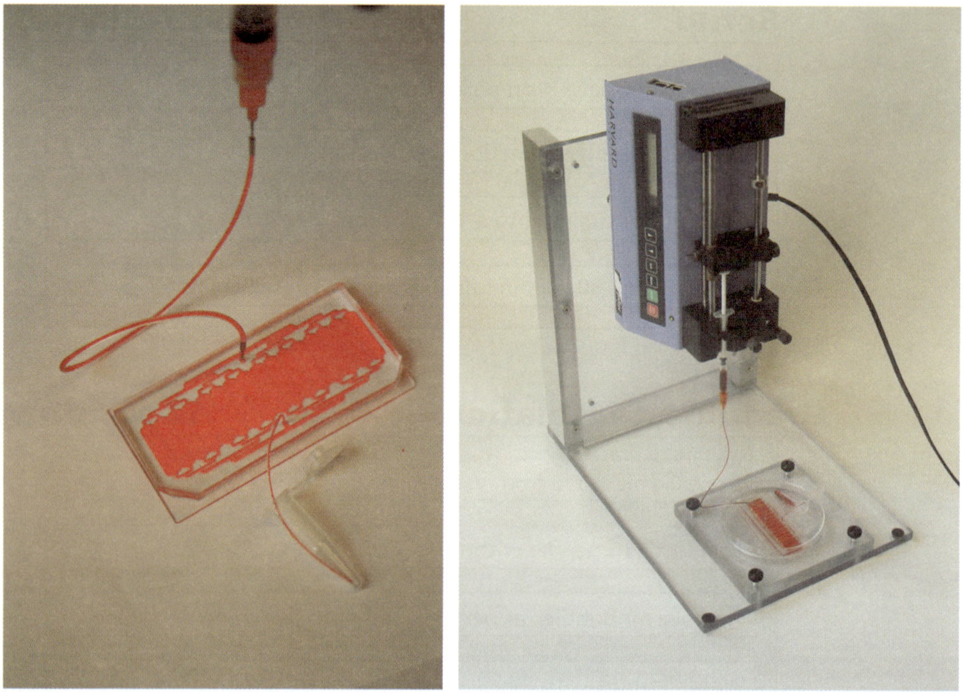

(a) (b)

**Figure 1.6** (a) Image of device connected to pump syringe with blood and outlet tubing set into waste tube, and (b) syringe pump set upright in stand with pump syringe connected to device.

collected at room temperature (18°C to 25°C) into one 2 mL lavender (EDTA) blood-collection tube (Becton-Dickinson Vacutainer, catalog no. 367841, or similar) using a standard Vacutainer collection system (Becton-Dickinson). The lavender top tube is gently inverted eight times to mix the blood with the contained anticoagulant. Time of draw is recorded and entered into a notebook. Blood should be processed within 1 hour of draw and held on a rocking platform at room temperature.

2. Take the Vacutainer containing the blood off the platform and gently invert it eight times to resuspend blood. Use a sterile, 1 mL syringe, and draw up 0.3 mL of whole blood from the Vacutainer tube. Cap the Vacutainer tube, and wipe the side of the syringe with a laboratory wipe.
3. Blood will be injected into the device using a syringe connected to the tubing connector of a pump syringe in Figure 1.2, which consists of 8" tubing connected on one end to an orange blunt-tip needle and on the other end to a short section of stainless steel tubing.
4. See step 3 in Section 1.4.4.1. Make certain that there is a droplet of PBS at the open inlet connection.
5. Carefully insert the tip of the blood-containing syringe into the blunt-tip needle that is attached to the tubing. As you do this, try not to introduce any air bubbles into the tip of the needle-syringe fitting. There may be some spill over of the blood and buffer. Wipe this with a laboratory wipe.

WARNING! The tubing connects to the inlet of the microfluidic cassette through a blunt-tip needle. Even though this needle is not sharp, the user should take care when handling it to avoid a needle-stick accident.

6. Check for air bubbles at the tip of the needle-syringe connection. The easiest way to remove bubbles at the connection of the needle and syringe is to:
   i. point the needle with the tip facing downwards
   ii. hold the top of the syringe with the left hand
   iii. tap with the right index finger at the connection
7. Load the 1 mL syringe onto the syringe pump. Carefully position the moving arm of the pump to butt against the plunger of the syringe. Gently push this arm until blood begins to flow through the tube connected to the needle on the syringe, but stop pushing before blood comes out of the tip of the end of the tubing.
8. Set up the syringe pump as follows:
   i. Select 1 mL syringe size (diameter 4.78 mm).
   ii. Enter infusion rate of 30 $\mu$L/min.
   iii. Begin flowing blood through the syringe on the syringe pump. Monitor the open stainless steel tube at end of the flexible tubing. When whole blood is flowing out of the end of the tubing, insert the stainless steel tip of the tubing into the inlet of the device.
9. Wipe away any small drops of blood that may have spilled around the tubing inlet with a laboratory wipe. Reapply a droplet of PBS at the inlet where the tubing connects to the device.
10. Using a laboratory timer, flow blood for 5 minutes. If blood does not flow through all the channels of the cassette, see the Troubleshooting Table. An image of blood running through the device can be seen in Figure 1.6.

### 1.4.6 Washing noncaptured cells with PBS

1. Stop the infusion pump by pressing the appropriate button to stop the flow of blood through the device.
2. Remove the 1 mL syringe on the syringe pump and set aside on the bench. Do not disconnect the needle with tubing from the syringe at this time. Load a 3 mL pump syringe containing nuclease-free PBS (Figure 1.2) onto the pump.
3. Position the moving arm of the pump to butt against the plunger of the 3 mL syringe. Do not change the settings on the infusion pump. Because of the larger-diameter syringe, the wash step will proceed at a flow rate proportional to the square of the ratio of the diameters of the 3 mL syringe to the 1 mL syringe (~3.3).
4. Begin flowing wash buffer through the syringe on the syringe pump and monitor the stainless steel tip at the end of the tubing. When droplets begin to form at this outlet, carefully remove the inlet tubing containing blood from the device inlet and dispose of the 1 mL syringe and tubing into a sharps biohazard container.
5. Connect the stainless steel tip of the wash buffer into the device inlet. Wipe away any small drops of blood that may have spilled around the tubing inlet with a laboratory wipe. Reapply a droplet of fluid at the inlet where the tubing connects to the device. Using a laboratory timer, let the device wash for 5 minutes.

6. At the end of 5 minutes, if blood visibly remains in any of the capture chambers of the microfluidic cassette, see the Troubleshooting Table.
7. Stop the syringe pump. Remove the 3 mL syringe on the syringe pump and set aside on the bench. Do not disconnect the needle with tubing from the syringe at this time.

At this point you will have cells bound to the surface of the microfluidic device. For the CD66b isolation protocol here, there will be approximately 250K to 400K cells in the device ready for postcapture processing.

### 1.4.7 Postcapture processing

The captured cells can now be processed for a variety of applications, including functional studies (oxidative activity, cytokine release, and so forth), enumeration, phenotyping (immunofluorescence, IHC, and histology), and molecular assays (FISH, genomics, proteomics, and so forth). Described next are protocols for fixing (Section 1.4.7.1) followed by immunofluorescence staining (Section 1.4.8), Giemsa staining (Section 1.4.9), or immediate lysis for genomics (Section 1.4.10).

#### 1.4.7.1 Formaldehyde fixing of cells for postcapture analysis

This procedure fixes the cells for postcapture analysis, including Wright-Giemsa or immunofluorescence staining. Immediately following washing with PBS, the cells are fixed with a 4% paraformaldehyde solution and refrigerated.

*Note:* For genomic or proteomic applications, cells should not be fixed. They should be lysed on-chip directly after capture for downstream processing.

WARNING! The fixing buffer contains a small amount (~4%) of paraformaldehyde, which is a known carcinogen. Do not allow this buffer to contact bare skin. Wear gloves and a lab coat when performing this procedure, and wash hands thoroughly afterwards. In case of contact with eyes, rinse immediately with water and seek medical attention. In the case of skin contact, wash with plenty of water.

1. See step 3 in Section 1.4.4.1. Make certain that there is a droplet of fluid at the inlet and the outlet connections.
2. Load 1 mL pump syringe (Figure 1.2) with fixing solution (BD Cytofix or equivalent) onto the syringe pump.
3. Position the moving arm of the pump to butt against the plunger of the 1 mL syringe. Set the infusion pump as for cell capture, step 6 in Section 1.4.5.
4. Let the device rinse with fixing buffer for 10 minutes.
5. Stop the flow of fixing buffer. Cap the microcentrifuge tube with the outflow (formaldehyde) and dispose of properly. Treat this as a biohazardous substance containing formaldehyde.
6. The device should sit at room temperature for 15 to 30 minutes to complete the fixation. To ensure reproducible results, fixation time should remain constant for each run.
7. Fill a 1 mL pump syringe containing 1% BSA, and load onto the syringe pump.
8. Let the device rinse with BSA (or FBS) for 10 minutes.

The device is not ready for staining and should be processed as soon as possible. If necessary, the device can be stored for at least 1 week as follows:

i. Remove the tubing from the inlet of the device and add approximately 1 mL of PBS to the top of the device.
ii. Place the top of the petri dish on the device, seal the edge with Parafilm, place the sealed petri dish into a Ziploc specimen bag, and place in a 4°C refrigerator.

### 1.4.8 Immunofluorescence staining

The following procedure is useful for determining overall cellular phenotype and consists of staining with monoclonal antibodies specific for the major leukocyte subpopulations, namely, lymphocytes, monocytes, and granulocytes, along with DAPI as a nuclear counterstain. The antibodies used in this procedure are mouse monoclonal, prelabeled with fluorescent tags, but primary-secondary staining also works in these chips, as do standard immunohistochemistry (IHC) protocols.

1. Make 300 $\mu$L total antibody staining solution for each device to be stained containing the following diluted in 1% BSA:
   i. 1:20 dilution of CD3-AF488 (Invitrogen MHCD0320)
   ii. 1:30 dilution of CD66-PE (BD, 551480)
   iii. 1:30 dilution of CD14-APC (BD, 340684)
2. See step 3 in Section 1.4.4.1. Make certain that there is a droplet of fluid at the inlet and the outlet connections.
3. Flush device with 1% BSA solution to block for nonspecific binding of antibodies. Let sit for 30 minutes at room temperature.
4. Flush antibody cocktail through the device, and let incubate covered with foil for 30 to 45 minutes. When you are working with small volumes of antibody solution, it is best to load the device manually with gel-loading pipette tips.
5. While devices are staining, prepare a 300 nM DAPI solution in PBS.
6. Rinse device with 5 to 10 volumes of 1% BSA using either a pipetteman or a syringe pump.
7. Inject 300 $\mu$L of DAPI solution into the device, and let sit 5 to 10 minutes.
8. Flush the device with PBS, and seal the inlet and outlet with PDMS plugs or with a piece of saline-filled Tygon tubing.

The devices are now ready to image using appropriate filters on a fluorescent microscope. In this protocol, granulocytes stain intense orange, lymphocytes are green, monocytes are deep red, and nuclei are blue. Purity is measured as total cells that are DAPI and CD66 positive versus total DAPI cells counted. A sample immunofluorescence image is shown in Figure 1.7. Eosinophils, a granulocyte subpopulation, can have highly autofluorescent granules. These will appear bright green, DAPI positive, and CD66 positive, but they should account for less than 10% of total granulocytes in a normal subject.

### 1.4.9 Giemsa staining protocol

Giemsa is one of the classic Romanowsky stains used in hematology to discriminate the different leukocyte subtypes in a blood smear. It consists of a set of azurophilic and

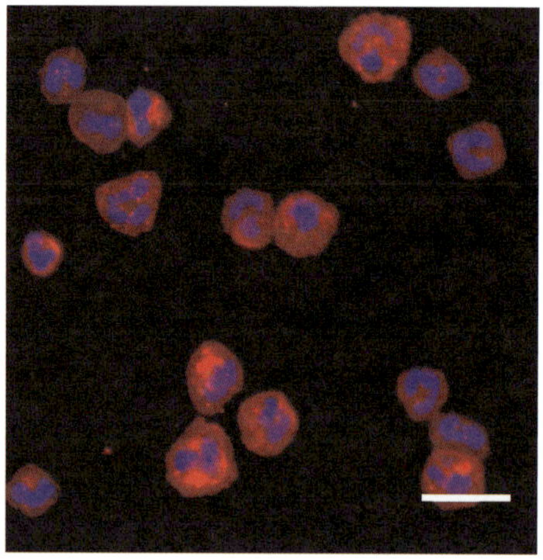

**Figure 1.7** Sample immunofluorescence image of captured granulocytes stained with DAPI, CD3-AF488, CD66b-PE, and CD14-APC. The white scale bar is 25 $\mu$m long.

basophilic dyes that differentially stain granules and nuclei of leukocytes. Identification of cell phenotype is made based on morphology and staining color.

1. See step 3 in Section 1.4.4.1. Make certain that there is a droplet of fluid at the inlet and the outlet connections
2. Fill a 3 mL pump syringe with 1 mL of 100% methanol.
3. Place new 1.5 mL centrifuge tube at the outlet of the device.
4. Set the infusion pump as for step 8 in Section 1.4.5. Place syringe with methanol onto the syringe pump, and begin the flow of methanol through the device. Using a laboratory timer, let the device fix with methanol for 4 minutes.
5. Dilute Giemsa stain (Sigma GS500) 1:4 with deionized water and load 2 mL into a 3 mL syringe, attach a 0.8 $\mu$m syringe filter (Millipore) to the end of the syringe, and prime the filter until a drop of fluid comes out at the tip of the filter.

   *Note:* Adding water to any Romanowsky stain (Wright, Giemsa, May-Grünwald) causes precipitation of some of the dye components. These precipitates, if unfiltered, will clog the channels of the microfluidic device and settle on the bottom of the chip.
6. Switch the tubing from the methanol needle to the syringe with Giemsa stain and syringe filter, and remove any air bubbles at the connection between the needle and syringe.
7. Place the syringe with Giemsa stain onto the syringe pump, and begin the flow of Giemsa stain through the device. Using a laboratory timer, stain the device for at least 5 minutes. It is helpful to examine the cells on a microscope to determine the level of staining.
8. Fill a 3 mL pump syringe (Figure 1.2) with 2 mL of deionized water.
9. Place the syringe with water onto the syringe pump and button, prime tubing with water, insert tubing into device inlet, and flush the device with water for 3 minutes.
10. Disconnect the water input needle into the device, and reinsert the waste tube into the inlet to seal the device.

11. Analyze on a microscope with a 40× or better objective.

    Typical cell staining is shown in Figure 1.8.

### 1.4.10 Cell lysis for genomic applications

One of the advantages of this cell isolation platform is the speed and efficiency of isolating cells. Because the isolation process is fast, few changes in transcription can occur, thus eliminating genomic transcripts that arise due to blood handling and minimizing effects due to endogenous RNA degradation pathways (e.g., apoptosis).

1. See step 3 in Section 1.4.4.1. Make certain that there is a droplet of fluid at the inlet and the outlet connections.
2. Using tweezers, remove the short piece of outlet Tygon tubing, and replace it with the short piece of Teflon tubing with an SS tip (Figure 1.2). When installing, plug the stainless steel tubing end into the device outlet. Dispose of the old outlet tubing in a biohazard container.
3. Place a QIAshredder column at the open end of the new Teflon outlet tubing. Bend the tube around into a *U* shape to allow the QIAshredder column to sit upright. If it is unstable, tape the QIAshredder column upright against the rim of the petri dish.
4. Using a standard 1 mL syringe with a blue blunt-tip needle that fits into the inlet of the device, draw up 0.35 mL of air, followed by 0.35 mL of RLT buffer. The plunger should be at the 0.7 mL mark, and the liquid RLT will rise to about the 0.25 mL mark.
5. Remove the stainless steel tip of the wash buffer tubing from the device inlet and set on the bench.
6. Insert the tip of the 1 mL syringe containing RLT buffer into the device inlet. Make sure the QIAshredder column is in place at the open end of the Teflon outflow tubing to collect the lysate. Slowly inject the RLT buffer over a 60-second period.

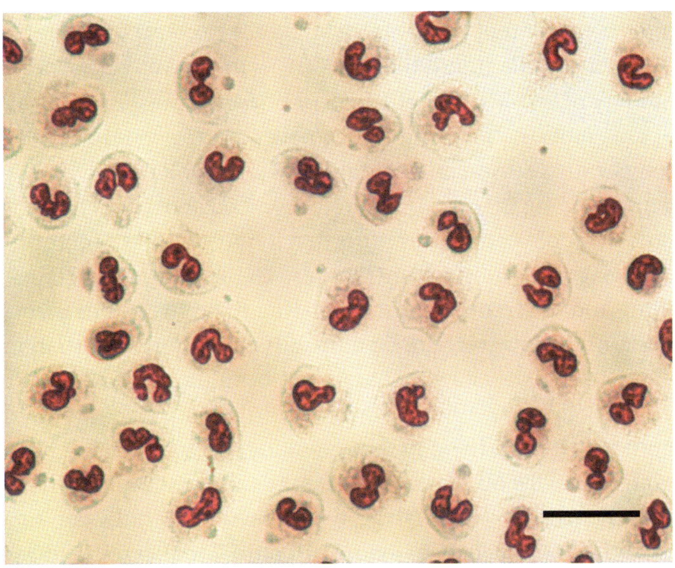

**Figure 1.8** Sample Giemsa-stained image of cells isolated on the microfluidic device imaged with a 40× objective.

Inject all the RLT as well as the air. Be sure that the QIAshredder column is collecting the outflow.

7. Cap the QIAshredder column, and place it into a benchtop microcentrifuge capable of about 15K rpm. Balance with a second QIAshredder column or tube of equal weight. Spin at maximum rpm (~15,000) for 2 minutes at room temperature. Discard the used microfluidics cassette in a Sharps biohazard container.
8. Using an RNase-free pipette, transfer the flow-through into a 2 mL collection tube (USA Scientific, catalog no. 1420-9705), and place into a –80°C freezer.

The RNA from the cell lysate can then be extracted for RT-PCR or microarrays using a standard Qiagen mini kit.

## 1.5 Data Acquisition, Anticipated Results, and Interpretation

Postcapture processing consists of a number of different approaches, depending upon the intended analysis. For device characterization, it is necessary to stain images with antibodies and count a significant portion, if not the entire chip, by fluorescence microscopy, usually at 10× or 20× magnification. Typically, this is done with an automated microscope and stage, but it can be done by hand with suitable filter sets and sufficient time. It is important to image both capture surfaces on the device. As mentioned above, typically total nucleated cells are counted in the DAPI channel, and this is compared with the number (and colocalization) of the positive fluorescent antibody marker. In the case of granulocytes, total number of granulocytes is typically 250K to 400K for 150 $\mu$L of blood, and the device shows a linear correlation between number of cells captured and input blood volume. Capture purity is routinely more than 98% in healthy volunteers, healthy blood that has been activated ex vivo with bacterial lipopolysaccharide (LPS), and severely burned patients. Depending on the size of the cells and the equipment available, it is possible to count on a fluorescence microscope a large number of cells (more than 1,000) as an assessment of purity and to count the total number of cells with bright-field or phase-contrast imaging at lower (4×) magnification. This reduces the number of images and provides a good estimate of capture parameters with a reasonable amount of work involved.

Giemsa staining is another method described above for characterizing cells. The primary advantage of Giemsa staining is that it only requires a bright-field microscope with moderate magnification to identify cells. The Romanowsky stains were developed for examining and differentially discriminating the different leukocytes in a blood smear. Giemsa staining, however, can be more difficult to interpret than immunofluorescence methods. Granulocytes have a characteristic multilobed nucleus and small granules distributed throughout the cytoplasm. As mentioned above, the staining of the granules discriminates between the different granulocyte types, namely, neutrophils, basophils, and eosinophils. CD66b captures neutrophils and eosinophils; the latter stain an intense red with Giemsa. Contaminating lymphocytes typically are smaller, with a round nucleus and a high nucleus-to-cytoplasm ratio. Monocytes tend to be large, with a kidney-shaped nucleus and pale cytoplasm. Accurate identification with Giemsa staining requires careful attention to staining procedure and special training by a hematologist or pathologist to get accurate, reproducible counts.

For genomic applications, RNA is typically purified using standard kits available from a wide range of manufacturers. RNA quantity is typically measured using a spectrophotometer or fluorescence assay. When dealing with small amounts of RNA, it is advantageous to use a device such as the NanoDrop spectrophotometer, which require only 1 to 2 $\mu$L of sample. RNA quality is typically assessed with an Agilent Bioanalyzer, a microfluidic capillary electrophoresis system that allows parallel processing of multiple samples. Granulocytes are terminally differentiated cells with much lower total RNA amounts compared to other cell types, but we typically purify 100 ng from 300K total cells. As a comparison, lymphocytes yield four to six times more RNA per cell captured.

## 1.6 Discussion and Commentary

Immunoaffinity capture is an excellent tool for cellular analysis over a wide range of samples. The shear stresses seen by cells in these devices are typically much lower than those seen under normal physiological blood flow; thus, fragile cells can be captured without damage. Another advantage to this device is the extreme rapidity of cell isolation. In FACS and MACS, bulk antibody or antibody-coated beads are diluted in blood and mixed via diffusion throughout the sample of interest, requiring 30 to 60 minutes to achieve optimal antibody-antigen binding interactions. In microfluidic immunoaffinity capture, the cells are presented to the antibody by flowing it across the device walls, thus removing the time-consuming incubation step.

While this technology is very powerful and general in its abilities to gently capture cells with high specificity, it has a number of fundamental limitations. Currently, any immunoaffinity technique is limited by the accessibility of the antigen. Thus, for MACS and microfluidic techniques, only cell-surface-bound ligands are available for capture. Biotin-avidin chemistry is used in this protocol to allow flexibility in applying commercially available biotinylated antibodies to the surface. Unfortunately, biotin-avidin binding is difficult to disrupt, as are antibody-antigen interactions. Thus, cells currently cannot be released once they have been captured. Another disadvantage to microfluidic cell isolation is the lack of infrastructure and skill base to prepare the microfluidic capture devices. This protocol aims to make the techniques for production and usage more accessible and to provide helpful tips in solving common problems that occur when rapid-prototyping and testing microfluidic designs in PDMS (see Troubleshooting Table).

## 1.7 Application Notes

The device described in this protocol has wide-ranging applications for processing many different, complex biological samples. The principle of immunoaffinity capture is not limited to granulocytes but can be extended to any cell or particle type typically found in biological specimens, including cells, proteins, platelets, and the like. This platform was developed to count CD4+ T lymphocytes but could be used to capture and enumerate any cell type in circulating blood, leading to many applications in clinical

**Troubleshooting Table**

| Problem | Possible Cause | Solution |
|---|---|---|
| 1. Device does not bond or device leaks during functionalization | Device or glass not close enough to UVO source | Bring closer to UVO source. |
| | UVO exposure too long | Decrease exposure time. |
| | UVO exposure too short | Increase exposure time. |
| | | Always test UVO exposure with scrap microscope slide and scrap PDMS pieces. |
| | Pressure too high when injecting solutions | Dispose of device; inject fluid slowly. |
| 2. Air enters device | Air in syringe | Flush device with ethanol and proceed. |
| | Dried-out device | Always keep device wet; flush with ethanol and proceed, or inject PBS through the device, cover the device with PBS, and place in a cold room for 24 to 48 hours. Any air should diffuse out of the PDMS, leaving a fluid-filled device. |
| | Introduction of air when plugging in tubing | Always keep a bead of buffer solution over inlets and outlets. |
| | Introduction of air originating at the syringe-needle connection | Always be sure to tap out air bubbles when connecting needle to syringe. |
| | Chambers drying out during storage | Store devices in petri dish containing 3 to 5 mL buffer and seal with Parafilm; if device is unused, inject PBS through the device, cover the device with PBS, and place in a cold room for 24 to 48 hours. Any air should diffuse out of the PDMS, leaving a fluid-filled device. |
| 3. Blood does not flow through all the capture chambers | Clogging of channels by cells or proteins | Use EDTA anticoagulated blood; design postbased filter to keep large clumps of cells away from channels. |
| | Small air bubbles occluding channels, blocking flow | Using tweezers, press down on top of channels only. Never press on chambers as this severely damages cells. This may restore flow through chambers but will not remove all bubbles from the device. |
| | Inlet or outlet tubing sealed by pressure to the bottom glass slide | Using tweezers, wiggle or partially pull out both the inlet and outlet tubing. |
| 4. Blood does not rinse out of device during PBS wash step | Complete clogging of channels by cells, proteins, or air bubbles | See solutions to #2. |
| | Partial clogging of channels by cells, proteins, or air bubbles | Run wash buffer an additional 5 to 10 minutes. |
| 5. Cells look damaged | Pushing down on the top of the PDMS capture chambers | Do not push down on capture chambers; design a device with posts to support the chamber walls. |
| 6. Cells are the wrong color (Giemsa) | Poorly buffered wash buffer | Try different wash buffer pH between 6.5 and 7.2 until the correct color is obtained |
| 7. There is significant immunofluorescence background | Antibodies binding nonspecifically to the cells or device chambers | Try a different blocking solution; decrease staining time; decrease antibody concentration. |
| 8. There is poor RNA quantity or quality | RNase contamination | Use RNase-free solutions; carefully clean all work surfaces during isolation; make sure the device is washed with nuclease-free PBS. |
| | DNA contamination | DNase-treat total RNA isolate. |
| | Low number of cells | Check cell capture before lysis. |

diagnostics, such as CD4/CD8 ratios in HIV patients, total leukocyte counts in leucopenia patients, cell subsets in blood-borne cancer patients, and the presence of cells in urine to monitor infection.

As briefly mentioned above, this device can be used to collect cell lysate for many downstream molecular diagnostics. The advantage of this approach is the speed in obtaining a homogenous cell population, which minimizes artifacts due to cellular activation or degradation. While not described explicitly, this device can be used to lyse cells for proteomics. Thus, this tool will have utility in next generation clinical diagnostic tools that rely on downstream molecular analysis of tissue or biofluid samples.

## 1.8 Summary Points

- Immunoaffinity capture is an effective way to rapidly isolate cells from complex fluids.
- Capture specificity can be tuned by choice of a wide range of antibodies specific to proteins of cells' surfaces and by shear stress at the walls of the capture surface.
- Capture specificity, efficiency, and total capture numbers can be varied by design parameters for parallel plate geometry.
- Isolated cells can be processed for many downstream applications, including phenotyping, enumeration, genomics, and proteomics.
- The method is very reproducible and easy for nonskilled personnel to perform.

## Acknowledgments

The authors gratefully acknowledge the exceptional help of Octavio Hurtado in developing protocols for SU-8 and PDMS device fabrication. Funding was provided through the BioMEMS Resource Center under NIH Grant P41EB002503-02, NIH "Inflammation and Host Response to Injury" U54GM062119-08, and a T32 training grant T32GM007035-32 (K. Kotz).

### References

[1] Johnson, M. M., and Lewandrowski, K. B., "Analysis of emergency department test ordering patterns in an urban academic medical center: Can the point-of-care option in a satellite laboratory provide sufficient menu to permit full service testing," *Point of Care*, Vol. 6, No. 2, 2007, pp. 134–138.

[2] Cheng, X., et al., "A microchip approach for practical label-free CD4+ T-cell counting of HIV-infected subjects in resource-poor settings," *J. Acquir. Immune. Defic. Syndr.*, Vol. 45, No. 3, 2007, pp. 257–261.

[3] Murthy, S. K., et al., "Effect of flow and surface conditions on human lymphocyte isolation using microfluidic chambers," *Langmuir*, Vol. 20, No. 26, 2004, pp. 11649–11655.

[4] Cheng, X., et al., "Cell detection and counting through cell lysate impedance spectroscopy in microfluidic devices," *Lab Chip*, Vol. 7, No. 6, 2007, pp. 746–755.

[5] Kotz, K., Cheng, X., and Toner, M., "Cell capture using a microfluidic device," *J. Vis. Exp.*, No. 8, 2007, p. 320.

[6] Wintrobe, M. M., and Greer, J. P., *Wintrobe's Clinical Hematology*, 11th ed., Philadelphia: Lippincott Williams & Wilkins, 2003.

[7] Elghetany, M. T., and Davis, B. H., "Impact of preanalytical variables on granulocytic surface antigen expression: A review," *Cytometry B Clin. Cytom.*, Vol. 65, No. 1, 2005, pp. 1–5.

[8] McDonald, J. C., et al., "Fabrication of microfluidic systems in poly(dimethylsiloxane)," *Electrophoresis*, Vol. 21, No. 1, 2000, pp. 27–40.

[9] Zola, H., et al., *Leukocyte and Stromal Cell Molecules: The CD Molecules*, New York: John Wiley & Sons, 2007.

[10] Shapiro, S. O., *Practical Flow Cytometry*, New York: John Wiley & Sons, 2003.

[11] Cheng, X., et al., "A microfluidic device for practical label-free CD4(+) T cell counting of HIV-infected subjects," *Lab Chip*, Vol. 7, No. 2, 2007, pp. 170–178.

[12] Sin, A., et al., "Enrichment using antibody-coated microfluidic chambers in shear flow: Model mixtures of human lymphocytes," *Biotechnol. Bioeng.*, Vol. 91, No. 7, 2005, pp. 816–826.

[13] Nagrath, S., et al., "Isolation of rare circulating tumour cells in cancer patients by microchip technology," *Nature*, Vol. 450, No. 7173, 2007, pp. 1235–1239.

CHAPTER

# 2

# Dynamic Gene-Expression Analysis in a Microfluidic Living Cell Array (mLCA)

Kevin R. King,[1,2] Rohit Jindal,[2,3,4] Yaakov Nahmias,[2,3,4] and Martin L. Yarmush[1,2,3,4]

[1]Harvard-MIT Division of Health Science and Technology, Boston, MA
[2]Shriners Hospital for Children, Boston, MA
[3]Center for Engineering in Medicine, Massachusetts General Hospital, Boston, MA
[4]Harvard Medical School, Boston, MA

## Abstract

Gene expression is a fundamental cellular process that allows for the dynamic regulation of protein production, adapting the cellular response to environmental changes. The ability to measure changes in gene expression is currently limited by the use of destructive techniques that measure only bulk properties of cell populations and by the inability to dynamically control the microenvironment of cells in culture. This chapter describes an integrated and highly scalable functional genomics platform called the microfluidic living cell array (mLCA) that combines a library of green fluorescent protein (GFP) transcriptional reporters with a high-throughput fluid-addressable cell array, enabling dynamic gene expression profiling in living cells. By creating a window to the temporal patterns of transcriptional regulation, the mLCA is poised to make important contributions to our understanding of transcription factor networks and dynamic processes such as development, wound healing, or disease.

Key terms gene expression dynamics
fluorescent reporter cells
systems biology

## 2.1 Introduction

Living cells dynamically respond to the changing microenvironment by altering their gene expression profile. Cell surface receptors continually sense the local microenvironment integrating information from numerous extracellular inputs via signal transduction cascades to modulate the activity of transcription factors, the principal regulators of gene expression (Figure 2.1). Transcription factors are proteins that bind to specific DNA sequences called *response elements*, where they modulate the expression of nearby genes. As transcription factors often regulate each other, their interaction takes the form of a network which controls the cellular gene expression profile. Gene expression involves *transcription*, in which DNA information is transferred to RNA, and *translation*, wherein messenger RNA (mRNA) serves as a template for protein synthesis. The relationship between the microenvironment and the transcriptional control of gene expression is fundamental to our understanding of dynamic processes such as stem cell differentiation [1], initiation and resolution of inflammation [2], and cancer [3].

One approach to study changes in gene expression is to screen the expression of hundreds or thousands of genes in a high-throughput format. Several technologies currently exist to study gene expression on the mRNA level (*transcription*). These include quantitative Reverse Transcriptase Polymerase Chain Reaction (qRT-PCR) [4, 5], which allows the quantification of numerous genes from a single cellular isolate in multiwell plates, and DNA microarrays, such as the Affymetrix GeneChip, which enables the study of tens of thousands of genes in a single experiment (Table 2.1). In qRT-PCR, isolated

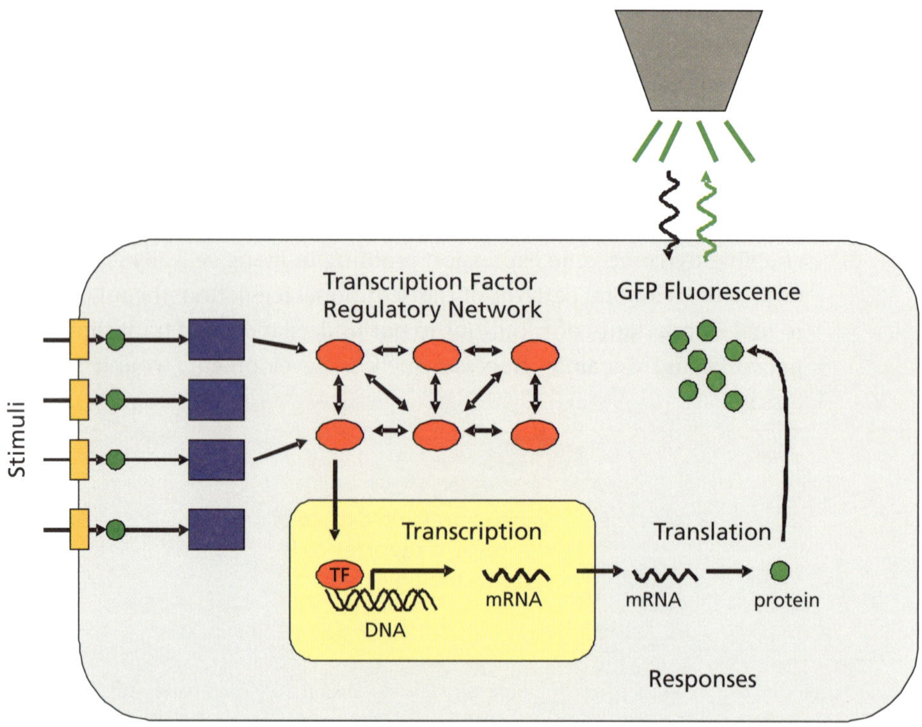


**Figure 2.1** Schematic of complex cell signaling: Multiple dynamic extracellular stimuli bind cellular receptors and activate a network of transcription factors that bind to DNA and induce expression of genes that are translated into proteins. In reporter cells, activation of the transcription factor of interest leads to production of GFP, which can be measured nondestructively using fluorescence microscopy.

**Table 2.1** Comparisons of Gene-Expression Measurement Techniques

| | RNA-Based Methods | Reporter-Based Methods |
|---|---|---|
| Low Throughput | Northern Blots, RT-PCR, qPCR<br>Pros: Direct high sensitivity measure of gene expression<br>Cons: Poor spatiotemporal resolution (destructive measure and spatial averaging) | Reporter Assay<br>Pros: High spatiotemporal resolution (nondestructive measurements with single cell resolution)<br>Cons: Indirect low sensitivity measure of gene expression |
| High Throughput | DNA Microarrays<br>Pros: 100,000's genes per experiment<br>Cons: Expensive (small number of stimulus conditions and time points) | Microfluidic Living Cell Array<br>Pros: Many dynamic stimulus conditions, many time points, single cell data<br>Cons: Requires stable reporter cell lines for each gene |

mRNA is converted to cDNA whose specific amplification is monitored in real time and compared to the amplification dynamics of a reference gene [6]. On the other hand, DNA microarrays use the hybridization of fluorescently labeled isolated mRNA on a micropatterned array to provide qualitative high-throughput data on tens of thousands of genes (Figure 2.2) [7, 8]. An alternative approach to screening the transcription of

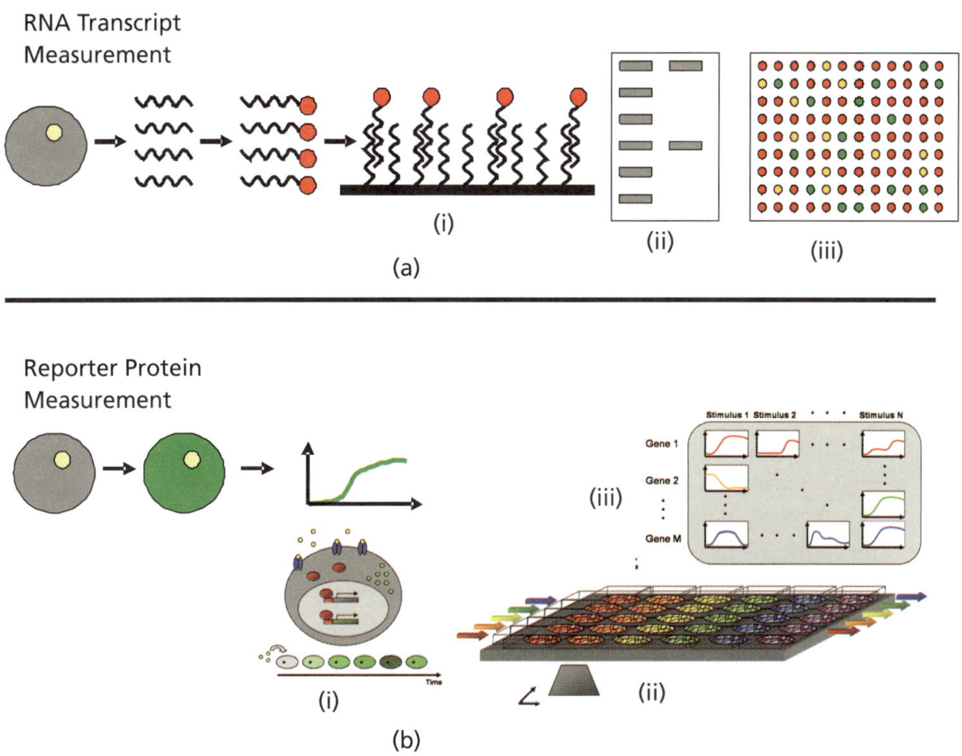


**Figure 2.2** Methods of gene-expression monitoring: (a) RNA-based methods: RNA extracted from lysed cell subjected to various methods for analyzing gene expression. (i) Glass slide spotted with various cDNA probes and exposed to labeled RNA. (ii) Separated RNA bands on a membrane exposed to labeled cDNA for a specific gene. (iii) RNA reverse-transcribed to cDNA and amplified with primers for different genes. (b) Reporter-based methods: (i) Stable reporter cell line exposed to stimuli expressing the fluorescent GFP protein with the same regulation as the gene of interest. (ii) Schematic illustrating various reporter cell lines exposed to multiple stimuli in an array format. (iii) 2-D matrix of dynamic stimulus-reporter gene response obtained from array described in (ii) [4].

thousands of genes is to probe the activity of the dozens of transcription factors that control their expression. Such a screen can potentially be carried out using an antibody array, such as Sigma-Aldrich's Panorama. However, although the techniques mentioned above can measure transcription factor levels on the mRNA or protein levels, these do not necessarily correlate with transcription factor activity, as it could be silenced or enhanced by multiple factors in the cell. For example, NFκB activity is regulated by the signal-induced degradation of its inhibitor IκB rather than by gene or protein expression. In an effort to produce quantitative data on transcription factor activity, several multiplex bead-based assays were recently developed by Luminex and its partners. By mixing a set of DNA response elements linked to color-coded beads with the cell lysate, the binding of transcription factors to DNA can be quantified using multiplex cytometry. However, regardless of the method of measurement, all of the techniques described above are destructive and therefore provide only static information from a single population of cells. While experiments can be repeated at different time points, the process requires a significant investment of labor at a prohibitive cost.

The need for dynamic measurements of gene expression in living cells has brought about the development of reporter plasmids. Reporters are easily detectable proteins designed to be regulated in the same fashion as the gene of interest. In this way, quantification of reporter protein levels can be used to infer the underlying gene expression pattern. The mechanics of this approach involves constructing plasmid DNA encoding the reporter protein, preceded by the appropriate regulatory sequence of the gene of interest, and introducing the plasmid DNA into cells using transfection or electroporation. Several types of reporter proteins have been developed: (1) secreted proteins such as alkaline phosphatase or secreted luciferase that can be measured nondestructively in cell culture supernatants [9]; (2) intracellular enzymes such as chloramphenicol acetyltransferase or firefly luciferase that require fixation or cell lysis for measurement [10]; and (3) fluorescent proteins such as GFP that can be measured nondestructively within individual living cells using flow cytometry or fluorescence microscopy. Of the available reporters, GFP and its destabilized variants are particularly well suited for dynamic gene expression measurement. The primary challenge of this approach is the need for reproducible plasmid transfection and the validation required to ensure that the reporter is modulated in the same fashion as the gene. In the mLCA platform described here, this challenge is overcome by creating a library of stable monoclonal GFP reporter cell lines and extensively validating their functional activity (Figure 2.2).

The second part of the mLCA is the microfluidic platform. Microfluidics represents a powerful technology that is able to dynamically regulate the external cellular microenvironment with high precision and reproducibility. Microscale fluidic systems are characterized by laminar flow and passive diffusive mixing and can be designed to deliver a range of different stimuli with predictable flow rates, concentrations, spatial distributions, and temporal profiles. They have been used to create a variety of chemical gradients for studying chemotaxis and to deliver stimuli to specific microdomains with subcellular resolution [11]. Microfluidic devices were used to generate both temporal and spatial temperature gradients [12], allowing differential regulation of development in anterior and posterior embryos. Building upon these demonstrations, we recently developed microfluidic circuits for delivering many different temporal regimens to downstream cell arrays for studying how cells decode their soluble microenvironment [13]. This approach, termed *flow-encoded switching* (FES), enables the simultaneous generation of many different stimulus patterns, including pulse trains of different widths, lengths, and frequencies. To demonstrate the power of this approach, FES devices were combined with one GFP reporter cell line and used to study the dynamics of gene expression in response to dynamic stimuli. As a whole, microfabricated culture systems offer high scalability, impressive integration, and novel functionalities, making them an ideal technology for constructing a high-throughput dynamic gene expression platform.

## 2.2 Materials

### 2.2.1 Reagents

The pEGFP-1 and pTRE-d2EGFP vectors used for reporter construction were obtained from Clontech (Mountain View, California). DMEM cell culture medium, fetal bovine serum (FBS), Penicillin-Streptomycin, phosphate buffered saline (PBS), Opti-MEM reduced serum medium, Geneticin (G418), calcein and cell tracker dyes, and LTX transfection reagent were acquired from Invitrogen (Carlsband, California). Cytokines such as TNF-$\alpha$, IL-1, IL-6, and IFN-$\gamma$ were purchased from R&D Systems (Minneapolis, Minnesota). SU8 photosensitive epoxy and PGMEA developing solution were obtained from Microlithography Corp. (Newton, Massachusetts). Poly(dimethylsiloxane) (PDMS, Sylgard 184) was acquired from Dow Corning (Midland, Michigan). Fibronectin and dexamethasone were purchased from Sigma (Sigma-Aldrich, St. Louis, Missouri). Reagents were stored under conditions recommended by suppliers.

### 2.2.2 Fabrication facilities

Mask designs were drawn on AutoCAD software. High-resolution mask printing on Mylar sheets was performed by Fineline Imaging (www.fineline-imaging.com). Microdevices were fabricated in the BioMEMS Resource Center clean room facility on 4-inch silicon single-sided polished wafers (MGH, Boston, Massachusetts). The following clean room fabrication equipment was used to create devices: March Instruments PX-250 for oxygen plasma treatment, Solitec 5110PD for spin coating SU-8, PMC 730A Digital Hot Plates for baking SU-8, Quintel Q2001 CT mask aligner, Kramer Olympus BX60 Microscope for wafer inspection, Blue M Oven for baking PDMS, and Kramer Combizoom-400 Microscope for aligning and bonding patterned PDMS layers. PDMS resin and curing agent were measured on a digital scale, mixed in weigh boats, and cast on the silicon mold in 15 cm plastic petri dishes. Glass slide substrates and plastic syringes were obtained from Fisher Scientific (Pittsburgh, Pennsylvania). Microfluidic inlet-outlet drills were made by blunting and beveling 18G needles obtained from Small Parts (Miramar, Florida).

### 2.2.3 Imaging equipment

Real-time imaging experiments were carried out using a Zeiss 200 M microscope (Carl Zeiss Inc., Thornwood, New York) equipped with temperature and $CO_2$ controller. Image capture and conversion were performed by Zeiss Axiovision software. Image analysis was performed using custom routines developed in MATLAB.

### 2.2.4 Perfusion components

Multichannel syringe pumps were purchased from Harvard Apparatus (Holliston, Massachusetts). Tygon tubing and syringe needles were obtained from Small Parts Inc. (Miami Lakes, Florida).

## 2.3 Methods

### 2.3.1 GFP reporter cell line construction

1. Reporter plasmids are constructed using traditional molecular cloning techniques. Briefly, response elements, or sequences of nucleotides known to have a high affinity for specific transcription factors (identified using the TRANSFAC database or primary literature as detailed in Table 2.2), are identified and used to create one plasmid per transcription factor. The basic design for each plasmid consists of three components: (1) three to four response element repeats with 5 bp intervals inserted upstream of (2) a minimal CMV promoter (to minimize transcription-factor-independent expression), regulating the expression of (3) a downstream fluorescent reporter protein with a 2-hour half-life, d2EGFP. Each reporter plasmid is also designed with a drug resistance gene to allow for the selection and purification of stably transfected cells. First, the CMV minimal promoter was digested from pTRE-d2EGFP and inserted between *Kpn I* and *Sma I* of the EGFP-1 plasmid (Clontech) to create the pCMVmin-EGFP1 plasmid. Response elements were then inserted before the CMVmin promoter between *Bgl* II and *Hin*d III on the multiple cloning sites of $\text{pCMV}_{\text{min}}$-EGFP-1. The EGFP gene was replaced with a destabilized variant with a 2-hour half-life (from pd2EGFP-1, Clontech), using *Bam*H I and *Not* I sites. The 2-hour EGFP variant was chosen to enable continuous monitoring of dynamic responses due to its short half-life. We note that commercial GFP reporter plasmids are currently available from SABiosciences (Frederick, Maryland).
2. Reporter plasmids are introduced into cells, such as H35 rat hepatoma, seeded at 25-50% confluency using LTX transfection reagent. Briefly 1 $\mu$g of DNA is mixed with 1 $\mu$L of LTX in Opti-MEM for 30 minutes to form DNA complexes. Once

**Table 2.2** GFP Reporter Sequences

| | | | | |
|---|---|---|---|---|
| NF-κB binding element/ **NF-κB** | GGGAMTNYCC | GGGAATTTCC | TNF-$\alpha$ | Proinflammatory, antiapoptotic |
| AP-1 binding element/ **AP-1** | TGASTMA | TGAGTCA | IL-1 | Proinflammatory, mitogenic |
| STAT3 binding element/ **STAT3** | TT(N)4–5AA | TTCCCGAA | IL-6 | Proinflammatory, anti-inflammatory |
| **ISRE**/IRF | SAAA(N)2–3AAASY | GAAACTGAAACT | IFN | Antiviral, innate immune activating |
| **GRE**/GR | AGAACANNNTGTTCT | AGAACAAAATGTTGT | Dexamethasone | Anti-inflammatory |
| **HSE**/HSF | CNNGAANNTTCNNG | CTAGAATGTTCTAG | 42°C | Cytoprotective |
| CMV-D4EGFP/D4G | — | — | Positive control | — |
| Nontransfected/NT | — | — | Negative control | — |

Note: Names in bold refer to stable monoclone reporter cell lines for each transcription factor.

formed, the culture medium is replaced with Opti-MEM and the complexes are delivered to the cells. Transfection is carried out for 4 hours, at which point the cells are washed and returned to standard culture medium for recovery.

3. Stable transfectants expressing the drug resistance gene are selected by culturing in the presence of geneticin (G418). A kill curve needs to be calculated for each new cell type transfected. Most cell lines die following 4 days of culture in concentrations ranging from 0.4 to 0.8 mg/mL G418. Selection is cell-type dependent but usually takes about 2 weeks. Once a stable transfectant has been selected, cells are cultured at low levels of G418 to maintain a positive selection pressure.
4. The stably transfected reporter cells are then stimulated with an inducer known to activate the transcription factor of interest. Stimulated cells are sorted by a Fluorescence-Activated Cell Sorter (FACS) to identify and purify cells with the highest fluorescence. The highly responsive cells are then expanded in culture for several days until GFP levels return to baseline. They are then sorted again in the absence of stimulation to purify those cells with the lowest background GFP expression.
5. The resulting cell population is then diluted and dispensed into 48 well plates such that an average of one cell is seeded into each well. The single cell cultures are then expanded to create highly responsive monoclonal reporter cell lines. Cells can also be cloned during the FACS sorting process.
6. This method is repeated for each transcription factor reporter plasmid to create a reporter cell-line library. The reporters described in this chapter were specifically selected for their roles in the regulation of inflammation and immune responses.

### 2.3.2 Microfluidic cell array fabrication

1. Microfluidic arrays are fabricated using planar microfabrication techniques based on conventional photolithography. After developing an initial design, each layer is drawn in two dimensions using a computer-aided design (CAD) tool such as AutoCAD, and electronic drawings are sent to a printing facility that prints the high-resolution pattern on Mylar transparency films with a minimum feature size of about 10 $\mu$m. These films are referred to as *photomasks*.
2. Silicon master molds are fabricated using conventional photolithography. Specifically, SU-8 photosensitive epoxy is spin-coated on silicon wafers at 1,000 rpm for 60 seconds, soft baked at 60°C for 5 minutes and 100°C for 15 minutes and returned to room temperature by placing the silicon substrate on a cooling block. Coated wafers are exposed through the photomasks with 365 nm UV light at 11 mW/cm$^2$ for 7 cycles of 5 second exposures and 5-second intervals (total of 35 seconds) to achieve a dose of around 400 mJ/cm$^2$ for selective photopolymerization, post-exposure baked at 60°C for 2 minutes and 100°C for 4 minutes, and developed by dissolving unexposed SU-8 in PGMEA solvent for about 10 minutes with agitation. SU-8 patterned silicon master molds are then rinsed in a fresh PGMEA developer and dried with compressed nitrogen.
3. Polymer replicas of the silicon masters are created by replica molding. To prevent sticking and tearing of polydimethylsiloxane (PDMS) films during the removal process, silicon masters are often exposed to trimethlylchlorosilane (TMCS) vapors for a few minutes prior to the application of PDMS. In this process, PDMS resin and curing agent are mixed at a ratio of 10:1, spin coated (for thin 50-$\mu$m PDMS layers) or

cast (for thick 0.5-cm PDMS layers) onto the silicon master, degassed and heated for 2–12 hours at 60°C. To a first approximation, the polymer is cured after 2 hours; however, the polymerization process and the mechanical properties of the material continue to evolve for at least 24 hours of heating. Once cured, the transparent PDMS replica is removed from the silicon support and inlets and outlets are formed by punching holes with a blunted and beveled 18G needle. Spin-coated PDMS films, due to their fragility, are not peeled, but are maintained on the silicon support until they are bonded to a thicker mechanically sturdy PDMS layer.

4. Assembly of the two-layer devices is conducted as follows (Figure 2.3). PDMS is spin-coated on silicon master mold #1 and cured at 60°C to form PDMS Layer#1, the *cell culture layer*. PDMS is then thickly cast on silicon master mold #2, cured, and peeled to form PDMS Layer #2, the *valve control layer*. Inlets and outlets are drilled in Layer #2 using 18G blunted and beveled needles to allow fluidic connection to the valve control channel network. After a brief oxygen plasma surface treatment, Layer #2 is manually aligned and bonded to the still silicon-supported PDMS Layer #1. The bonded layers are heated for 10 minutes at 100°C to complete the bonding. The bonded PDMS layers can then be peeled from the underlying Layer #1 silicon support. Note that in order to correct for the contraction of PDMS after its release

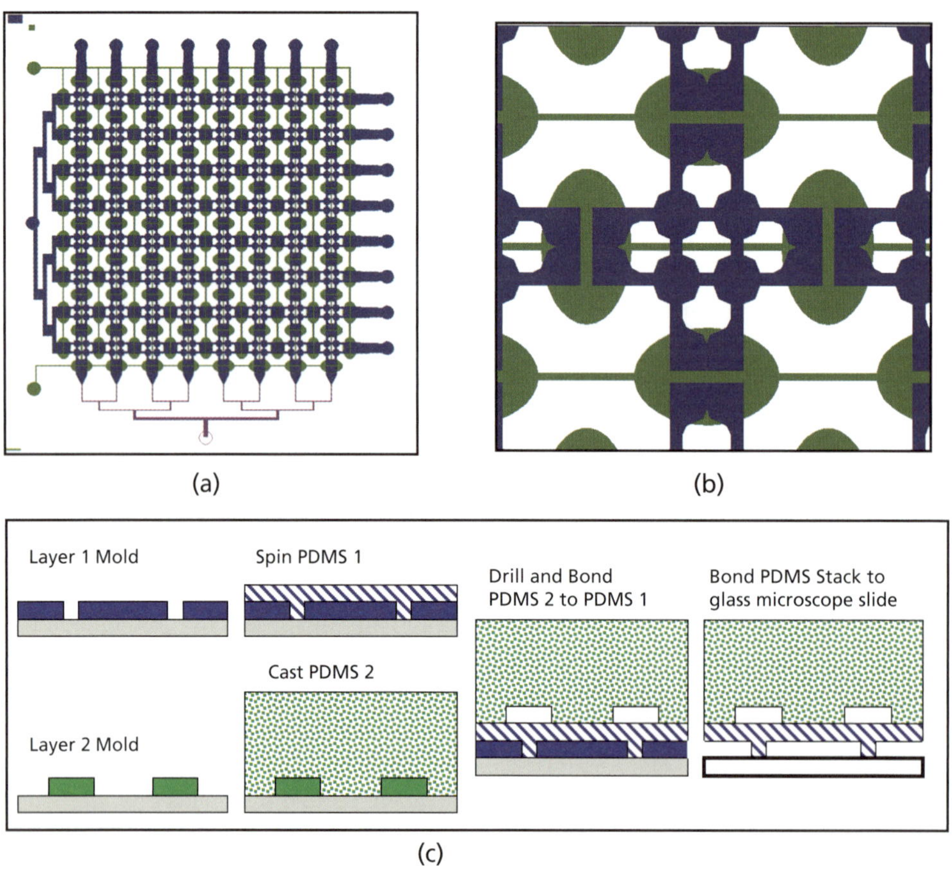


**Figure 2.3** Microfluidic living cell array (mLCA) fabrication and assembly: (a) high-level layout of the mLCA, (b) close-up of four cell culture chambers (blue circles) and their isolating microvalves (green ellipses), and (c) process for fabricating the mLCA shown in cross section.

from the silicon master, Layer #2 photomask needs to be printed at 101.8% of the target dimensions to align with Layer #1.

5. At this point, additional cell culture inlets and outlets are drilled to provide connections to the Layer #1 channel network. The multilayer PDMS stack and a glass microscope slide are treated with oxygen plasma and bonded. During this bonding step, valves are held open by connecting tubing and a syringe and drawing back on the syringes to create negative pressure in the valve control channels. This avoids permanent bonding of valves in the closed position. After bonding, the valves are cycled rapidly to ensure that permanent bonding is avoided. All bonding is performed by an oxygen plasma treatment of the two bonding surfaces for 15–30 seconds at low power (50W).

### 2.3.3 Microfluidic array pretreatment and seeding

1. Once fabricated, devices are sterilized by autoclave treatment. Tygon tubing is inserted (tight-fit) into the PDMS inlets with tweezers, and blunted needles are attached to the distal ends of the tubing allowing Luer-Lock syringe access to the system.
2. Layer #1 of the array is filled with a 0.1% fibronectin solution by closing all but one inlet and pressurizing the interconnected channel network using the fluid-filled fibronectin-containing syringe. Using this approach, air trapped in the network is actively driven through the walls of the gas-permeable PDMS device. Once the layer #1 channels have been completely degassed and there are no bubbles remaining, valves are closed and the device is incubated for 1–2 hours to allow fibronectin t to the glass bottom of the microchannel facilitating cell attachment.
3. Each cell line is maintained in a separate culture. Before seeding the microfluidic array, each cell line is trypsinized, counted with a hemocytometer, resuspended to a final density of $5 \times 10^6$ cells/mL, and triturated with a small-tipped pipette to ensure that all cell aggregates are broken and only single cells are injected into the microscale channels of the array.
4. Tubing from each cell type inlet is immersed in each reporter cell suspension, making sure to connect fluid menisci to avoid introducing bubbles. A syringe is connected to the seeding outlet, the stimulation valves are closed, the seeding valves are opened, and negative pressure is drawn from the seeding syringe, simultaneously pulling reporter cell suspensions through their respective channel rows and out the common seeding outlet. While the cells are flowing, the seeding valves are abruptly closed, and all flow abruptly comes to a halt. Seeding inlet and outlet tubings are clamped and immersed in culture medium-containing reservoir to avoid evaporation and introduction of air into the device. Most cell lines are typically attached and well spread on the fibronectin-coated glass surface within 1 hour. Cells are then cultured overnight with valves open and no perfusion for stabilization.
5. Note that the type of extracellular matrix, seeding density, and incubation times are cell-type dependent and need to be determined experimentally. Shear forces should be kept below 0.01 Pa for most cell types. Cell viability following seeding should be assessed for wild-type cells using a Live-Dead fluorescent assay (Invitrogen), and for H35 cells it was greater than 95%.

### 2.3.4 Stimulation and reporter imaging

1. Once cells are stably seeded in the microchambers, the device is mounted on an automated microscope with an incubated stage that controls both temperature and $CO_2$. To further minimize fluctuations in pH, all culture medium is supplemented with 30 mM HEPES. Imaging locations are selected using the Axiovision Mark and Find software (Carl Ziess), and the auto-focus routine is calibrated.
2. Time-lapse phase and fluorescence imaging are initiated prior to stimulation in order to characterize initial reporter fluorescence levels. Molecular stimuli are prepared at the desired concentrations in HEPES containing medium and loaded into syringes that are mounted in a multichannel syringe pump. Seeding valves are clamped closed and stimulation channels are drawn open by repeatedly applying positive and then negative pressure to using the Layer #2 control lines. Molecular stimulation is initiated by advancing the pump at a high flow rate to prime the tubing and to create a bolus of flow, thereby synchronizing the start of stimulation. The flow rate is then gradually decreased to the desired steady-state flow rate to provide controlled long-term delivery of each stimulus into its respective channel for the duration of the experiment. An alternative stimulation strategy involves loading stimuli into reservoirs and drawing them through the array by applying negative pressure to a single syringe connected to the common stimulation outlet. It is important to be cautious using this technique, as the approach requires sufficient flow rate to avoid gravity-driven flow from one reservoir to another.
3. Care should be taken not to photobleach the reporters during the procedure. Intervals of 30 to 90 minutes provide sufficient dynamic data without affecting the reporters.

## 2.4 Data Acquisition, Anticipated Results, and Interpretation

Each experiment in the 256 chamber proof-of-principle device, shown in Figure 2.4, generates about 5,000 images per day [14]. Fluorescence images are captured on a Zeiss 200 Axiovert microscope using an AxioCAM MRm digital camera, and are quantified as grayscale images using custom analysis routines written in MATLAB. Each image is divided by an image with uniform fluorescence to correct for spatial variations in fluorescence excitation. The intensity histogram of each image is then combined with a user-defined threshold parameter to automatically determine and subtract a background fluorescence level. The remaining fluorescence is integrated and assumed to be the result of cellular GFP expression. The resulting quantified time courses are then organized by location in the array and visualized collectively as a heat map, shown in Figure 2.5. Numerous analysis techniques have been developed to categorize and synthesize such multiparameter dynamic cell response datasets [15, 16]. To highlight temporal aspects of the responses, correct for differences in cell number, and facilitate comparisons between locations, each response was normalized between its maximum and minimum fluorescence levels according to $\Phi_{ij}(t) = [F_{ij}(t) - F_{ij\_min}]/[F_{ij\_max} - F_{ij\_min}]$, where $\Phi_{ij}(t)$ is the normalized fluorescence of row $i$ and column $j$ at time $t$, $F_{ij}(t)$ is the post-processing image fluorescence, and $F_{ij\_max}$ and $F_{ij\_min}$ are the maximum and minimum post-processing fluorescence values for the $ij$ array location, respectively. In the study

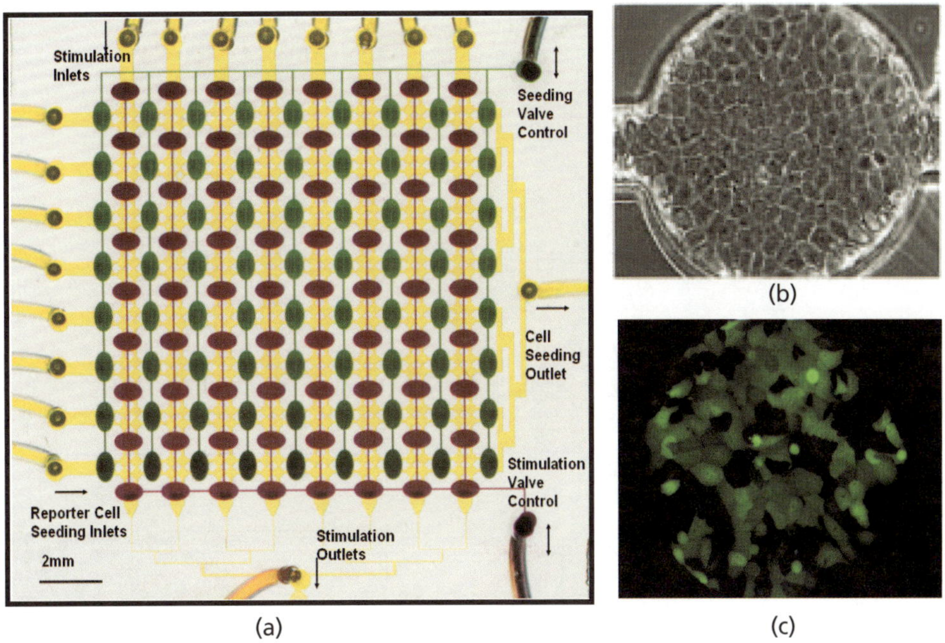


**Figure 2.4** Characterization of the mLCA. (a) Image of dye-filled mLCA. Cell culture chambers in layer 1 are filled with yellow dye, while layer 2 seeding and stimulation channel networks are filled with green and red dye [4]. (b) Phase-contrast image of confluent reporter cells in an mLCA cell culture chamber. (c) Example of GFP reporter fluorescence imaged in one cell chamber.

shown here, 3–4 cell culture chambers containing the same stimulus-reporter pair are measured at each time point and averaged, and standard deviations are calculated.

The proof-of-principle experiments described here demonstrate the promise of the mLCA technology. Nevertheless, several differences between cells cultured under microfluidics and those cultured in conventional tissue culture techniques need to be considered. Thus far, the dynamic responses of GFP reporter cells observed in the mLCA have proven to be similar to those seen in conventional culture. However, it is likely that certain aspects of cellular function will be altered by flow, and these effects remain an active area of continuing investigation. The most obvious differences between conventional and microfluidic culture is the large surface area-to-volume ratios and the potential for perfusion-related artifacts. While it is straightforward to develop simplified mathematical models to simulate the effects of different channel geometries and fluid flow rates, such effects have yet to be thoroughly examined experimentally. Another difference between conventional and microfluidic cultures is the chemical composition of the cell culture surface. Whereas conventional cell culture is usually performed on tissue culture plastic, microfluidic cultures are carried out on fibronectin-coated glass surfaces. Additional studies will be necessary to fully elucidate the mass transport and mechanical shear effects of continuous perfusion in order to understand the differences between conventional and microfluidic cell culture and optimize designs to take advantage of these differences.

Finally, when using GFP reporters, one must carefully consider the assumption that GFP levels reflect the dynamics of the activity of a transcription factor or the transcrip-

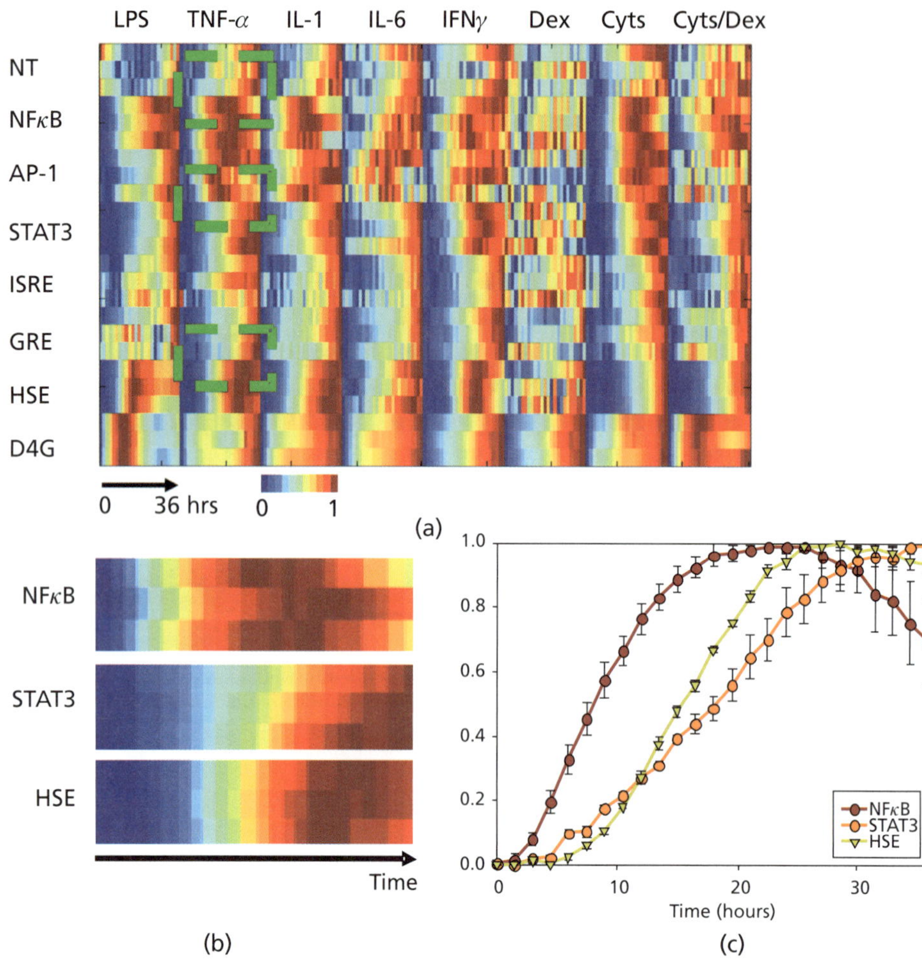


**Figure 2.5** Experimental results from high-throughput dynamic reporter profiling. (a) Heat map of dynamic responses from one time-lapse mLCA experiment [4]. (b) Close-up of TNF$\alpha$-induced transcription factor responses showing early NF-$\kappa$B and delayed heat shock (HSE) activation. (c) Plot of TNF$\alpha$-induced transcription factor activation dynamics [4].

tion of a gene of interest, for example, when GFP mRNA stability or post-translational processing is affected by a given stimulation. Some of these concerns are addressed in our design by the presence of a cell line constitutively expressing d2EGFP. Normalizing one signal against the other should provide for a more sensitive measurement of transcriptional activity. It is important to note that the CMV promoter is probably not ideal for this design, and a weaker promoter, such as hTK, will be more a more sensitive indicator of nonspecific events.

## 2.5 Discussion

This chapter describes methods for constructing a library of GFP reporters, fabricating a microfluidic living cell array, and combining them to create a high-throughput experimental platform for profiling transcriptional activity dynamics. First, we discussed the generation of a library of highly responsive stable monoclonal reporter cell lines that can be readily interrogated using fluorescence microscopy and fluorescence cytometry. Compared to existing gene expression measurement techniques, only reporter cells are uniquely suited for nondestructive monitoring of gene expression dynamics in living cells with single cell resolution. In addition to their stability, important features of GFP reporter construction include their high inducibility by classic stimuli, low background GFP expression, and monoclonality. These attributes allow each cell in a population to be interpreted as a genetic equivalent, such that variations between cells can be attributed to aspects of cell physiology and environment rather than differences in the number of reporter copies or the location of integration.

This chapter also describes methods for constructing a highly integrated microfluidically addressable living cell array. The array is fabricated by replica molding transparent PDMS elastomer from microfabricated silicon masters. Once fabricated, the silicon masters can be reused numerous times to create PDMS polymer replicas. Polymer devices can then be cast, cured, bonded, sterilized, coated, and seeded in approximately 24 hours, such that cell-seeded arrays are ready for experimentation the day after fabrication. As long as cells are well triturated prior to seeding, aggregation does not appear to complicate seeding in the microscale channels. Instead, cell suspensions flow smoothly through the channels and underneath open valves. When the valves are suddenly closed, the cells immediately stop and sediment to the bottom where they attach and spread on the fibronectin-coated surface. Since all cell lines can be seeded simultaneously, future versions of the device can be scaled to accommodate many more reporter cell lines. Similarly, the number of molecular stimuli can also be scaled without significantly increasing the device complexity or experimental setup.

It might be argued that the experiments described here can be performed using multiwell plates and automated high-throughput screening technologies. However, unlike multiwell plates, the microfluidic platform is not limited to constant stimulation, but instead allows for dynamic control of the cell microenvironment. For example, the chemical milieu surrounding the cell can be varied in time, just as hormone levels fluctuate in the blood or inflammatory mediators are transiently released after detection of a pathogen. This was recently demonstrated in a novel microfluidic circuit design that allowed many different stimulus time courses to be controlled from a single pressure input [13]. These and other such dynamic stimulus control circuits can be readily integrated upstream of the valve-controlled multireporter array. In addition to enabling dynamic stimulus delivery, microfluidics are also considerably less expensive than high-throughput screening facilities and require significantly less reagent volumes, which is particularly important for long duration dynamic experiments.

The microfluidic platform allows tremendous integration and scalability. The proof of concept device described here uses two manifolds of 144 integrated valves to control delivery of eight different stimuli to eight different reporter cell lines. Each stimulus-reporter pair is imaged in four separate wells to control for variations in seeding density as well as imaging artifacts. Each experiment yields one phase image and one fluorescence image for each location of the 256-element array at each time point. When time-lapse experiments are performed, hourly sampling for 24 hours yields over 5,000 data points per experiment.

Initial experiments were performed by summing fluorescence across the entire well; however, since the data is in the form of images, they can be retrospectively mined to extract single cell data. Furthermore, future devices can be designed to integrate single cell array technologies into the microfluidic array in order to expand the number of data points to more than 1 million single cell measurements per day. Single-cell analysis is particularly powerful because GFP reporter measurements can be correlated with other cell parameters such as cell motion, shape, neighbors, and initial cell state. This would facilitate investigation of biological processes where population heterogeneity plays an important role. At present, the factors that contribute to cell population variability are unknown but are thought to include local variations in the microenvironment, differences in cell history, and the stochastic nature of gene expression [18]. Fluctuations in gene expression are thought to generate diversity in cell phenotypes, even across genetically identical cell populations exposed to the same environment [19]. Early studies in bacteria provided the basic experimental and theoretical framework for investigating cell heterogeneity [20–22], and more recently, significant population heterogeneity was demonstrated in mammalian cells. For example, monitoring p53-Mdm2 dynamics at an individual cell level revealed that expression of p53 occurred as a series of discrete pulses [23] rather than as damped oscillations suggested by population level studies [24]. Another dynamic system that demonstrates oscillatory behavior is NF-$\kappa$B signaling. Pop-

**Troubleshooting Table**

| Problem | Explanation | Potential Solution |
|---|---|---|
| The GFP reporter responses are heterogeneous | Intrapopulation heterogeneity can be due to polyclonality of the reporter as well as cell cycle asynchrony, variations in cell shape, and differences in local cell microenvironments | Ensure populations are clonal (created from single cells) |
| Reporters are GFP+ after 7 days of no stimuli | We have found that inducibility can change with extensive passaging. | Thaw a new vial from reporter frozen stocks. |
| GFP reporters no longer respond to classic stimuli | We have found that inducibility can change with extensive passaging. | Thaw a new vial from reporter frozen stocks. |
| Spin-coated PDMS layers are sticking to the microfabricated silicon master | The nonadhesive nature of the master surface has been compromised | Reapply nonadhesive trichloromethylsilane before each PDMS spin. |
| PDMS-to-PDMS or PDMS-to-glass bond is not strong (peels off with applied pressure) | It is likely surfaces were touched and removed briefly prior to final bonding position. Alternatively, bond did not strengthen because of insufficient time and temperature. | Restart with new devices. We have not had success rebonding. |
| After bonding, a few valves are stuck to the glass surface and cannot be actuated | PDMS valves were allowed to contact plasma treated glass too early or too long | Hold valves in open position for 5 minutes after bonding. Then rapidly cycle them >20 times and return to open position. |
| Bubbles are forming in device during experiments. | Either air was introduced in through inlet or outlet tubing, devices were not completely degassed before cell seeding, excessive negative pressure was applied to the fluid-filled channel network, or solutions were excessively aerated prior to injection. | Remove all bubbles at fibronectin coating step. Allow fibronectin to incubate without bubbles so that all device walls hydrate adequately. Incubating all tubing with fibronectin solution reduces the chance of introducing air bubbles from an inlet. Ensure there are no bubbles in valve control line. |
| Some phase image of cells are not in focus | Autofocus routine failed to find optimal z-level because starting z-position was too far from optimal | Check that starting z-position for autofocus routine is closed to the optimal focus position. |
| Auto-threshold on image analysis routine fails to find a threshold that identifies cells. | Illumination was not uniform because fluorescence bulb for GFP excitation was not aligned. | Align fluorescence blub so that reporter cell field is uniformly illuminated. |

ulation level studies lead to the initial identification of oscillatory behavior in the temporal response of NF-$\kappa$B [25]; however, time-lapse fluorescence imaging at the single cell level established that oscillations were asynchronous [26]. Subsequent studies identified NF-$\kappa$B signaling heterogeneities in physiologically relevant macrophage cells [27]. Future experiments that employ microfluidics in concert with fluorescence imaging at single-cell levels are well poised to make significant contributions towards identifying the factors that lead to population level heterogeneities.

## 2.6 Application Notes

Results from the mLCA experiments can be used to gain insight into the architecture and dynamic operation of transcriptional networks. For example, the data from our proof-of-principle experiments helped identify a delayed heat shock response to stimulation with the inflammatory cytokine, TNF-$\alpha$ [Figure 2.5(b)]. While early activation of NF-$\kappa$B is a classic cell response to stimulation with TNF-$\alpha$, delayed TNF-induced activation of heat shock factor is not well understood. These temporally distinct responses to a single cytokine stimulus are particularly interesting in the context of apoptosis, because heat shock and NF-$\kappa$B regulate both pro- and anti-apoptotic genes to determine cell fate [17]. It is dynamic phenomena such as this that the mLCA is ideally suited to uncover.

The ability to study spatial and temporal gene expression patterns in response to environmental changes has been elusive due to the limitations of conventional experimental tools. The development of well-characterized stable monoclonal reporter cell lines and a high-throughput microfluidic experimental system represents a powerful addition to the functional genomics toolkit and will provide an important window into spatiotemporal patterns of gene expression and the organization of complex transcription factor regulatory networks.

## 2.7 Summary Points

- A library of stable monoclonal GFP reporters was constructed to monitor dynamic gene expression in living cells.
- A valve-controlled microfluidic array can be fabricated to allow seeding of each reporter cell line in rows and delivery of different molecular stimuli in columns.
- Time-lapse fluorescence microscopy and automated image analysis currently allow the entire reporter library to be exposed to eight different stimuli while being monitored across 24 to 48 hours, yielding approximately 5,000 to 10,000 single time-point measurements per experiment.
- The scalable nature of this functional genomics platform will allow dynamic response profiling of increasing numbers of stimulus-reporter combinations.

## Acknowledgments

We would like to acknowledge NIH Grants GM065474, AI063795, K01 DK080241, and BioMEMS Resource Center Grant P41 EB002503. K.R.K was supported by a postdoctoral fellowship from Shriners Burns Hospitals. The authors would like to thank Sihong Wang, Mehmet Toner, Pohun Chris Chen, Cindy Zia, Deanna Thompson, Ken Wieder, Arul Jayaraman, Daniel Irimia, and Octavio Hurtado for helpful discussions and technical support.

## References

[1] Iwasaki, H., et al., "The order of expression of transcription factors directs hierarchical specification of hematopoietic lineages," *Genes Dev.*, Vol. 20, No. 21, 2006, pp. 3010–3021.

[2] Lawrence, T., et al., "Possible new role for NF-kappaB in the resolution of inflammation," *Nat. Med.*, Vol. 7, No. 12, 2001, pp. 1291–1297.

[3] de Bivort, B., Huang, S., and Bar-Yam, Y., "Empirical multiscale networks of cellular regulation," *PLoS Comput. Biol.*, Vol. 3, No. 10, 2007, pp. 1968–1978.

[4] Alwine, J. C., Kemp, D. J., and Stark, G. R., "Method for detection of specific RNAs in agarose gels by transfer to diazobenzyloxymethyl-paper and hybridization with DNA probes," *Proc. Natl. Acad. Sci. USA*, Vol. 74, No. 12, 1977, pp. 5350–5354.

[5] Liang, P., and Pardee, A. B., "Differential display of eukaryotic messenger RNA by means of the polymerase chain reaction," *Science*, Vol. 257, No. 5072, 1992, pp. 967–971.

[6] Gibson, U. E., Heid, C. A., and Williams, P. M., "A novel method for real time quantitative RT-PCR," *Genome Res.*, Vol. 6, No. 10, 1996, pp. 995–1001.

[7] Schena, M., et al., "Quantitative monitoring of gene expression patterns with a complementary DNA microarray," *Science*, Vol. 270, No. 5235, 1995, pp. 467–470.

[8] Duggan, D. J., et al., "Expression profiling using cDNA microarrays," *Nat. Genet.*, Vol. 21, No. 1 Suppl., 1999, pp. 10–14.

[9] Berger, J., et al., "Secreted placental alkaline phosphatase: a powerful new quantitative indicator of gene expression in eukaryotic cells," *Gene*, Vol. 66, No. 1, 1988, pp. 1–10.

[10] Castano, J. P., Kineman, R. D., and Frawley, L. S., "Dynamic monitoring and quantification of gene expression in single, living cells: a molecular basis for secretory cell heterogeneity," *Mol. Endocrinol.*, Vol. 10, No. 5, 1996, pp. 599–605.

[11] Takayama, S., et al., "Subcellular positioning of small molecules," *Nature*, Vol. 411, No. 6841, 2001, p. 1016.

[12] Lucchetta, E. M., et al., "Dynamics of Drosophila embryonic patterning network perturbed in space and time using microfluidics," *Nature*, Vol. 434, No. 7037, 2005, pp. 1134–1138.

[13] King, K. R., et al., "Microfluidic flow-encoded switching for parallel control of dynamic cellular microenvironments," *Lab Chip*, Vol. 8, No. 1, 2008, pp. 107–116.

[14] King, K. R., et al., "A high-throughput microfluidic real-time gene expression living cell array," *Lab Chip*, Vol. 7, No. 1, 2007, pp. 77–85.

[15] Gaudet, S., et al., "A compendium of signals and responses triggered by prodeath and prosurvival cytokines," *Mol. Cell Proteomics*, Vol. 4, No. 10, 2005, pp. 1569–1590.

[16] Miller-Jensen, K., et al., "Common effector processing mediates cell-specific responses to stimuli," *Nature*, 2007.

[17] Beere, H. M., "Death versus survival: functional interaction between the apoptotic and stress-inducible heat shock protein pathways," *J. Clin. Invest.*, Vol. 115, No. 10, 2005, pp. 2633–2639.

[18] Elowitz, M. B., et al., "Stochastic gene expression in a single cell," *Science*, Vol. 297, No. 5584, 2002, pp. 1183–1186.

[19] Kaern, M., et al., "Stochasticity in gene expression: from theories to phenotypes," *Nat. Rev. Genet.*, Vol. 6, No. 6, 2005, pp. 451–464.

[20] McAdams, H. H., and Arkin A., "Stochastic mechanisms in gene expression," *Proc. Natl. Acad. Sci. USA*, Vol. 94, No. 3, 1997, pp. 814–819.

[21] Ozbudak, E. M., et al., "Regulation of noise in the expression of a single gene," *Nat. Genet.*, Vol. 31, No. 1, 2002, pp. 69–73.

[22] Swain, P. S., Elowitz, M. B., and Siggia, E. D., "Intrinsic and extrinsic contributions to stochasticity in gene expression," *Proc. Natl. Acad. Sci. USA*, Vol. 99, No. 20, 2002, pp. 12795–12800.

[23] Lahav, G., et al., "Dynamics of the p53-Mdm2 feedback loop in individual cells," *Nat. Genet.*, Vol. 36, No. 2, 2004, pp. 147–150.

[24] Lev Bar-Or, R., et al., "Generation of oscillations by the p53-Mdm2 feedback loop: a theoretical and experimental study," *Proc. Natl. Acad. Sci. USA*, Vol. 97, No. 21, 2000, pp. 11250–11255.

[25] Hoffmann, A., et al., "The IkappaB-NF-kappaB signaling module: temporal control and selective gene activation," *Science*, Vol. 298, No. 5596, 2002, pp. 1241–1245.

[26] Nelson, D. E., et al., "Oscillations in NF-kappaB signaling control the dynamics of gene expression," *Science*, Vol. 306, No. 5696, 2004, pp. 704–708.

[27] Ramsey, S., et al., "Transcriptional noise and cellular heterogeneity in mammalian macrophages," *Philos. Trans. R Soc. Lond. B Biol. Sci.*, Vol. 361, No. 1467, 2006, pp. 495–506.

CHAPTER

# 3

# Micromechanical Control of Cell-Cell Interactions

Elliot E. Hui,[1] Salman R. Khetani,[1] and Sangeeta N. Bhatia[1,2,3]

[1]Division of Health Sciences and Technology (Harvard-MIT), Electrical Engineering and Computer Science, Massachusetts Institute of Technology, Cambridge, MA 02139
[2]Division of Medicine, Brigham and Women's Hospital, Boston, MA 02115
[3]Howard Hughes Medical Institute

## Abstract

Cell-cell interactions play a critical role in determining cellular fate and function but have been challenging to manipulate using conventional tools. Micromechanical reconfigurable culture (MRC) is a method in which adherent cells are cultured on microfabricated plates that can be positioned and rearranged with micrometer accuracy. In this way, dynamic modulation of cell-cell interactions between multiple cell populations is possible. In addition, contact-dependent interactions can be decoupled from diffusible signals, and populations can be rapidly separated and recombined to change the composition of a mixed culture. This chapter describes the fabrication and use of the MRC device in practical detail, including examples of different experimental configurations and consideration of troubleshooting techniques.

Key terms
microfabrication
MEMS
dynamic substrate
cell-cell interaction
microenvironment
cell patterning
cell separation
juxtacrine
paracrine

## 3.1 Introduction

### 3.1.1 Cell-cell interactions

Cell-cell interactions form a key component of the cellular microenvironment and play a critical role in regulating functional behavior and driving cell-fate processes. Classic examples include the interactions between endothelial cells and smooth muscle cells in blood vessels [1], hepatocytes and stroma in the liver [2], progenitor cells and the various embryonic layers [3], stem cells and their surrounding niche [4], and tumor-stromal interactions in cancer [5].

Geometric factors play an important role in determining the extent of signal propagation. Certain types of cell-cell signaling require cells to be in direct physical contact. Examples include the binding of membrane-associated ligands, diffusion of signaling molecules through gap junctions, and integrin-mediated force transduction. On the other hand, secreted paracrine factors can diffuse over many cell lengths, but signal intensity can be modulated by factors such as distance, obstructions in the intervening space, and competitive consumption by other cells. Spatial organization thus plays a fundamental role in determining intercellular communication within a multicellular community.

In this chapter, we present micromechanical reconfigurable culture (MRC). This method employs a microfabricated mechanism to manipulate the spatial organization of multiple cell populations in culture. In this way, both contact-dependent and paracrine cell-cell interactions between distinct cell populations can be dynamically manipulated.

### 3.1.2 Conventional cocultivation models

The conventional methods for studying cell-cell interactions in vitro are the use of monolayer cocultures, the transfer of conditioned media, or the employment of membrane inserts. Each of these methods offers relative advantages and disadvantages with regard to the types of signals that are active, the ease of separating back to pure populations, and the ability to manipulate the composition of the culture dynamically (Table 3.1).

Monolayer cocultivation of multiple cell types is the most straightforward way of forming a mixed culture. Different populations are combined and plated together into a single well; hence, contact-dependent signaling and soluble signaling are both possible between the different cell populations. However, once the cells are plated, their posi-

**Table 3.1** Comparison of Reconfigurable Culture and Conventional Methods for In Vitro Cocultivation

| | Monolayer Culture | Media Transfer | Transwell Insert | Reconfigurable Culture |
|---|---|---|---|---|
| Soluble Signals | ✓ | ✓ | ✓ | ✓ |
| Contact Signals | ✓ | × | × | ✓ |
| Separable Populations | × | ✓ | ✓ | ✓ |
| Dynamic Manipulation | × | ✓ | ✓ | ✓ |

tioning is largely static and not amenable to reconfiguration. Similarly, it is not trivial to separate the cells back into pure populations. In certain circumstances, two populations may have enough of a difference in substrate adhesion that a carefully timed trypsinization step can release just one cell type. In general, however, the cells must be all released, dissociated into a single-cell suspension (not always easy), and then separated by fluorescence-activated cell sorting (FACS).

In order to decouple soluble factors from contact-mediated signals, researchers will often culture different populations in separate wells and then transfer conditioned media from one well to the other. Another option is to culture two populations in the same well, but with the second population suspended above the first by using a membrane insert [6]. The membrane allows secreted factors to diffuse across freely, but the two populations do not form cell-cell contacts with each other as they are separated by a few millimeters. In both of these configurations, the two populations are easily separable and can also be quickly reconfigured, for example, by replacing the insert with another containing different cells.

### 3.1.3 Micromechanical reconfigurable culture

The reconfigurable culture method described in this chapter offers key advantages over conventional approaches. Cocultures formed using this system can interact through both soluble and contact-mediated signals, yet can still be quickly separated or reconfigured. In addition, contact-mediated signals can be modulated independently of soluble factors.

The device is illustrated in Figure 3.1. Two sets of fingers are arranged as interlocking combs, and adherent cells are cultured on the top surface of the fingers. Different cell populations can be placed on each comb so that as the parts slide together and apart, cell-cell interactions are modulated between the two populations. In the contact configuration, the parts are pushed together and fit precisely enough that the two populations can form direct cell-cell contact across the interface. In the gap configuration, the populations are separated by 80 $\mu$m, and only soluble factor interactions are possible. The parts are interchangeable so that one population can be removed and replaced while the second population is held fixed (Figure 3.2).

The integrated latching mechanism provides precise positioning in both the contact and gap modes. The mechanism is self-centering and incorporates a 20:1 mechanical

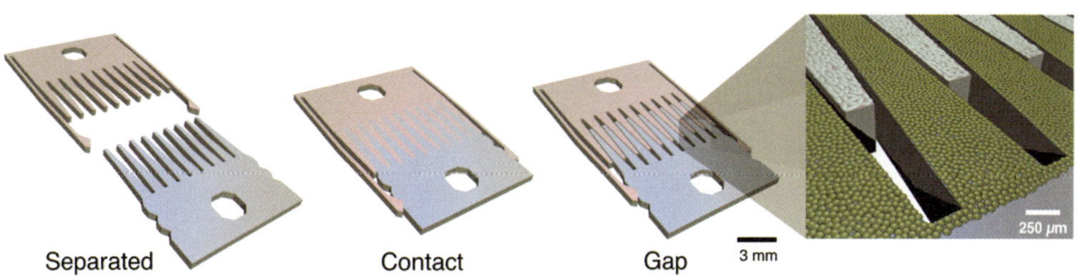


**Figure 3.1** Micromechanical substrates enable precise and reconfigurable cell positioning. The MRC device consists of two silicon combs that can be separated, locked together with the fingers in contact, or locked together with an 80 $\mu$m gap in between the fingers. The integrated latching mechanism is self-centering and allows the device to be manually actuated with an accuracy of microns. Cells are meant to adhere to the top surface of the fingers. (Reprinted with permission from [7] © 2007 National Academy of Sciences, USA.)

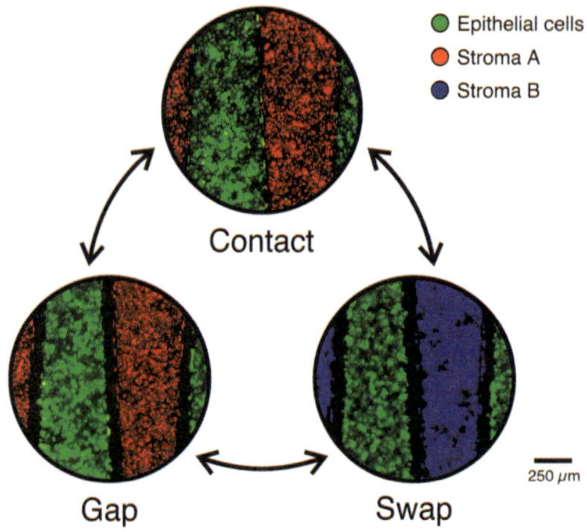


**Figure 3.2** Reconfigurable cell culture. Distinct cell populations can be combined in coculture and configured to allow direct cell-cell interactions between the populations (contact) or to allow only soluble factor exchange (gap). Additionally, one cell population can be held constant while the other population is removed and replaced (swap). Manipulations are reversible and can be combined sequentially, allowing a very large number of different experimental permutations. (Reprinted with permission from [7] © 2007 National Academy of Sciences, USA.)

transmission ratio (sliding the parts 1.6 $\mu$m changes the separation by only 80 $\mu$m). It is thus possible to actuate the device accurately by using only manually operated tweezers. Additional micromanipulation machinery is unnecessary.

### 3.1.4 Application examples

This section presents a number of examples to illustrate the different uses for reconfigurable culture. Every capability described here has been demonstrated, at minimum, by a successful proof-of-concept experiment.

#### 3.1.4.1 Cell patterning

The most straightforward application of reconfigurable culture is simply to form well-defined patterns of cells. Cell patterning has proven to be a powerful biological tool for understanding how geometric relationships affect tissue function [8]. A number of methods exist for micropatterning cells, including photolithography, microcontact printing, and microfluidic stenciling [9]. Most approaches work by patterning adhesive regions on a substrate and then seeding cells uniformly and allowing the cells to attach selectively to the adhesive regions. To form patterned cocultures, a second cell type is added after patterning of the first cell type is complete. It can be helpful to employ adhesion chemistries that are selective for particular cell types, if these are available. Still, both cell types are seeded over the entire substrate, and it is nearly impossible to avoid some level of cross-contamination between the different populations.

With reconfigurable culture, the device elements are separated into isolated wells for cell seeding so that each part receives a pure population of the desired cell type (Figure 3.3). Cell patterning is independent of the underlying surface chemistry; in fact, both cell types can be cultured on the same surface chemistry if desired.

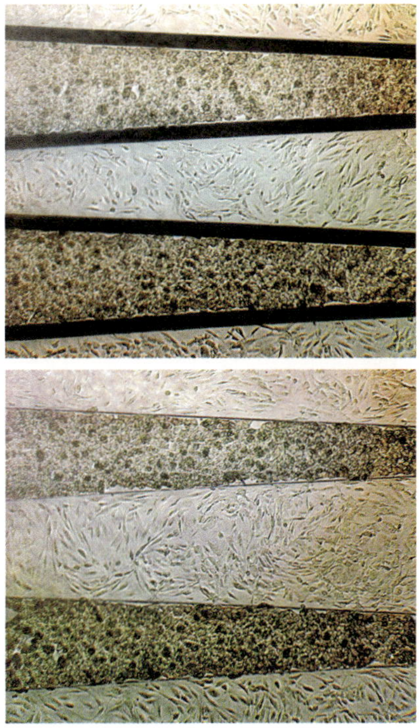

**Figure 3.3** Patterning of two separate cell populations can be achieved with no cross-contamination. Cells are seeded onto individual comb parts and allowed to attach and spread before pairs are assembled to form patterned cocultures in gap mode (top) or contact mode (bottom). In order to facilitate good cell adhesion, the surface is coated with plasma-treated polystyrene to create a surface comparable to standard tissue-culture plastic. (Reprinted with permission from [7] © 2007 National Academy of Sciences, USA.)

#### 3.1.4.2 Decoupling of contact-mediated and soluble signals

Comparing the function of cocultures in the gap mode versus the contact mode is an excellent way to study the relative importance of contact-dependent and secreted factors (Figure 3.4). The experiment is well controlled, as the two configurations are exactly identical, except for the 80 $\mu$m change in separation. In comparison, the use of membrane inserts introduces more variation because the two cell populations are cultured on different substrates. Also, the distance between the cells when using either media transfer or membrane inserts is much greater than with the MRC device in gap mode, and so factors that depend on high concentrations or short diffusion distances may not be effective.

It should be noted that the contact configuration enables a host of different types of signaling, including binding of membrane ligands, ECM deposition, physical forces, or possibly extremely short-range diffusible factors. Additional studies are thus required to differentiate between these different possible modes of signaling.

#### 3.1.4.3 Range of soluble signaling

The MRC device allows the effective range of diffusible factors to be studied. Using the standard design as detailed above, there is only one possible separation setting in gap mode. However, the cell pattern that is established separates the two populations more

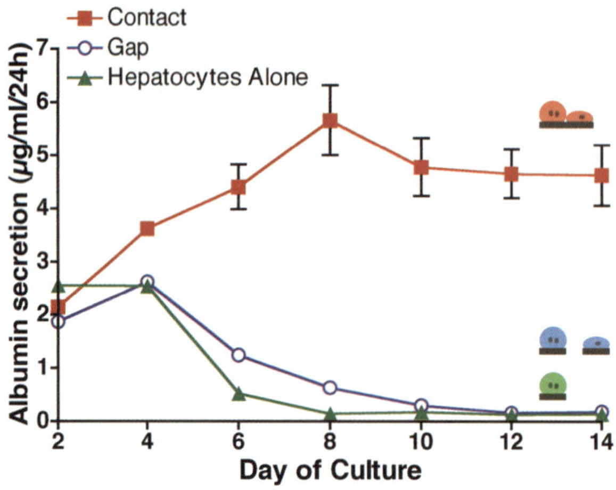


**Figure 3.4** Decoupling of contact-mediated and soluble signals. A coculture of hepatocytes and supportive stroma maintains a liver-specific phenotype (albumin secretion) only with the MRC device configured in contact mode. When the cells are separated in gap mode, function is lost in a manner similar to when the stroma is absent altogether. (Reprinted with permission from [7] © 2007 National Academy of Sciences, USA.)

in some regions than others, and an in situ assay can be used to monitor phenotypic changes as a function of separation distance (Figure 3.5). In effect, the geometry of the device and the positioning of the two cell populations allow soluble gradients to be established and studied. In addition, the separation distance in gap mode could be modified by changing the device design, if required.

An important point to note in the experimental example cited in Figure 3.5 is that certain soluble factors in this model system were not effective beyond a range of about 325 $\mu$m. This shows that there are situations in which the separation distance achieved via membrane inserts or media transfer might be too large to allow effective soluble factor signaling. The fact that the MRC device separates cell populations by only 80 $\mu$m can thus be a significant advantage.

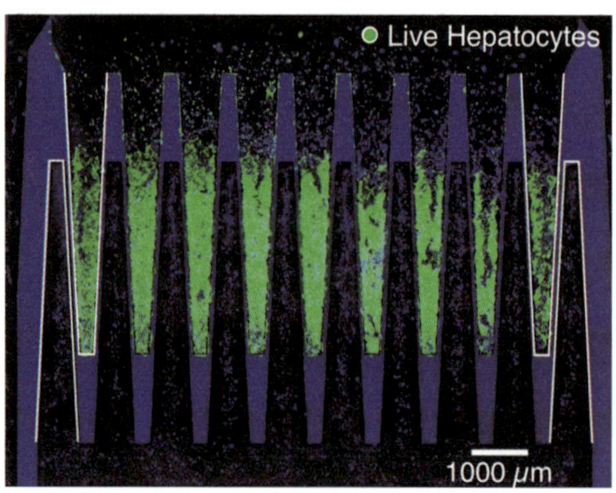


**Figure 3.5** Investigating the range of soluble signaling. Viability of hepatocytes (top row of fingers) is determined through in situ Calcein staining. It is observed that viability is maintained (2-week culture) only by the hepatocytes within close proximity (<325 $\mu$m) to supportive stroma (bottom row of fingers). (Reprinted with permission from [7] © 2007 National Academy of Sciences, USA.)

#### 3.1.4.4 Temporal control of cell-cell signaling

One of the most unique aspects of reconfigurable culture is that it allows cell-cell signaling between two populations to be dynamically modulated. For example, by switching back and forth between the contact and gap modes, contact-dependent signals can be turned on and off, while other factors such as soluble signaling and cell-substrate adhesion are held constant (Figure 3.6).

Using conventional methods, the same level of dynamic control is only possible if a particular signaling molecule has been identified and can be blocked or enhanced. With reconfigurable culture, all signaling factors that are active between two populations can be modulated, even if the factors are unknown.

#### 3.1.4.5 Deconvolution of overlapping cell processes

Reconfigurable culture allows novel manipulations of cell cultures that effectively isolate the effect of specific biological processes. For example, when two cell populations are plated together into a well, many overlapping processes occur over the first couple of hours. Cell attachment and spreading proceed concurrently with the evolution of cell-cell interactions. Using reconfigurable culture, attachment and spreading can be decoupled from the initiation of cell-cell signaling. Cells can be seeded beforehand and allowed to reach a quiescent state before being combined to form cocultures. There is thus greater certainty that measured changes arise specifically from interactions between the two populations, especially in the case of early-time-point measurements.

#### 3.1.4.6 Separation of cell populations

In reconfigurable cultures, the separation of cocultures back into their original components is trivial, simply requiring that the two parts of the device be pulled apart manu-

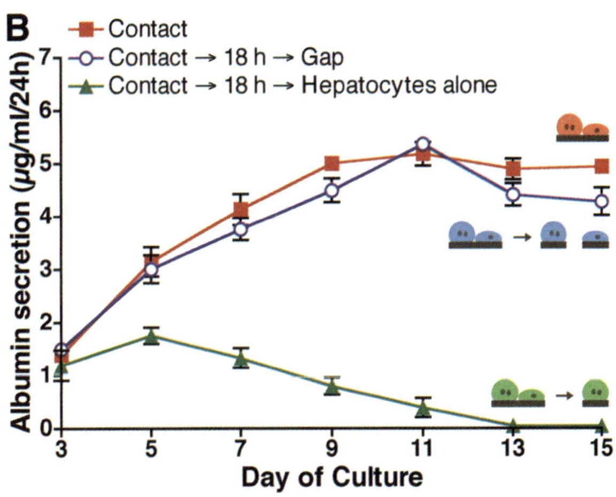


**Figure 3.6** Temporal control of cell-cell signaling. A coculture of hepatocytes and supportive stroma requires contact for only 18 hours in order to maintain function over 2 weeks at levels comparable to continuous contact. Significantly, after the initial 18 hours of contact, the support cells must remain in the culture in gap mode in order to maintain function, indicating that sustained soluble exchange is required. (Reprinted with permission from [7] © 2007 National Academy of Sciences, USA.)

ally. There are a couple of caveats, however. With highly migratory cell types, considerable cross-contamination can occur as cells cross over to adjoining fingers. In addition, as the parts are separated, the cells at the contact interface are susceptible to being pulled off of their underlying substrate by cell-cell adhesions to neighbors on the opposing comb. (This can be counteracted to some extent by chemically blocking cell-cell junctions prior to separating the combs, for example, by using a calcium chelator like EDTA.) Still, even though separation is not always perfect, reconfigurable culture offers very quick, fairly efficient segregation. Notably, separation does not depend upon any marker or label to distinguish the two populations.

#### 3.1.4.7 Population-specific stimulation and interrogation

After a mixed culture has been formed, it can be challenging to address a specific subpopulation selectively, either to introduce a treatment or to take a measurement. With reconfigurable culture, a coculture can be temporarily separated and then reassembled after stimulation or interrogation is complete (Figure 3.7).

#### 3.1.4.8 Investigation of reciprocal signaling and cell-cell crosstalk

The ability to reconfigure the composition of cultures by removing one subpopulation and replacing it with another enables a variety of manipulations that can be used to examine directionality and crosstalk in cell-cell signaling. For example, crosstalk could be investigated by conditioning population A with population B, before removing B and replacing it with C. Similarly, reciprocal signaling could be investigated by using B to condition A in contact mode, then culturing A in gap mode with B′, where B′ is the same cell type as B, except that it is naive and has not been exposed to contact with A. An example of such an experiment is shown in Figure 3.8.

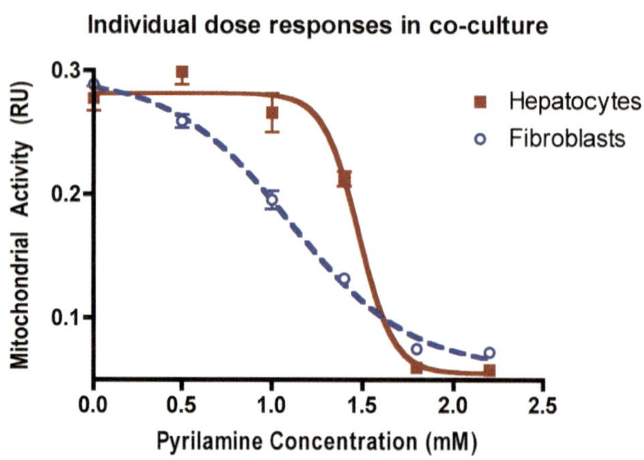


**Figure 3.7** Separability enables population-specific interrogation. Hepatocytes and supportive fibroblasts are cocultured in order to maintain liver-specific phenotype. The cocultures are exposed to a drug compound for a period before a bulk assay is utilized for precise quantitation of cell viability. Through the use of reconfigurable culture, the populations can be quickly separated immediately prior to performing the viability assay. In this case, the fibroblasts appear to be more susceptible to the compound than do the hepatocytes.

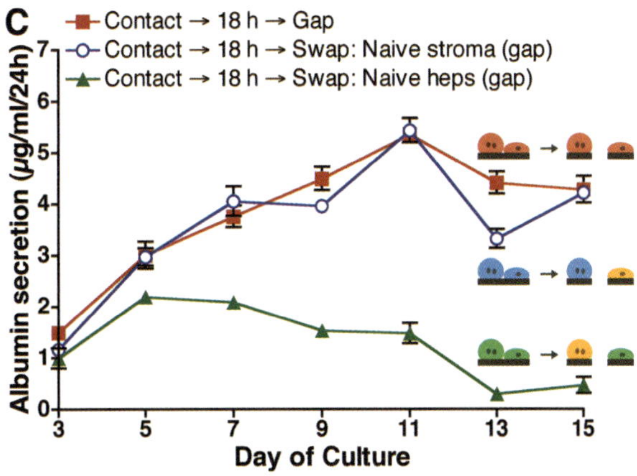


**Figure 3.8** Investigation of reciprocal signaling. Conditioned hepatocytes are cocultured with naive stroma (circles); likewise, conditioned stroma are cocultured with naive hepatocytes (triangles). It is observed that conditioned hepatocytes are required to maintain function comparable to the positive control (squares). This demonstrates that during hepatocyte-stroma contact, only stroma-to-hepatocyte signaling is important. The reciprocal signaling path from hepatocyte to stroma is shown to be noncritical for the measured phenotypic behavior. (Reprinted with permission from [7] © 2007 National Academy of Sciences, USA.)

## 3.2 Experimental Design

### 3.2.1 Experimental variables

The reconfigurable culture method is most appropriate when studying interactions between two or more different cell types. It should already be established through conventional coculture methods that a measurable effect is observed when the cell populations are combined. Reconfigurable culture can then be applied to explore this interaction in greater detail.

The possible experimental variations can be sorted into a few general categories, which are summarized below. Each experimental condition generally requires multiple replicates. For bulk assays, each replicate requires a separate MRC device. For in situ assays, often each finger can be regarded as a replicate, in which case a complete set of replicates can require only one MRC device.

- *Configuration:* the physical arrangement of the MRC device
  - *Isolation:* a single population cultured alone (It is still important to culture on the device, even though only one comb is utilized, in order to keep conditions consistent with the other configurations.)
  - *Gap:* two populations with soluble signaling only
  - *Contact:* two populations with full cell-cell interactions
- *Timing:* the duration of a certain configuration
- *History:* a specified sequence of configurations and timing, which can include removal and/or addition of individual cell populations

### 3.2.2 Readout

Cellular assays can be performed largely according to standard protocols. One limitation of the MRC system is the minimal number of cells per device. The exact number of cells at confluence will vary according to cell type, but a typical number is 40,000 cells per comb (hepatocytes), counting only the cells that sit directly on the fingers. If this is insufficient for a particular assay, multiple replicates may need to be pooled together. The comb pairs are typically cultured in 12-well plates using 750 $\mu$L of media per well. Secreted factors may thus be considerably diluted, so it is important to check that the molecules of interest will be available at detectable concentrations.

## 3.3 Materials

### 3.3.1 Reagents/supplies

- Flat-tipped metal tweezers (0S2AP-XD, SPI Supplies, West Chester, Pennsylvania)
- Molded PTFE tweezers (PFA01-AB, SPI Supplies, West Chester, Pennsylvania)
- Polystyrene (50,000 MW, Polysciences, Warrington, Pennsylvania)
- BISCO HT-6240 silicone sheets (Rogers, Carol Stream, Illinois) or Polydimethylsiloxane (Sylgard 184, Dow Corning, Midland, Michigan)
- Toluene
- Sulfuric acid
- Hydrogen peroxide
- Phosphate-buffered saline (PBS)
- Ethanol
- 12-well culture plates
- 50 mL polypropylene conical tubes
- Glass pipettes
- Glass Pasteur pipettes and rubber bulbs
- Pyrex glass dishes
- Graduated beakers
- Deionized water
- Adherent mammalian cells (i.e., epithelial cells, endothelial cells, fibroblasts)
- Appropriate culture medium for cells of interest
- Appropriate extracellular matrix (ECM) to support cells of interest
- Fluoromount-G (Southern Biotechnology, Birmingham, Alabama)
- Glass slides
- Microscope cover glass

### 3.3.2 Facilities/equipment

- *Silicon reconfigurable culture devices.* These may be obtained by contacting the authors of this chapter. Fabrication of these devices is accomplished in an advanced microfabrication facility by single-mask Bosch-process plasma etching through a thermally oxidized double-side-polished silicon wafer [7, 10]. The details of this process are beyond the scope of this chapter.
- *Upright reflecting microscope.* Compound objectives are preferred since many stereomicroscopes do not have adequate resolution for imaging cells. However, a

working distance of greater than 20 mm is preferred so that devices can be inspected in culture plates with the lid in place in order to preserve sterility. The solution is to employ ultralong working-distance objectives (Mitutoyo, Kawasaki, Japan) or a macroscope instrument (Nikon AZ100, Olympus MVX10, Leica MacroFluo). The silicon comb devices are opaque and thus generally incompatible with inverted biological microscopes.

- *Spin coater.* This is available in microfabrication facilities or from vendors such as Laurell Technologies or Specialty Coating Systems.
- *Oxygen plasma system.* This is available in microfabrication facilities or from vendors such as Technics and SPI Supplies.

## 3.4 Methods

To perform an experiment using the reconfigurable culture system, (1) the devices are coated in order to promote cell adhesion, (2) cells are seeded onto individual parts for culture, (3) the devices are assembled and reconfigured to achieve different experimental conditions, and (4) standard assays are performed. At the conclusion of an experiment, the devices may be cleaned and recoated for reuse. The parts are cleaned with a strong acid, and the surface coating is completely removed and replaced, so there should be no contamination passed from one experiment to the next.

### 3.4.1 Device handling and actuation

The silicon MRC devices may be manipulated manually with tweezers and should be handled with care. The most fragile elements are the latching spring arms, which can be rather easily broken. The device parts with arms are referred to as female; the parts which lock into the female parts are referred to as male.

The device parts may be firmly grasped in the rear by using flat-tipped metal tweezers as shown in Figure 3.9. The parts may also be grasped by the edges using larger plastic tweezers as shown in Figure 3.10(a), though the grip is less secure. It is important to hold the female parts at the rear corners and not to put pressure on the arms [Figure 3.10(b)]. Plastic PTFE tweezers should always be used when handling the parts in acid.

Place a device pair onto a flat surface before sliding together to lock into the gap or contact configurations. Most commonly, the devices are utilized inside 12-well culture plates. The bottom of each well is adequately flat, but sometimes it curves up slightly at the edges, so try to position the parts in the middle of the well before locking together in order to achieve good planar alignment. Slide the parts by using the holes at the rear of each comb. When inserting a device pair into a culture well, place the female part first and slide it back to the edge of the well in order to provide as much room as possible for inserting the male part (Figure 3.11).

One final point is very important. When disassembling a comb pair, do not simply slide the parts apart horizontally as this often results in breaking an arm. Instead, slide the parts to gap mode and then lift one part up vertically while holding the other part down (Figure 3.12). If the above procedures are followed properly, the devices can be manipulated and reused more than 20 times without significant damage or wear. In the case that one arm does get broken, the female part can still be employed in contact

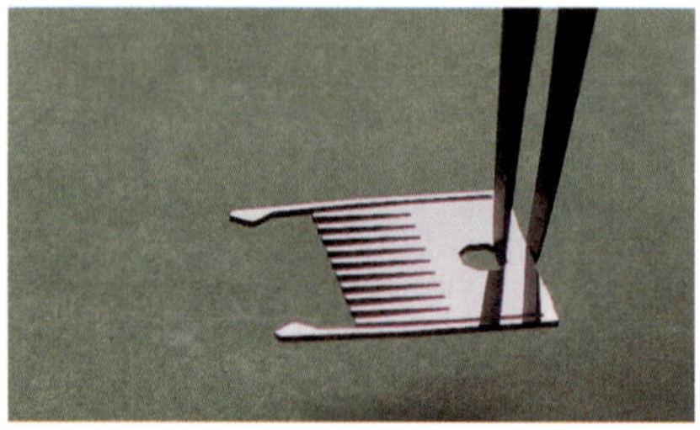

**Figure 3.9** Use flat-tipped metal tweezers to pick up and transfer device parts. Hold the parts at the rear end using the hole.

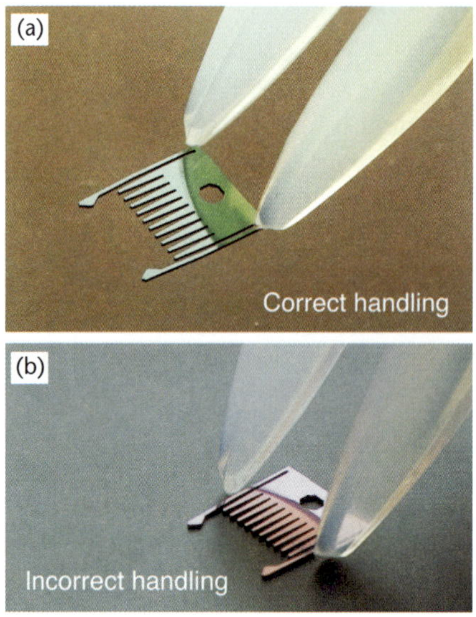


**Figure 3.10** Molded PTFE tweezers are necessary when metal tweezers are not chemically compatible with a process. (a) Grasp female parts at the edges of the rear corners (the base of the spring arms). (b) Grabbing the part directly by the spring arms is not recommended and will likely result in breakage.

mode, but gap mode is not possible. If the second arm is also broken, the female part is no longer usable.

### 3.4.2 Preparing devices for cell culture

Adhesion of cells on the untreated device surface has generally been found to be inadequate, even when the surface is coated with adsorbed ECM proteins. The best solution that has been found is to coat the device with polystyrene, followed by oxygen plasma treatment, resulting in a surface that is comparable to standard tissue-culture plastic [11, 12].

#### 3.4.2.1 Removal of cells and polystyrene

1. If adhered cells are present, remove by soaking in bleach for 10 minutes and then rinsing three times in water.

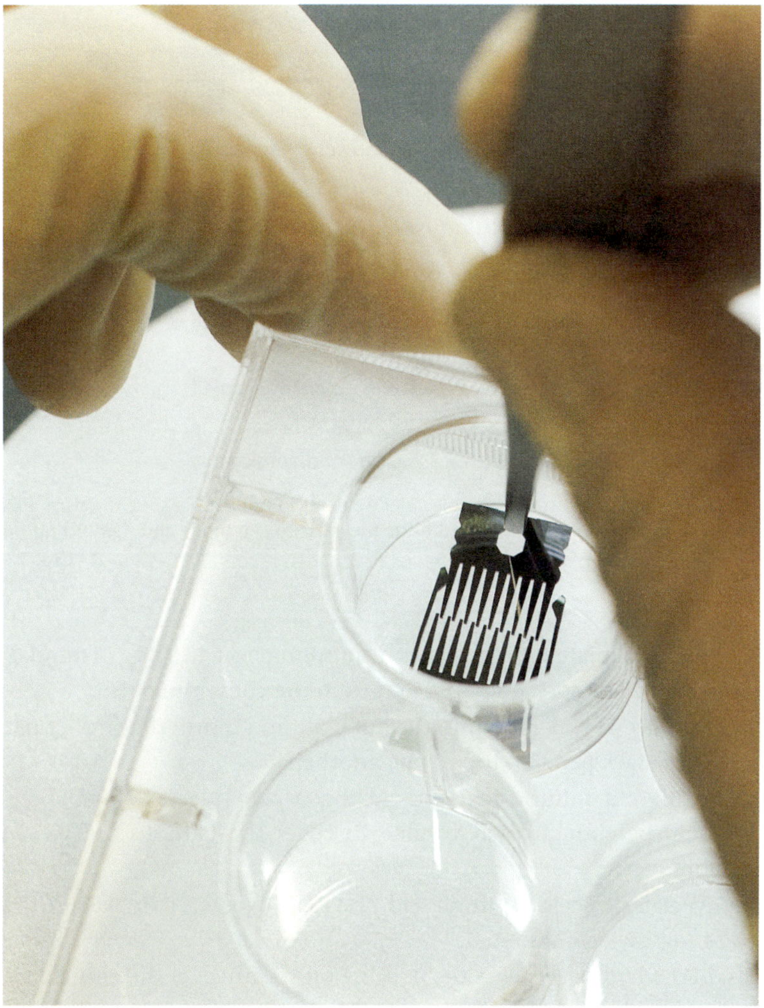

**Figure 3.11** To assemble a device pair in a 12-well plate, first insert the female part and slide it back to the edge of the well in order to maximize the space available for the male part to be inserted. In order to protect cells from being scraped off, avoid letting the male fingertips touch the top surface of the female fingers.

2. If a previous polystyrene coating is present, strip in toluene for 2 hours in a covered dish with gentle agitation on an orbital shaker. Toluene vapor is toxic and flammable, so this step must be performed in a chemical fume hood. Also, toluene and water are immiscible, so the parts must be completely dry prior to immersion in toluene. Use a glass pipette to dispense toluene because a polystyrene pipette will be dissolved by the toluene.

#### 3.4.2.2 Piranha cleaning

This step involves strongly corrosive chemicals and vapors. Wear proper protective equipment, including nitrile gloves and safety goggles, and use a chemical fume hood. Do not use metal tweezers as they will react with the acid.

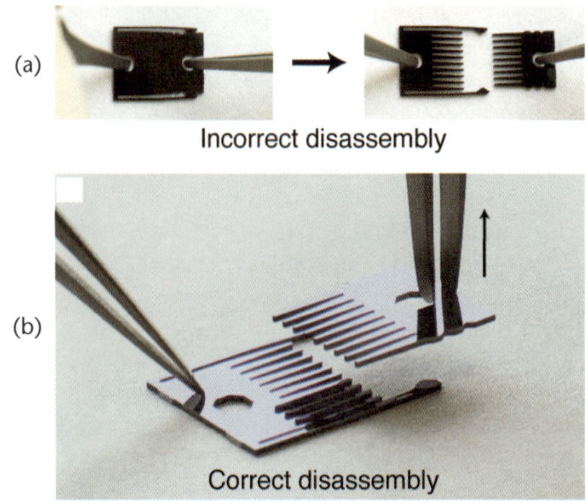


**Figure 3.12** (a) In order to disassemble a device pair, do not simply slide the parts apart as it is easy to break an arm in this fashion. (b) Instead, slide the parts to gap mode and then lift off one part vertically while holding down the other part.

1. Cover a hot plate generously with aluminum foil, keeping in mind that tiny droplets of acid will be spread up to a foot from the containing dish.
2. Place a deep Pyrex dish onto the hot plate and transfer the device parts into the dish. Do not stack parts on top of one another.
3. Measure out sulfuric acid and hydrogen peroxide at a volume ratio of 2:1. There should be enough total volume to cover the parts to a depth of at least a few millimeters.
4. Cover the devices in sulfuric acid. Use PTFE tweezers to make sure that all of the parts are completely immersed.
5. Add the hydrogen peroxide, pouring uniformly over the entire dish. Use caution as the chemicals will undergo an exothermic reaction.
6. Using the hot plate, heat the solution past 120°C. The solution will begin to bubble quite visibly when the proper temperature is reached.
7. Allow the reaction to proceed until the hydrogen peroxide is completely consumed, at which point the bubbling comes to a halt. This takes about 15 to 30 minutes.
8. Turn off the hot plate, and let the dish cool until it can be handled comfortably.
9. Use PTFE tweezers to transfer the device parts from the acid mixture into a dish of ultrapure water.
10. Rinse the parts in a continuous flow of ultrapure water for at least 10 minutes. Alternatively, transfer the parts through three sequential dishes of ultrapure water, agitating after each transfer.
11. Store the parts in a covered dish of ultrapure water until just before polystyrene coating.

#### 3.4.2.3 Polystyrene coating

For this procedure, be careful to avoid dust contamination. A clean room environment is preferred; however, this is not essential. Since toulene is a solvent for polystyrene, do not use polystyrene pipettes or labware.

1. Dissolve polystyrene in toluene at 100 mg/mL. Measure out the toluene using a glass pipette. Vortex in a 50 mL polypropylene conical tube for about 30 minutes, or until fully dissolved. Prepare 2 mL of solution for every 10 parts.
2. Air-dry the device parts and then dehydrate at 120°C using a hot plate or oven. Allow the parts to cool for 10 minutes to room temperature.
3. Prepare silicone chucks. Take a sheet of HT-6240 silicone, and use a blade to cut out circles large enough to cover the spin coater vacuum chuck. Alternatively, mix PDMS 10:1 base polymer to curing agent ratio and degas for 1 hour in a desiccator. Pour into 4" petri dishes and cure at 85°C for 2 hours.
4. Place a silicone chuck onto the vacuum chuck of the spin coater. Mount device parts onto the silicone chuck as shown in Figure 3.13.
5. Use a glass Pasteur pipette to dispense polystyrene solution generously onto each device part. Make sure that each part is completely covered, but work quickly since the silicone will begin to absorb toluene and swell. If too much time elapses before beginning the spin (about 1 minute), the parts will lose adhesion to the silicone chuck.
6. Spin at 1,800 rpm for 30 seconds.
7. Carefully remove the silicone chuck from the vacuum chuck. Carefully remove the parts from the silicone sheet. Bending back the silicone sheet can facilitate detachment. Be careful not to break the fragile arms on the female parts. Also, try not to touch the surface of the parts in the regions where you plan to culture cells as the polystyrene coating will scratch easily.
8. Place the parts on aluminum foil and bake in an oven at 65°C for 15 minutes. Increase the temperature to 95°C and bake for an additional 15 minutes. Cover the parts to protect from dust as the polystyrene will be sticky during this step.

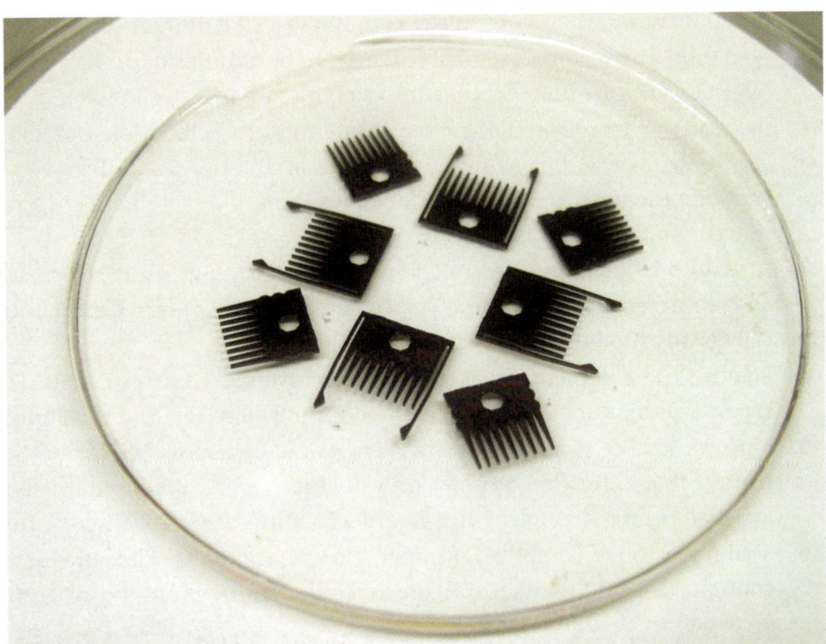

**Figure 3.13** To mount the device parts for spin-coating of polystyrene, place a fresh sheet of silicone rubber onto the spin coater vacuum chuck and then tack the parts onto the silicone.

9. After baking, some parts may be stuck to the underlying aluminum foil. Carefully peel the aluminum foil away from the parts and remove using tweezers.
10. Plasma-treat the polystyrene-coated parts for 1 minute. A range of process parameters is effective, for example, $O_2$ gas at 200 mTorr and 200W.
11. Wait for at least 3 hours after plasma treatment to allow the surface to stabilize [11].
12. At this point, the devices are ready for cell culture. They may be stored dry in this state for months before use.

### 3.4.3 Cell seeding

Once surface preparation is complete, cell cultures and assays can be performed by using largely standard techniques. For experiments requiring cell-cell contact between populations on opposing combs, it is important to obtain confluent cell monolayers that extend right to the edge of the fingers. If cells are not highly proliferative, the following steps should be followed to obtain confluent monolayers. This procedure should be performed in a biosafety cabinet following aseptic technique.

1. Load the device parts into 12-well plates. Individual male parts can fit into 24-well plates if lower fluid volume is needed.
2. If polystyrene-covered devices are being used, any desired ECM proteins may now be adsorbed. For example, primary hepatocyte culture requires collagen type 1 (adsorb 50 $\mu$g/mL at 37°C for 45 minutes), but no ECM is needed for 3T3 fibroblasts.
3. Pair every part with a matching complement, and lock the pairs into contact. This creates a continuous flat surface to facilitate uniform cell seeding. Cell adhesion on the complements is not required, so untreated parts may be used. Check the alignment of the pairs with an upright reflecting microscope.
4. Soak in 70% ethanol for a minimum of 10 minutes to sterilize. Rinse twice in ultrapure water, then once in the appropriate cell culture media.
5. Seed cells. The seeding protocol will need to be optimized for each specific cell type. Typical values are 500,000 cells/mL, 1 mL per well (12-well plate).
6. Incubate for 1 hour, shaking every 15 minutes to redistribute the unattached cells evenly. Be careful not to jostle the plates at all after the cells have settled to the surface; otherwise, seeding will not be uniform. If a plate happens to get bumped, shake it up to resuspend the cells.
7. After 1 hour, if the attached cell density is not sufficient, aspirate the cell suspension and repeat seeding with a fresh suspension for another hour. Repeat until the desired cell density has been achieved.
8. Remove the complements. Transfer the active parts to fresh wells. For many cell types, overnight incubation is required for full adhesion and spreading.
9. If necessary, a cell scraper can be used to manually remove cells from the rear of each part so that cells are confined only to the fingers. Wait until cells have firmly attached before attempting this in order to minimize inadvertent removal of cells.
10. Seeding is now complete. Arrange the device parts into the desired starting configuration and then reconfigure as required throughout the experiment. Use 750 $\mu$L of culture medium per well.
11. Due to the geometry of the device, many cell types are quite prone to detachment. Therefore, when changing media or transferring parts from well to well, it is very

important to minimize the amount of time that the cells are not covered with fluid: 1 second or less is best. For media changes, use two pipettes simultaneously; draw out fluid with one hand and immediately replace using the other hand.

### 3.4.4 Assay preparation

Samples may be mounted for high-resolution microscopy. Fix and stain following standard procedures. Lightly blot the device and place on a glass slide with cells face up. Cover the device in Fluoromount-G, and overlay with a coverslip. Store horizontally at 4°C in the dark. Once the mounting is set (anywhere from 15 minutes to overnight, depending on volume of mounting medium used), the slide can turned over and imaged on an inverted epifluorescent microscope. Fluoromount-G is water soluble, so mounting is reversible. Simply immerse the slides in water until the coverslips and devices detach (hours to days).

For experiments requiring extraction of protein, RNA, or DNA, culture the cells targeted for extraction on the male part of each device pair. Prepare a 24-well plate on ice, loading PBS into the first three rows and filling the bottom row with 150 $\mu$L per well of the appropriate reagent for lysing or extraction. After culturing to the desired time point, wash each part by dipping sequentially in three wells of PBS, blot lightly, and place into the extraction reagent, tilting the plate so that the fingers are immersed. Use a cell scraper to facilitate cell removal from the device part; if the head of the scraper does not fit into the well, simply cut it down to size using a razor blade. Complete the extraction protocol following standard procedures.

## 3.5 Discussion

To summarize, micromechanical reconfigurable culture enables clean patterning of multiple cell types and dynamic reconfiguration of these patterns. In particular, the device discussed in this chapter is designed to modulate contact between two populations while keeping constant the exchange of diffusible factors. In the noncontact configuration, the cell populations are separated by 80 $\mu$m, which is much closer proximity than is achieved using membrane inserts or conditioned media transfer. This close proximity is significant because some soluble factors have very limited effective range [7]. Further, as cultures are reconfigured, each cell population remains adhered to its own individual substrate. Populations can thus be quickly separated and recombined independent of cell adhesion dynamics.

The most intimidating aspect of this system for many life scientists seems to be the physical handling of the device. The only advice that can be offered here is practice makes perfect. New users should accept that breaking a few parts is simply a part of the learning process as one is still developing a feel for the device. At any rate, precise and careful handling is certainly required, but operation of this system is no more challenging than, say, mouse surgery.

The most common source of trouble with the MRC device has been found to be cell detachment. While protocols for cell culture and assays can be utilized more or less according to standard procedure, small adjustments may have to be made to account for the fact that cells often seem to detach from the MRC device more easily than from a

**Troubleshooting Table**

| Problem | Potential Solution |
|---|---|
| During initial seeding, cells detach at the first media change. | Vary the duration of time between cell seeding and the first media change. |
| After cell seeding, the cell monolayer is not sufficiently confluent. | Seed additional cells; it is best if supplemental seeding takes place prior to cell spreading. |
| Cells detach during media change. | Add fresh media immediately (<1 second) after removal of old media. Use both hands: one hand to remove and the other to add media. |
| During cell fixing and staining, cells detach from the edges of the comb fingers. | Keep combs parts locked in contact with complementary pair during washing steps. |
| Adjoining combs parts are not coplanar or properly aligned. | Disassemble comb pairs and reassemble into desired configuration. |
| Polystyrene coating on comb parts detaches prematurely. | Clean parts thoroughly in heated piranha solution immediately prior to coating. |

standard tissue-culture well. In general, try to be gentle during wash steps, and minimize the amount of time that parts are not covered with fluid.

Although cell patterning is perfect when device pairs are first assembled, migratory cells can cross over the interface between adjoining fingers in contact. The rate of cross-contamination varies greatly as a function of the two cell populations. The tendency to cross-migrate can be counteracted by seeding combs to full confluence, by growth-arresting the cell populations, or simply by reducing the duration of time spent in contact.

The ability to modulate interactions between different cell populations with temporal control is a powerful tool yet to be fully exploited. While early experiments have primarily been confined to liver models [7], this tool is broadly applicable to adherent cell types in general and should prove useful for advancing our understanding of intercellular communication in tissue.

## 3.6 Summary Points

- Cocultures formed using micromechanical reconfigurable culture can interact through both soluble and contact-mediated signaling factors; contact-mediated signals can be modulated independently of soluble factors. The composition of the culture can be changed dynamically, and cocultivated populations can be separated quickly.
- To perform an experiment, devices are coated in order to promote cell adhesion, cells are seeded onto individual parts for culture, devices are assembled and reconfigured to achieve different experimental conditions, and standard assays are performed. At the conclusion of an experiment, the devices may be cleaned and recoated for reuse.
- While reconfigurable culture appears to be a stark departure from traditional tools, individual cell populations encounter a culture environment almost identical to standard cell culture; hence, established culture models should adapt readily to this system.

- The silicon devices are opaque and thus generally incompatible with inverted biological microscopes. Recommended imaging alternatives are listed in Section 3.3.2. In the future, transparent devices may become available.
- It is important to practice handling and actuating the devices extensively. Train without cells, in media, in 12-well plates, and inside of a biosafety cabinet. Only when operation becomes smooth and routine should cell culture be attempted. The tweezers listed in Section 3.3.1 (part numbers included) are optimal for the handling of these devices.

## Acknowledgments

This work was supported by the National Science Foundation Faculty Early Career Development Program, National Institutes of Health–National Institute of Diabetes and Digestive and Kidney Diseases, and the David and Lucile Packard Foundation. E. E. H. was supported by a Ruth L. Kirschstein National Research Service Award.

### References

[1] Furchgott, R. F., "Role of endothelium in responses of vascular smooth muscle," *Circ. Res.*, Vol. 53, No. 5, 1983, pp. 557–573.

[2] Bhatia, S. N., et al., "Effect of cell-cell interactions in preservation of cellular phenotype: Cocultivation of hepatocytes and nonparenchymal cells," *FASEB J.*, Vol. 13, No. 14, 1999, pp. 1883–1900.

[3] Lemaigre, F., and Zaret, K. S., "Liver development update: New embryo models, cell lineage control, and morphogenesis," *Current Opinion in Genetics & Development*, Vol. 14, No. 5, 2004, pp. 582–590.

[4] Moore, K. A., and Lemischka, I. R., "Stem cells and their niches," *Science*, Vol. 311, No. 5769, 2006, pp. 1880–1885.

[5] Hanahan, D., and Weinberg, R. A., "The hallmarks of cancer," *Cell*, Vol. 100, No. 1, 2000, pp. 57–70.

[6] Sheridan, S. D., et al., "Microporous membrane growth substrates for embryonic stem cell culture and differentiation," in J. P. Mather (ed.), *Stem Cell Culture,* Methods in Cell Biology 86, Cambridge, MA: Elsevier, 2008, pp. 29–57.

[7] Hui, E. E., and Bhatia, S. N., "Micromechanical control of cell-cell interactions," *Proc. Natl. Acad. Sci. USA*, Vol. 104, No. 14, 2007, pp. 5722–5726.

[8] Liu, W. F., and Chen, C. S., "Cellular and multicellular form and function," *Adv. Drug Deliv. Rev.*, Vol. 59, No. 13, 2007, pp. 1319–1328.

[9] Falconnet, D., et al., "Surface engineering approaches to micropattern surfaces for cell-based assays," *Biomaterials*, Vol. 27, No. 16, 2006, pp. 3044–3063.

[10] Ayón, A., et al., "Characterization of a time multiplexed inductively coupled plasma etcher," *J. Electrochem. Soc.*, Vol. 146, No. 1999, p. 339.

[11] Blagovic, K., et al., "Patterned polystyrene substrates for integrated cell culture devices," *Proc. Biomed. Eng. Soc.*, St. Louis, MO, October 1–4, 2008.

[12] Chakraborty, M., Chowdhury, D., and Chattopadhyay, A., "Spin-coating of polystyrene thin films as an advanced undergraduate experiment," *J. Chem. Educ.*, Vol. 80, No. 7, 2003, pp. 806–809.

### Related sources

A video demonstration of the methods described in this chapter can be found in the *Journal of Visualized Experiments* at www.jove.com/index/Details.stp?ID=268.

CHAPTER

# 4

# Mechanotransduction and the Study of Cellular Forces

Michael T. Yang and Christopher S. Chen

Department of Bioengineering, University of Pennsylvania, Philadelphia, PA 19104

## Abstract

Cells exert traction forces on their surrounding extracellular matrix (ECM). These forces play a critical role in regulating processes such as proliferation, differentiation, migration, and apoptosis. Traction forces arise from interactions between myosin motor proteins and the actin cytoskeleton, which is anchored to the ECM through cell-matrix contacts known as focal adhesions. In this chapter, we provide an overview of tools for measuring forces at these adhesions. Furthermore, we describe one of these tools, the microfabricated post array sensor (mPADs), in practical detail, followed by a discussion of its applications and limitations.

Key terms

microfabrication
MEMS
soft lithography
traction forces
mechanotransduction

## 4.1 Introduction

### 4.1.1 Cellular forces: Functions and underlying mechanisms

In addition to the obvious use of muscle contraction to generate forces, a close examination of numerous cells and tissues reveals a fundamental role for intracellularly generated forces in mediating many functions of nonmuscle cells. In a cell undergoing cytokinesis, a contractile ring pinches the cytoplasm and cell membrane, spawning two daughter cells [1]. During gastrulation, embryonic cells self-sort into the three germ layers based on differential levels of intercellular adhesion and cell-cortex tension [2]. In healing wounds, fibroblasts and endothelial cells exert traction forces against the extracellular matrix (ECM) as they crawl into the area of the wound. Once at the wound site, these cells deposit new collagen matrix and form new blood vessels, respectively. After the provisional tissue has been formed, keratinocytes migrate over the wound to create a barrier to the environment, while myofibroblasts contract the wound [3].

Although intracellular forces regulate such diverse biological processes in nonmuscle cells, the same proteins implicated in muscle contraction, actin and myosin, are responsible for generating these forces. Traction forces, in particular, have garnered intense interest due to a wide body of evidence that implicates these forces in mechanotransduction, or the conversion of such forces into biochemical signals associated with regulating behaviors from cell proliferation to apoptosis to differentiation [4–8]. Adherent cells exert traction forces on the ECM through adhesion plaques that form at cell-ECM contacts, commonly termed focal adhesions (FAs) (Figure 4.1) [9, 10]. Biochemical analysis of these FAs has revealed not only numerous signaling proteins but also many scaffolding proteins that anchor the actin cytoskeleton to the adhesion sites [11–14]. In culture, cells form actin-myosin stress fibers that are seen to terminate at these FAs [15, 16].

A number of studies have indirectly probed the role of traction forces in cellular function. Perturbations of signaling proteins that regulate myosin activity, such as Rho GTPases and myosin light chain kinases, have shown that stimulating contractility leads to stress fiber and FA formation and that inhibiting contractility yields the opposite

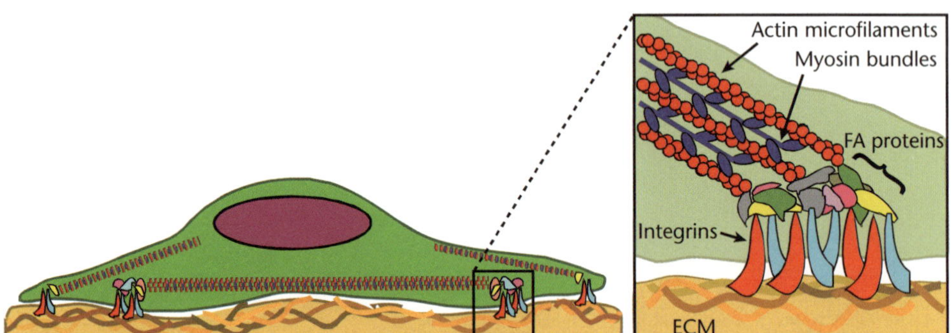


**Figure 4.1** Forces at focal adhesions. Cells adhere to the extracellular matrix through specialized structures known as focal adhesions. These structures form in a tension-dependent process. First, integrins bind to the ECM and cluster together, leading to the recruitment of cytoplasmic focal adhesion proteins, many of which are able to bind actin microfilaments. Nascent focal adhesions, known as focal complexes, are small and transient. However, contraction of the actin microfilaments by myosin produces force at the focal adhesions, leading to growth and stability of the adhesions. On a deformable substrate, these intracellular forces can be visualized as traction forces.

result [16]. Local application of external forces to FAs using bound beads or pipettes elicited increased adhesion assembly and strengthening [17, 18]. Importantly, these same manipulations of the contractility system also have been found to modulate rates of cell proliferation, differentiation of progenitor cells to different lineage, and many other cellular functions [19–21]. While these observations reinforce the role of traction forces in mechanotransduction, tools needed to be developed to measure these forces directly in adherent cells.

### 4.1.2 Techniques for studying traction forces

In the 1980s, Harris pioneered the first approach to observing cellular traction forces with silicone film substrata. These deformable substrates are essentially ultrathin silicone films that have been crosslinked on top of a viscous silicone fluid [22–25]. The silicone films are so soft that cells cultured on them are able to induce wrinkles that propagate in an accordion-like pattern underneath their centroid [Figure 4.2(a)]. Such wrinkles were the first visual confirmation of traction forces in cells. However, silicone films have many drawbacks. Not only is it difficult to accurately quantify forces from wrinkles, but silicone films are notoriously difficult to handle for biochemical and immunofluorescent assays with cells.

Use of polyacrylamide (PAA) gels to measure cellular forces was a marked improvement over silicone film substrata. These versatile gels can be functionalized with adhesive ligands and their stiffness can be precisely tuned by varying the concentration of crosslinker used [30]. PAA gels are prestressed on rigid surfaces such that wrinkling does not occur. However, cells cultured on these substrates are still able to generate in-plane deformations. To visualize these deformations and measure traction forces, Pelham and Wang developed traction force microscopy (TFM) [Figure 4.2(b)]. In TFM, fluorescent beads are embedded near the surface of PAA gels to sample the displacement field of the gel during deformation [31]. Cells are observed to deform the PAA gel principally in the

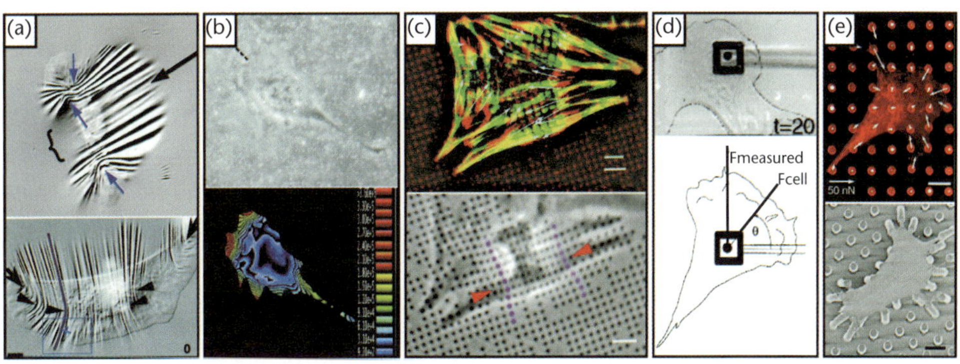


**Figure 4.2** Techniques to study cellular traction forces. (a) Contraction of thin silicone films by cells induces microscopic wrinkles in the films. The curvature and length of the wrinkles provide a qualitative measure of the direction and magnitude of force. (Reprinted with permission from [24].) (b, c) Deformation of nonwrinkling elastic substrates by cells can be visualized from the displacement of randomly embedded fluorescent beads or patterned fluorescent dots. The displacement field provided by these markers is used to calculate a map of traction stresses across the entire cell body. (Reprinted with permission from [26, 27].) (d) Deflection of a horizontal cantilever measures the force generated at a single site underneath a cell. (Reprinted with permission from [28] © 1997 National Academy of Sciences, USA.) (e) Deflection of vertical microposts in an array measures forces generated at multiple sites across a cell. (Reprinted with permission from [29] © 2003 National Academy of Sciences, USA.)

plane of the surface, with maximum bead displacements of approximately 1 $\mu$m [32]. Because small deformations can be tracked with the beads, one can treat the PAA gel as an incompressible, linearly elastic material of semi-infinite thickness (bead displacements are small compared to gel thickness) and thereby relate the traction field $F(r)$ to the displacement field $u(r)$ in the following integral transform:

$$u_i(r) = \int dr' G_{ij}(r - r') F_j(r') \tag{4.1}$$

where $G(r - r')$ is the tensorial Green's function, which provides the displacement at any point $r$ on the substrate induced by a traction at point $r'$. Given $u(r)$, which is measured from the movements of the fluorescent markers, and $G(r - r')$, which can be derived from Boussinesq's theory for an elastic half-space, (4.1) must be inverted to solve for $F(r)$. This is an ill-posed problem requiring regularization schemes to achieve stable, unique solutions. Such schemes include restricting traction forces to sites of adhesion and placing boundaries on the deformation field of the cell [32–34].

TFM has grown in sophistication due to a number of computational and experimental advances. The earliest algorithm, known as the boundary element method, integrates (4.1) over a computational mesh with fixed boundary conditions [32]. This method is computationally intensive, requiring the construction and inversion of very large matrices. Fourier-transform traction cytometry vastly improves computational speed by transforming the displacement field into Fourier space, multiplying it with the inverse Green's function, and then transforming the product back into real space to obtain the traction field [33]. Despite these advances, the major challenge of TFM has been to develop methods to more accurately estimate the displacement field from the experimental data. Several imaging methods such as particle-image velocimetry, particle-tracking velocimetry, and correlation-based particle-tracking velocimetry have been devised to extract this information from the particle displacements [35, 36]. Recently, two differently colored nanobeads have been embedded in PAA gels to provide a denser and more precise sampling of the displacement field [35].

A similar approach to TFM employs microfabricated bead markers to regularize the placement of the points being tracked, giving a better measurement of the displacement field. Balaban et al. photopatterned equally spaced photoresist markers on a wafer and embedded these markers on the surface of a PDMS substrate [26]. The use of a regular pattern not only permits easier visualization of substrate distortions but leads to more uniform force measurement [Figure 4.2(c)]. This approach has been used to measure traction forces at known positions of focal adhesions by culturing cells transfected with fluorescently tagged adhesion proteins.

Other researchers have departed from the thin gel systems entirely, using microfabrication approaches to create deformable beams. Galbraith and Sheetz utilized microfabrication to etch horizontal cantilevers out of silicon [28]. Cells cultured on these silicon devices are able to attach to adhesive pads on the cantilevers and exert traction forces, which in turn causes these cantilevers to bend in plane [Figure 4.2(d)]. However, horizontal cantilevers can only measure force along one direction and only at a few locations on the device.

To address these limitations, we developed the microfabricated post array detectors (mPADs) [29]. This substrate consists of an array of closely spaced posts that act as vertical cantilevers. Cells cultured on this array are able to spread across the tips of multiple posts, which are coated with adhesive ECM proteins. Contractile forces generated by these cells are then exerted on the underlying posts, causing them to deflect [Figure 4.2(e)]. For small deflections, these posts can be treated as linearly elastic beams subjected to a pure bending force at a point moment. The applied force $F$ is directly proportional to the post deflection $x$ as described by (4.2).

$$F = \left(\frac{3EI}{L^3}\right)x \tag{4.2}$$

In this equation, $E$ is the elastic modulus, $I$ is the area moment of inertia, and $L$ is the height of the post. This elegant method not only simplifies calculation of forces but allows the control of substrate stiffness through manipulations in post geometry rather than substrate chemistry. In this chapter, we provide a detailed protocol for using the mPADs system to measure traction forces. Sections 4.2 and 4.3 describe the materials and methods for fabricating the mPADs masters and using mPADs substrates for cell culture and traction force measurement. Section 4.4 discusses the applications and pitfalls of the mPADs as well as highlights biological insights gained from using this tool. Lastly, Section 4.5 provides tips for troubleshooting and summarizes important points for using the mPADs effectively.

## 4.2 Materials

### 4.2.1 Reagents and supplies

- 3″ to 6″ diameter *n*-type <100> silicon wafers (Silicon Quest International, Santa Clara, California)
- SPR700-10 photoresist (Rohm and Haas Company, Philadelphia, Pennsylvania)
- MF-701 developer solution (Rohm and Haas Company, Philadelphia, Pennsylvania)
- (tridecafluoro-1,1,2,2-tetrahydrooctyl)-1-trichlorosilane (United Chemical Technologies, Bristol, Pennsylvania)
- Polydimethylsiloxane (Sylgard 184, Dow Corning, Midland, Michigan)
- Aluminum weighing dishes (Heathrow Scientific, Vernon Hills, Illinois)
- Human fibronectin (BD Biosciences, San Jose, California)
- DiI (chemical name: 1,1′dioleyl-3,3,3′,3′-tetramethylindocarbocyanine methane-sulfonate) (Invitrogen, Carlsbad, California)
- Pluronics F127 (BASF, Ludwigshafen, Germany)
- Mammalian cells (i.e., fibroblasts, endothelial cells)
- Fluoromount-G (Electron Microscopy Sciences, Hatfield, Pennsylvania)
- Primary and secondary antibodies for immunofluorescent staining
- Paraformaldehyde
- Triton X-100
- Phosphate-buffered saline
- Ethanol
- Deionized water

- Petri dishes
- Microscope cover glass
- Glass slides

### 4.2.2 Facilities, equipment, and software

- AutoCAD (Autodesk, San Rafael, California) or L-Edit (Tanner Research Inc., Monrovia, California).
- MATLAB (Mathworks, Natick, Massachusetts).
- Automatic or manual spin coater: These are commonly found in academic microfabrication facilities but can also be purchased from vendors such as Laurell Technologies Corporation or Specialty Coating Systems.
- Wafer stepper (projection mask aligner): We used a Nikon NSR2005i9 (Nikon Precision Inc., Belmont, California) for projection photolithography. This instrument is usually found in advanced microfabrication facilities. If access to a stepper is not possible, a contact mask aligner will suffice (see Section 4.3.1.1).
- Deep reactive ion etching (DRIE) tool: We used an ST Systems Multiple ICP tool (Surface Technology Systems, Newport, United Kingdom). This instrument is usually found in advanced microfabrication facilities. Access can also be obtained through a foundry service such as MEMS Exchange.
- Scanning electron microscope: This can be found in electron microscopy core facilities and some microfabrication facilities.
- Critical point drier (Tousimis Research Corporation, Rockville, Maryland).
- Plasma cleaner (SPI Supplies, West Chester, Pennsylvania).
- UV-ozone cleaner (Jelight, Irvine, California).
- Epifluorescent microscope such as a Zeiss Axiovert 200M or Nikon TE-2000.

## 4.3 Methods

### 4.3.1 Microfabrication of micropost arrays

The mPADs are devices on which arrays of elastomeric posts have been cast out of the silicone rubber polydimethylsiloxane (PDMS). In order to use the mPADs to measure cellular forces, these arrays must be closely spaced and have excellent uniformity among the posts. To fabricate such a device, it is necessary to employ tools developed in the semiconductor industry for fabricating integrated circuits (ICs) and microelectromechanical systems (MEMS). First, a master template of the micropost array must be produced. The microfabrication processes used to generate this master are highly dependent on the intended biological application of the force sensor. In this section we will discuss the different options for master fabrication and describe in detail the Silicon-LIGA process for generating a high-resolution mPADs master (Figure 4.3). After creating a master, the second step is to replicate it in PDMS to create the final device.

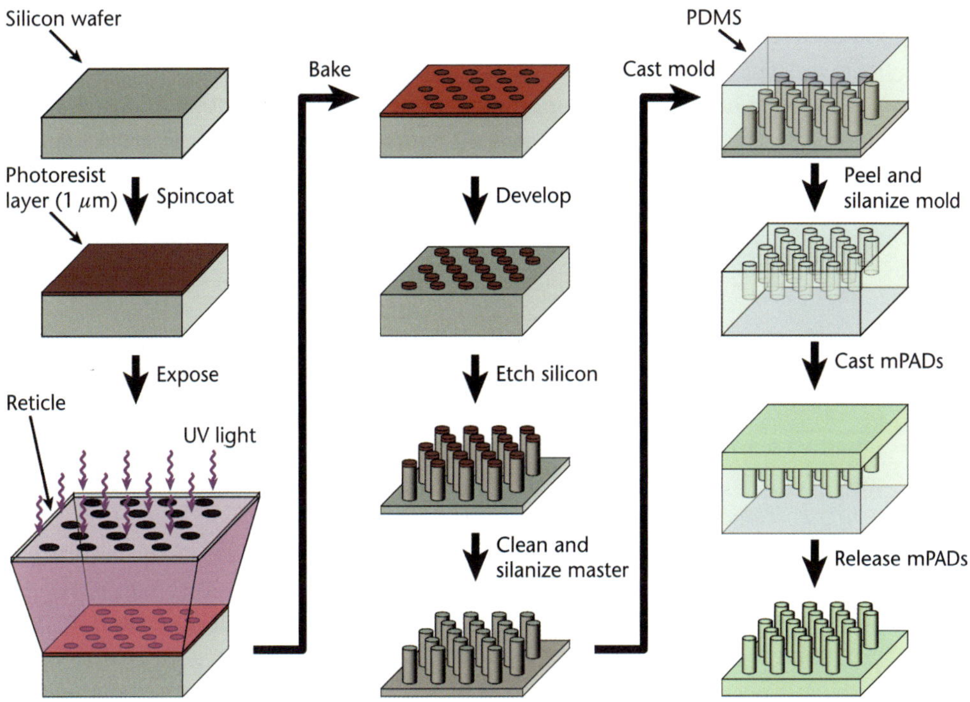


**Figure 4.3** Process flow for fabricating mPADs. The master is first patterned with projection photolithography and then etched with deep reactive ion etching. Double casting the master with PDMS yields the mPADs substrate.

#### 4.3.1.1 Rationale for different master-fabrication processes

One must consider three critical design parameters when fabricating the mPADs master. These are the center-to-center spacing, diameter, and height of the posts. The first parameter determines the array density and therefore affects the spatial resolution of reported traction forces. In our experience, more closely spaced arrays are better suited to measure forces in small cells, such as epithelial cells. However, farther-spaced arrays may be sufficient for large cells such as mesenchymal stem cells or even monolayers of cells. The second parameter is affected by the first parameter. As center-to-center spacings are decreased, so must post diameter be decreased to ensure that there is a sufficient gap between neighboring posts for deflections. Lastly, post height, along with diameter, determines the spring constant of the post. Once again, the intended biological application must be considered as some cell types are more contractile than others [37].

Choosing geometric parameters for the mPADs is important as, generally speaking, the smaller the features, the more costly they are to fabricate. The most cost-effective method is to fabricate mPADs masters with a process called UV-LIGA. In this process, high-aspect ratio microstructures are fabricated out of a photosensitive polymer, also known as photoresist. These structures are patterned using photolithography, a technique developed by the IC industry to generate nano- to microscale features with extremely high fidelity. The basic premise of photolithography is to use a photomask, which consists of opaque and transparent regions, to selectively allow ultraviolet (UV) light to pass through, and react with, photoresist that has been coated on a flat substrate such as a silicon wafer. Depending on the tone of photoresist, UV light can render exposed regions either soluble (positive tone) or insoluble (negative tone) in a developer solution. The most commonly used photoresist for UV-LIGA is SU-8 [38].

UV-LIGA has the lowest resolution of the different methods we discuss here. Contact photolithography, where the photomask is pressed against the photoresist-coated wafer, is used to pattern SU-8 and has a theoretical resolution $R$ of

$$R = \frac{3}{2}\sqrt{\frac{\lambda z}{2}} \tag{4.3}$$

where $\lambda$ is wavelength of light, and $z$ is the photoresist thickness. SU-8 maximally absorbs UV light with a wavelength of 365 nm. Thus, the thickness of the photoresist determines the minimum feature size. In practice, the theoretical resolution is rarely obtained due to defects such as uneven photoresist coating and mask damage. We have used UV-LIGA to generate masters consisting of SU-8 posts with diameters, heights, and center-to-center spacings ranging from 3 to 4 $\mu$m, 7 to 12 $\mu$m, and 6 to 9 $\mu$m, respectively. The protocol for fabricating these *standard-resolution* mPADs masters has been described in detail elsewhere [39]. Moreover, by modifying the SU-8 protocol with a contrast-enhancement step, others have fabricated posts with diameters as small as 2 $\mu$m [40]. UV-LIGA is cost-effective in that for equipment only a spin coater and a contact mask aligner are required, both of which are readily available in even the most basic microfabrication facilities.

However, to fabricate higher-resolution mPADs, another process known as Silicon-LIGA is used. Here, the master consists of posts that have been etched out of silicon. To create these structures, a thin layer of photoresist must first be patterned on the wafer. Projection photolithography, an advanced form of photolithography, possesses a number of advantages over contact photolithography for this step. In projection photolithography, UV light passes through a special photomask called a reticle, and the resultant image is focused through an objective lens onto a photoresist-coated silicon wafer. In practice, projection photolithography can generate patterns with features as small as 0.5 $\mu$m. Some types of projection photolithography systems, known as steppers, are able to expose small sections of the wafer at a time and therefore produce repeats of the reticle pattern across the entire wafer. This is particularly useful for patterns with periodic features, such as the mPADs. Moreover, steppers are often equipped with a reduction lens that scales down the reticle pattern 5 or 10 times when projecting an image onto the wafer. Thus, to pattern 0.5 $\mu$m features requires a reticle with a minimum feature size of 2.5 $\mu$m. Taken together, these equipment capabilities allow us to decrease the pattern density and increase the minimum feature sizes on the reticle, which translates into lower reticle production costs.

The second step in Silicon-LIGA is to etch the bare regions of silicon on a patterned wafer. Deep reactive ion etching (DRIE) is an attractive solution for fabricating silicon microstructures with very high aspect ratios [41, 42]. The operating principle of this technique, known as the Bosch process, is to iteratively passivate and etch the wafer to achieve a very directional etch [43]. During the passivation step, the wafer is uniformly coated with a chemically inert polymer, typically $C_4F_8$, which acts as a protective layer. In the etching step, the polymer is removed by a nearly isotropic plasma etch. For silicon, the gas source for plasma etching is usually $SF_6$. An inductively coupled plasma (ICP) source creates plasmas with high electron density, low pressure, and low energy.

Such plasmas achieve high etch rates and produce very directional etching. By cycling between passivation and etching, lateral etching is minimized and high-aspect-ratio structures can be fabricated. DRIE etches silicon about 100 times more quickly than photoresist. Thus, to etch short depths (i.e., less than 20 $\mu$m), as is desired for the mPADs masters, an approximately 1 $\mu$m layer of photoresist is sufficient as a masking layer.

DRIE presents a number of challenges unique to the application of the final device. Despite being a very directional process, minor lateral etching occurs during each etch cycle, resulting in scalloping along the sidewalls. In the case of the mPADs masters, large scallops along the sidewalls of silicon posts or holes can significantly affect the stiffness of PDMS posts cast from these structures [44]. Another common challenge is aspect-ratio-dependent etching or microloading [45]. Etch rate is diffusion limited and inversely correlated with aspect ratio. As a result, small, high-aspect-ratio structures, such as those found on high-resolution mPADs masters, etch more slowly than larger structures. However, these problems can be overcome by optimizing the gas flow rates, cycle times, and ICP power settings. Moreover, careful design of the patterns on the reticle can ensure that differential etch rates do not affect the performance of the final device.

We have used Silicon-LIGA to fabricate arrays of silicon posts with diameters, heights, and center-to-center spacings ranging from 0.67 to 2 $\mu$m, 3 to 12 $\mu$m, and 1.5 to 4.5 $\mu$m, respectively. Moreover, similar devices composed of holes instead of posts have been fabricated with the same process [46, 47]. The main disadvantage of Silicon-LIGA is that it requires access to advanced microfabrication facilities and is therefore more costly. However, it is much less expensive than X-ray LIGA, which is the gold standard in building high-aspect-ratio microstructures [48]. Other tools, such as electron-beam lithography and focused ion beam, are capable of making *nanofabricated* post arrays with nanometer dimensions. However, not only are nanoposts nearly impossible to visualize, but they are impractical as these tools have to serially write each feature in a pattern. For further reading, several excellent resources describe microfabrication processes in greater detail [49–51].

#### 4.3.1.2 Reticle design

The first step in fabricating the mPADs master is to design the reticle. Reticles are fused quartz plates on which a thin chromium layer has been patterned according to the specifications in the design. Reticle design is carried out with computer-aided design (CAD) software, such as AutoCAD or L-Edit. The mPADs pattern depends on the user's desired level of resolution. In practice, we have designed arrays of circles with diameters of 0.75, 1, 1.5, and 2 $\mu$m, with center-to-center spacings of 1.5, 2, 2.5, 3, 4, and 4.5 $\mu$m. We and others have found that a center-to-center spacing twice as large as the post diameter offers excellent spatial resolution [47]. To maximize post density in the arrays, a hexagonal lattice should be used (as opposed to a square lattice), which shortens the spacing between diagonal posts. Moreover, to factor in the reduction lens equipped on steppers, the features on the reticle should be 5 or 10 times larger than they would be on the wafer. A useful feature of advanced steppers is electronic shutters, which can limit UV exposure to user-defined sections of a reticle. This permits the design of multiple arrays with different dimensions on the same reticle. However, when using a reticle containing multiple patterns, it is ill advised to pattern arrays with different dimensions on the same wafer. As previously described, this can lead to nonuniform etch depths across different arrays due to microloading effects. To facilitate microcontact printing on the mPADs, as described in Section 4.3.2.1, large flat structures should be fabricated adjacent to the arrays. These flat structures support the weight of stamps used to transfer fibronectin onto the easily deformed posts. We recommend dividing the mPADs array into multiple 2 × 2 mm arrays separated by rectangular flat structures. The last design criterion to address is whether to make the circles opaque or transparent. To fabricate silicon *posts* using *positive tone* photoresist, the circles should form the opaque regions on the reticle. Once the reticle design is complete, it can be outsourced to a commercial mask-writing service for reticle production (Microtronics, Newton, Pennsylvania).

#### 4.3.1.3 Projection photolithography of micropost arrays

Once the reticle has been fabricated, the next step is to coat wafers with photoresist for projection photolithography. To minimize contamination from wafer handling, an automated spin coater should be used to coat a batch of test grade, *n*-type <100> silicon wafers with a 1-$\mu$m-thick layer of SPR700-10 photoresist. These systems have preprogrammed routines for generating resist coatings, so the parameters described here may need to be optimized for different spin coaters. The wafer is dehydrated at 175°C for 30 seconds, briefly cooled, and then transferred to a chamber where photoresist is dispensed onto its surface. The wafer is held in place by a vacuum chuck and spun at 4,600 rpm for 30 seconds, leaving a uniform coating 1 $\mu$m in thickness. The wafer is then transferred to a hot plate, where it is soft-baked at 95°C for 60 seconds. This baking process evaporates any residual solvent in the photoresist and promotes its adhesion to the wafer.

Once the wafers have been coated, they are loaded into the wafer stepper for UV exposure. This instrument is also completely automated. The only parameters input by the user are the Cartesian coordinates on the reticle containing the pattern of interest, the step size for patterning each section of the wafer, and the exposure time. Moreover, the stepper has test routines for optimizing the exposure time. After all the wafers have been exposed, they are transferred back to the automated spin coater, which doubles as an automated developer. First, the wafers are postexposure-baked at 115°C for 60 seconds, briefly cooled, and then transferred to a developer chamber, where MF-701 developer solution is dispensed. The wafer is spun during development to remove the solubilized resist and then rinsed with deionized (DI) water. After drying, the wafers are fully patterned and ready to be etched.

#### 4.3.1.4 Deep reactive ion etching of silicon microposts

An ST Systems Multiple ICP tool is used to etch silicon wafers. A wafer is placed into the loading chamber, after which it is loaded into the plasma chamber and etched according to a user-defined recipe. Process parameters for passivation (85 sccm $C_4F_8$, 7s duration, 0W platen, 600W coil, 16 mTorr) and etching (130 sccm $SF_6$, 9s duration, 12W platen, 600W coil, 33 mTorr) have been developed but may need to be adjusted for different machines. To achieve the desired etch depth, the etch rate must be characterized. This is done by etching a test wafer using a known number of cycles, dicing the wafer in half, and imaging the cross-sectional profile with a scanning electron microscope (SEM).

#### 4.3.1.5 Device packaging

After DRIE, the mPADs master is fully fabricated. However, dicing a finished wafer with a die saw can produce smaller chip-sized dies, which are much easier to handle during the mPADs casting process. Before this is done, the wafer should be coated with a thick layer of photoresist to protect the surface from debris during the dicing process. Individual dies can then be cleaned in a strongly oxidizing solution, such as piranha solution, to remove the photoresist and any organic residue. Special care should be taken when handling piranha as it can be explosive. To facilitate casting with PDMS in subsequent steps, the mPADs master should be silanized with (tridecafluoro-1,1,2,2-tetrahydrooctyl)-1-trichlorosilane. This is done by placing the mPADs master in a desiccator with a few drops of silane placed on a glass slide. Evacuating the desiccator overnight then allows the silane to vaporize and covalently bind to the silicon dioxide surface of the wafer. Handle trichlorosilanes with extreme care as they readily react with water vapor to generate toxic hydrochloric gas.

#### 4.3.1.6 Soft lithography of micropost array substrates

PDMS is an optically transparent, biologically inert, silicon-based polymer whose applications in micromolding are well-known [52]. To cast PDMS replicas of an mPADs master with silicon posts, we use a double-casting procedure. In the first casting, a negative mold (i.e., holes) of the master is generated. This mold is then used for a second casting to create a positive replica of the mPADs. To generate the negative molds, PDMS is mixed at a 10:1 base-polymer-to-curing-agent ratio and degassed for 1 hour in a desiccator. The mPADs master is placed in an aluminum weighing dish and covered with an approximately 1 cm layer of PDMS. The dish is then heated at 110°C for 10 to 15 minutes to polymerize the PDMS into an elastic solid. After allowing the aluminum dish to cool to room temperature, the PDMS can be peeled away from the master to yield a negative mold. This casting process can be repeated many times to create a large batch of negative molds.

In order to cast from the negative molds, they must be treated with the same fluorosilane used to passivate the master in order to ensure that PDMS in the second casting will not permanently bond with the molds. First, the molds are placed in a plasma cleaner for 1.5 minutes to oxidize their surfaces. Next, the molds are placed in a desiccator with a few drops of fluorosilane, and the chamber is evacuated overnight. To do the second casting, PDMS is mixed and degassed as previously described and applied as a thin layer to the negative molds. Precleaned glass coverslips are then placed on top of these molds to provide the substrates with rigid support. These molds are then cured at 110°C for 20 hours to fully crosslink the PDMS. Peeling away the molds then yields the final mPADs substrate.

#### 4.3.1.7 Characterization of micropost spring constant

As previously described, for small deflections, a post can be modeled as a slender beam undergoing pure bending with minimal shearing. For a cylindrical post, (4.2) can be reexpressed as

$$F = \left( \frac{3\pi ED^4}{64L^3} \right) x \tag{4.4}$$

where $D$ is the diameter, and all other variables are as previously described. In this equation, $3\pi ED^4/64L^3$ represents the spring constant $K$ of the post and is assumed to be conserved across a uniform array. Thus, to accurately calculate $K$, it is critical to obtain accurate measurements for the diameter and height. These can be obtained from top-down and cross-sectional SEM images of the mPADs. If the diameter varies along the length of the post due to scalloping or suboptimal etching parameters, $K$ can still be calculated via numerical integration in a software package such as MATLAB. Moreover, $K$ can also be determined through experimental calibration with a calibrated glass microneedle. This technique has been described elsewhere [39]. In practice, experimental values of $K$ agree closely with theoretical values of $K$ [29, 40, 46]. Therefore, for higher-resolution mPADs where the posts are very small and difficult to visualize during experimental calibration, the theoretical spring constant alone is sufficient.

### 4.3.2 Analysis of traction forces with micropost arrays

The mPADs must undergo a series of surface treatments before cells can be cultured on the substrates and their traction forces measured (Figure 4.4). Most importantly, cell adhesion must be restricted to only the tips of the posts. This is done by microcontact-printing ECM onto the tips of the posts and then treating the sidewalls and base with a nonadhesive surfactant. Furthermore, labeling the posts with a fluorescent dye facilitates measurement of the traction forces from the post deflections. Cells are then cultured on the micropost arrays, after which they can be fixed and stained for

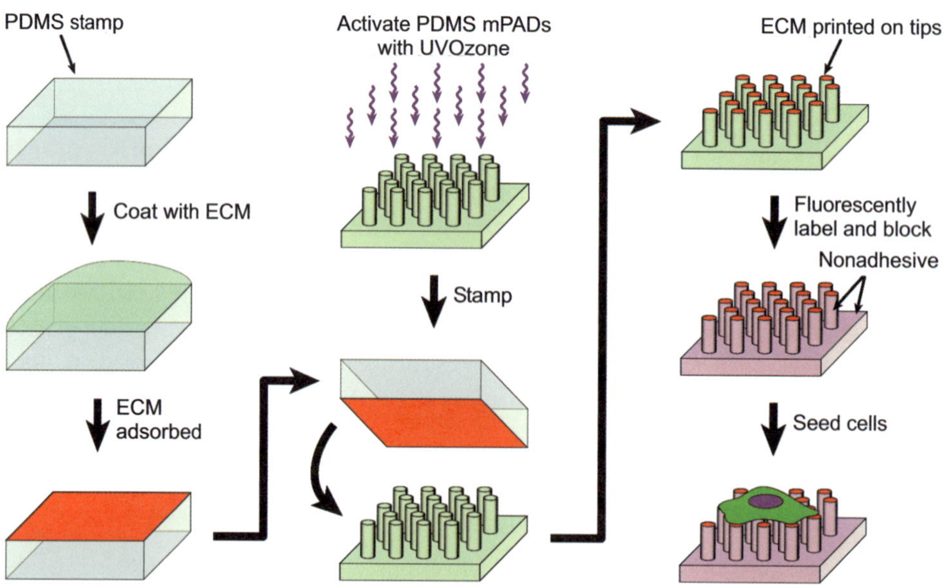


**Figure 4.4** Preparation of the mPADs for seeding cells. PDMS stamps are inked with ECM proteins that absorb out of solution. The mPADs are treated with UV-ozone to render the surface hydrophilic compared to untreated PDMS. Placing the coated stamp in contact with the mPADs transfers the ECM to the tips of the posts. Subsequent washing and incubation steps fluorescently label the posts and render the unprinted surfaces nonadhesive. The final step is to seed cells on the mPADs.

fluorescent microscopy. Images of posts under deflection are analyzed in MATLAB to determine traction forces exerted by cells.

#### 4.3.2.1 Substrate preparation

In this protocol, we describe microcontact printing with flat PDMS stamps, which transfers ECM over the entire array. First, PDMS stamps are generated by preparing a 30:1 base-polymer-to-curing-agent mixture, degassing it for 1 hour, pouring it into a 150 mm petri dish, and finally curing the PDMS at 60°C for at least 2 hours. Stamps can then be cut out of the petri dish at approximately the same size as the mPADs substrate. In a biosafety cabinet, human fibronectin is diluted in DI water to a final concentration of 50 $\mu$g/mL. The flat surfaces of the stamps are then covered with 50 to 100 $\mu$L of fibronectin solution for 1 hour to allow the protein to adsorb. Excess fibronectin is rinsed off in DI water, and the stamps are dried with nitrogen. The mPADs are rendered hydrophilic with UV-ozone treatment for 7 minutes in a UV-ozone cleaner. The flat stamps are then placed on top of the mPADs and pressed gently to transfer fibronectin from the hydrophobic PDMS stamp to the hydrophilic posts.

After microcontact-printing fibronectin on the posts, the mPADs undergo a sequence of sterilizing and washing steps. First, the mPADs are submerged in 100% ethanol to wet the posts and then sterilized in 70% ethanol. The substrates are then washed in successive dishes of DI water to remove the ethanol. The mPADs are then labeled with a fluorescent lipophilic dye (DiI, 1,1'dioleyl-3,3,3',3'-tetramethylindocarbocyanine methanesulfonate) at 5 $\mu$g/mL for 1 hour. Next, the mPADs are washed in successive dishes of DI water to remove the excess DiI and then immersed in 0.2% Pluronics F127 for 30 minutes to prevent protein adsorption and cell adhesion to the regions that were not stamped with fibronectin.

After the substrates are washed in successive dishes of phosphate buffered saline (PBS) to remove excess Pluronics, the mPADs are transferred to a standard tissue-culture dish filled with culture medium. Cells are then seeded at 1,000 to 10,000 cells/cm$^2$ using normal tissue-culture procedures [53]. If too many cells have been seeded, then the substrates can be transferred to fresh dishes of media after monitoring cell attachment. Cells are typically allowed to attach and spread on the mPADs for at least 10 hours before experiments are performed.

#### 4.3.2.2 Staining and microscopy of micropost arrays

Before microscopic images of the mPADs for analysis are acquired, the cells can be fixed and stained to label specific structures such as actin filaments and focal adhesions. Substrates are typically fixed with 4% paraformaldehyde in PBS, permeabilized with 0.2% Triton X-100 in PBS, and stained with antibodies or other fluorescent reagents. The substrates are mounted on glass slides and overlaid with a glass coverslip prior to image acquisition.

To analyze the traction forces for a cell, two fluorescence images are required: an image of the top of the posts and an image of the base of the posts. Though not required, a third image of the cell is helpful to identify which posts are being deflected by the cell. High-magnification oil-immersion objectives (25×, 40×, or 60×) are used for collecting these images, with the choice of objective depending on the size of the cell being imaged.

#### 4.3.2.3 Image analysis of micropost deflections

Traction force analysis is performed with an image-analysis algorithm developed with the MATLAB image-processing toolbox [37]. This algorithm imports the acquired top and base images and performs localized thresholding to determine the centroids of the fluorescent posts at both planes. The difference in positions (in pixels) between the top and base images is converted to traction forces by multiplying the displacements (in micrometers) by the spring constant (in nano-Newtons per micrometer). Moreover, vector plots can be generated from the displacements to visualize the distribution of subcellular forces. Another useful measure is strain energy, which is obtained by multiplying the squared displacements by one-half $K$. Our MATLAB code is available upon request.

## 4.4 Discussion

### 4.4.1 Applications and enhancements of the micropost arrays

#### 4.4.1.1 Micropatterning single cells and multicellular aggregates

Changes in cell shape and structure appear to have dramatic effects on many cellular functions, such as proliferation, migration, differentiation, and apoptosis, and have been investigated extensively with flat substrates on which ECM has been micropatterned to constrain cell shape [19, 20, 54, 55]. Traction forces appear to play a critical role in transducing these changes in cell shape into biological effects. Cell adhesion on the mPADs can be confined to a subset of ECM-coated posts by using patterned stamps with raised microstructures to transfer fibronectin. These stamps are cast from photolithographically patterned wafers that contain different geometric features (i.e., squares, circles, triangles, teardrops). Using this technique, increasing the degree of cell spreading increased the average traction force per post [Figure 4.5(a)] [29]. This increase in traction force in turn was found to be involved in regulating cell behaviors (e.g., proliferation and differentiation) [20, 56]. Another study, in which large, multicellular aggregates were patterned on the mPADs using stamps with mesoscale features, revealed that the contractile activity of individual cells can be transmitted across cell-to-cell adhesions, leading to increased traction forces at the edges (breaks) within a monolayer [Figure 4.5(b)] [46, 57]. Interestingly, this emergent pattern of regions of high mechanical stress, as dictated by the shapes of the multicellular structures, leads to patterns of proliferation localized to these regions of stress.

#### 4.4.1.2 Magnetic actuation of microposts to investigate localized force application

Several studies have shown that traction forces and externally applied forces can separately elicit focal adhesion assembly and reinforcement [17, 18, 26]. To examine the interplay between these two sources of force in regulating cellular function, we modified the mPADs to apply forces to individual focal adhesions and to sense traction forces simultaneously. The mPADs are well suited for this application as each post deflects

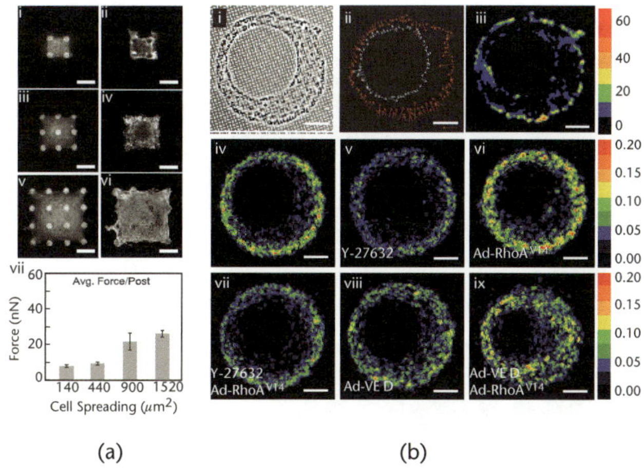


**Figure 4.5** Micropatterning cells on the mPADs. (a) Single cells are patterned on (i, ii) 2 × 2 posts, (iii, iv) 3 × 3 posts, and (v, vi) 4 × 4 posts. The average force per post correlates positively with cell area (vii). (Reprinted with permission from [29] © 2003 National Academy of Sciences, USA.) (b) (i) Monolayers of cells are cultured in an asymmetric annular pattern. (ii, iii) The forces are asymmetrically distributed along the inner and outer edges of the annulus-shaped monolayer. (iv) A gradient of proliferation is observed at the outer edge of these monolayers. (v) Inhibiting actomyosin contractility with the drug Y-27632 significantly reduces the gradient of proliferation. (vi, vii) Upregulating contractility with constitutively active RhoA enhances the gradient, although this effect is abrogated with simultaneous addition of the contractility inhibitor. (viii, ix) The use of a dominant negative mutant of vascular endothelial cadherin (VEΔ) decouples the transmission of forces between cells, inducing uniform rather than patterned proliferation. (Reprinted with permission from [57] © 2005 National Academy of Sciences, USA.)

independently of its neighbor. In the modified mPADs substrate, called the mag-mPADs, magnetic nanowires are sparsely seeded into the mPADs negative molds during substrate casting [58]. The result is an mPADs substrate in which a subset of "magnetic" posts contains embedded nanowires [Figure 4.6(a)]. Applying a uniform horizontal magnetic field imparts a torque on these nanowires, causing the magnetic posts to deflect [Figure 4.6(b–d)]. By using a sparse density of nanowires, cells will probabilistically attach to only one magnetic post among many nonmagnetic posts, thereby allowing the measurement of global changes in traction force in response to localized force application. With the mag-mPADs, it was observed that locally applied force induces only local FA growth, while neighboring adhesions on nonactuated posts are unchanged [Figure 4.6(e–h)]. However, the effect of locally applied force on traction forces is more complex, with discrete sites along the cell periphery experiencing either rapid or gradual losses in force in response to the applied force. Thus, it appears that a complex interaction exists between applied forces and cell-generated forces.

### 4.4.2 Potential pitfalls of micropost arrays

#### 4.4.2.1 Accuracy of the slender beam approximation

The linear approximation of a post deflection assumes that cells exert force at a point moment at the top of a post. However, it is possible that cells apply different types of force, such as shear, axial loading, torsion, and point moments. Among these different forces, only shear and point moments affect post bending. To examine which type of

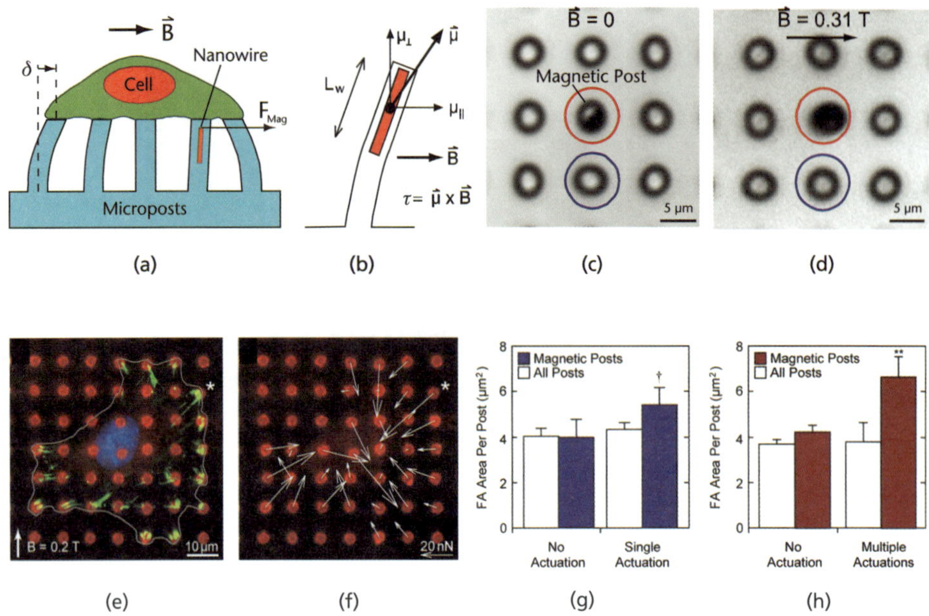


**Figure 4.6** Magnetic actuation of individual posts in the micropost array. (a) External force $F_{mag}$ is applied to an adherent cell through magnetic posts containing cobalt nanowires that bend in the presence of a magnetic field. (b) The magnetic torque, $\tau$, on a magnetic post depends on the applied field, $\vec{\beta}$, the nanowire length, $L_W$, and the dipole moment, $\vec{\mu}$. (c, d) Phase-contrast images of a magnetic post under no field and a 0.31T field. (e) Representative immunofluorescent image of FAs (green), posts (red), and nucleus (blue) after force actuation. The direction and magnitude of the field are displayed. The cell is outlined, and the location of the magnetic post is marked by the asterisk. (f) Vectors (white arrows), indicating the direction and magnitude of force, are plotted for each post for the same cell. The magnetic post is marked by the asterisk. (g) FAs associated with magnetic posts are slightly larger after a single application of force compared to FAs at nonmagnetic posts. (h) Repeated application of force enhances growth of FAs associated with magnetic posts. (Reprinted with permission from [58] © 2007 National Academy of Sciences, USA.)

force is predominantly applied to posts by cells, a 3-D reconstruction of a deflected post was compared to theoretical calculations of post deflection, assuming either a shear load or a point moment at the tip [Figure 4.7(a)] [37]. Contrary to our approximation, the deflection profile of the post closely follows the predicted bending pattern of a post subjected to a shear load at its tip [Figure 4.7(b)]. However, the shear component does not significantly alter the spring constant until the height-to-diameter ratio becomes sufficiently small. For example, for a post with a height-to-diameter ratio of 3:1, the spring constant, assuming shear, is only 7% smaller than the spring constant neglecting shear.

Another potential drawback of the mPADs is that forces generated by strongly contractile cells may deflect posts beyond the range where the linear approximation is valid. To investigate the accuracy of this approximation for a range of post deflections, a finite element model (FEM) of a PDMS post was constructed [Figure 4.7(c)] [37]. Compared to the FEM results, the linear approximation slightly underestimates the force for a given deflection. Moreover, the approximation is accurate to within 10% for forces exceeding 100 nN, typically the upper limit of forces measured at FAs [Figure 4.7(d)]. Ideally, a batch of mPADs masters covering a range of post heights will provide the sensitivity to measure forces across all cell types.

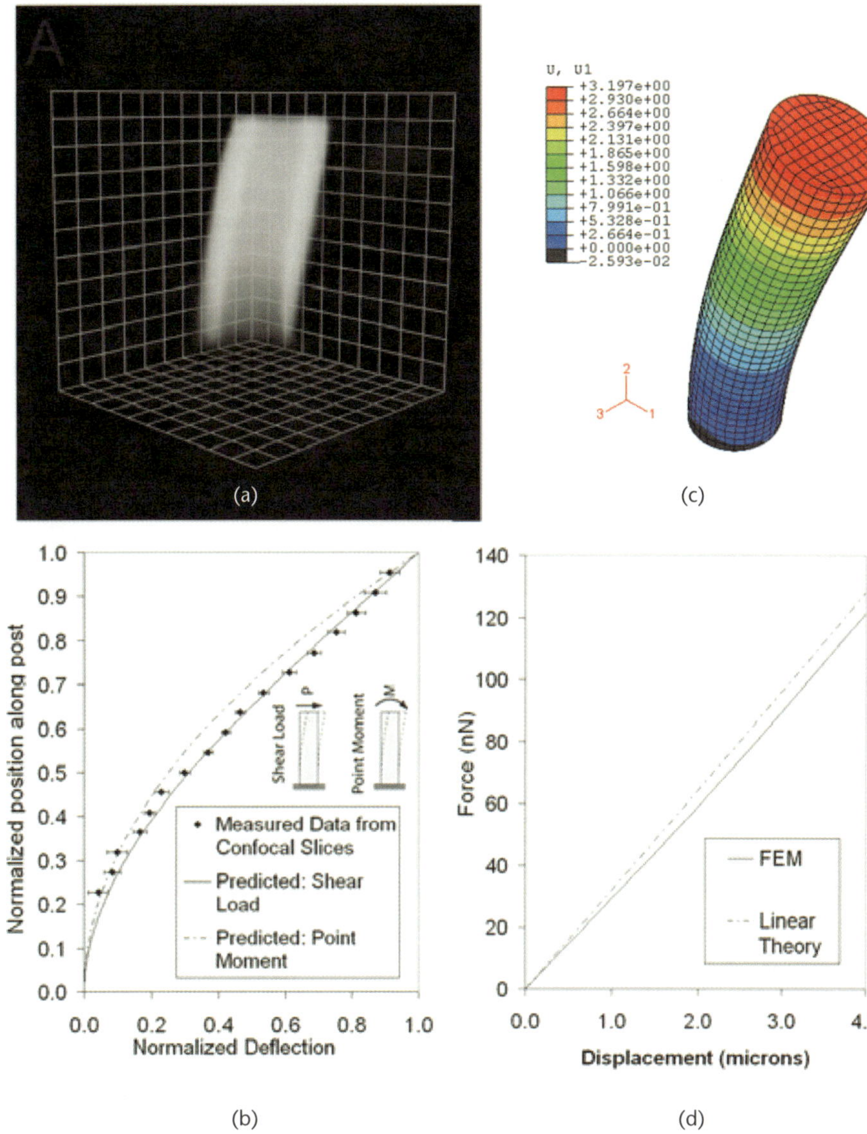


**Figure 4.7** Accuracy of the linear approximation for determining traction forces. (a) A volocity 3-D reconstruction of a deflected post ($D = 3\ \mu$m, $L = 11\ \mu$m) with confocal image slices taken at 0.5 $\mu$m intervals. (b) Normalized deflection versus normalized length along a post for two cases: deflection under shear load (solid) and deflection due to a point moment (dashed). The measured centroid positions from the confocal images are shown as black diamonds. (c) An FEM model of a shear force applied to the top of a post ($D = 3\ \mu$m, $L = 11\ \mu$m). (d) Applied force as a function of deflection for linear theory (dashed) and FEM analysis (solid). (Adapted from [37].)

#### 4.4.2.2 Scaling down micropost array geometry

A major criticism of the mPADs is that the discrete surface topology of the arrays presents an unnatural environment for culturing cells. However, this raises the question of what is considered a natural environment. In the in vivo microenvironment, cells are exposed to gradients of soluble cytokines and surrounded by ECM with heterogeneous composition, mechanics, and 3-D microstructure. None of the in vitro systems in which cells are laid down on an artificial surface can reproduce that degree of complexity. As such, the mPADs arrays are in no way inferior to more conventional continuous substrates for studying cellular behavior. In fact, cells cultured on *standard-resolution* mPADs form similar stress fibers and FAs as those on rigid substrates, and they generate traction forces comparable to those measured on PAA gels [37]. One aspect, however, in which the mPADs could be improved upon is the spatial resolution of force measurement. On the *standard-resolution* mPADs, the large spacing between posts limits the number of attachments cells can form. Decreasing post spacing to increase post density is the obvious solution to improving the spatial resolution of traction force measurement. However, altering micropost array geometry can have unpredictable effects on cell behavior.

To investigate these concerns, we cultured cells on six different mPADs geometries, with post diameters from 0.67 to 1.3 $\mu$m and spacings from 1.5 to 4.5 $\mu$m [Figure 4.8(a–f)] [59]. Cells cultured on posts spaced less than 4.5 $\mu$m apart display similar morphologies to those found on flat surfaces, suggesting that decreasing the spacing between posts beyond some optimal spacing does not yield significant improvements in cell spreading [Figure 4.8(g, h)]. Moreover, we observed no statistical difference in total strain energy per cell on the different array geometries [Figure 4.8(i)]. Thus, strain energy per post correlates inversely with post density [Figure 4.8(j)]. In our studies, we did not use a wide range of post diameters to examine the role of adhesive area per post on traction forces. However, based on data that point to a positive correlation between FA size and traction force, it is likely that smaller-diameter posts will experience lower forces [26, 29]. Due to the potentially confounding effects of micropost array geometry, it may be difficult to reconcile traction force measurements generated on mPADs with different geometries. Moreover, because PAA gels do not restrict areas of adhesion, it is also unlikely that traction forces from mPADs and PAA gels of "equivalent" stiffness can be directly compared [60].

### 4.4.3 Biological insights from using micropost arrays

In our group and others, the mPADs have been used to investigate a wide range of biological problems. In a study comparing the functions of two isoforms of nonmuscle myosin II, denoted NMM-IIA and NMM-IIB, Sheetz et al. observed that 60% and 30% of the traction force generated by cells were contributed by NMM-IIA and NMM-IIB, respectively [61]. They further demonstrated that depletion of NMM-IIA, but not NMM-IIB, dramatically decreased retrograde F-actin flow and cell-spreading rate. Moreover, the distributions of NMM-IIA and NMM-IIB were generally concentrated within different regions of a cell. Taken together, these findings provide insight into how different isoforms of nonmuscle myosin II generate contractile force.

Pirone et al. examined the role of focal adhesion kinase (FAK), a cytoplasmic signaling protein recruited to FAs, in proliferation [56]. Different mutants of FAK were introduced into endothelial cells to upregulate and downregulate different functions of FAK. They observed that cells expressing a C-terminal fragment of FAK (FRNK), generated greater traction forces than control cells expressing GFP. In contrast, there was no difference between control cells and cells expressing wild-type FAK. Interestingly, cells expressing a FAK mutant lacking a certain phosphotyrosine residue (Y397F) generated

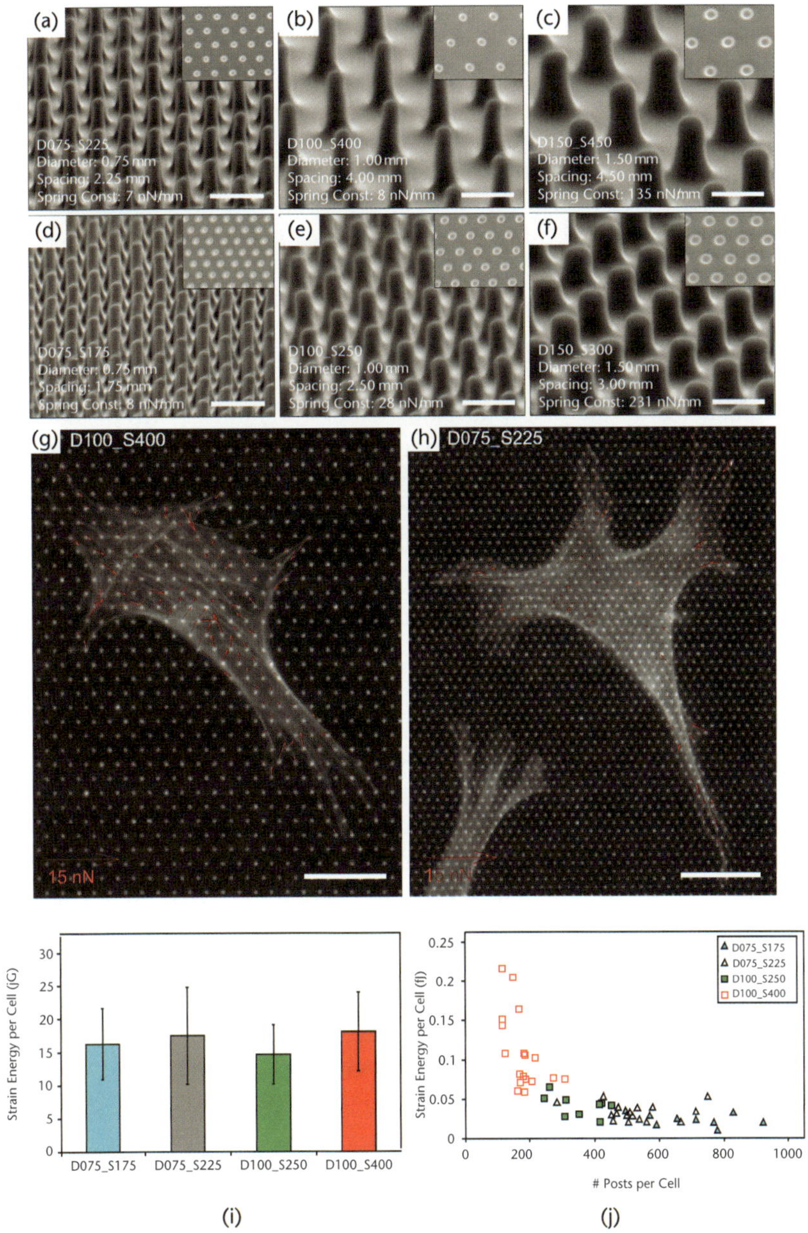


**Figure 4.8** Scaling down the mPADs: (a–f) Scanning electron micrographs of six mPADs with different geometries, with top-down images inset. Post diameter, spacing, and spring constant are shown. Scale bars are 3 $\mu$m. (g, h) Representative immunofluorescent images of cells on two of the mPADs geometries. Scale bars are 20 $\mu$m. (i) Plot of the strain energy per cell for four different mPADs geometries. (j) Plot of the strain energy per post (per cell) as a function of the number of posts attached to each cell. (Reprinted with permission from [59].)

lower forces than control cells. These observations, along with proliferation studies, support a novel role for FAK in growth control in which loss of FAK signaling increases Rho-mediated cytoskeletal tension, which results in a loss of adhesion-dependent control of cell proliferation.

In another approach to investigating contractility, Deshpande et al. used the mPADs to frame a theoretical model of contractility that accounts for the dynamic reorganization of the cytoskeleton [62]. In this model, cells comprise representative volume elements containing a fine network of actin filaments. Initiating an activation signal (i.e., a soluble contractility agonist) then triggers the formation of stress fibers in these elements. In each element, stress fibers can form in any direction in a tension-dependent manner. As a proof of principle, the model was applied to square cells attached to four posts at the corners. By varying the post stiffness, the model predicted that cells on stiffer posts would generate stress fibers with greater intensity and, concomitantly, larger forces. Moreover, in square cells, stress fibers originated from each corner post and were oriented toward the center of the cell, thus demonstrating the effect of cell shape and boundary conditions on the development of structural anisotropy. As experimental validations with the mPADs are done, future refinements to this model can be made, such as coupling stress fiber formation with focal adhesion formation.

### 4.4.4 Future innovations for studying cellular forces

Tools for investigating cellular traction forces can be taken in many directions. The most obvious direction taken thus far has been to combine different techniques to examine the response of intracellular force to external force. Integration of magnetic nanowires into the mPADs was the first example of a hybrid device for simultaneously applying and measuring forces at FAs. Recently, a second such device was developed to measure traction forces in cells subjected to global, equibiaxial stretch [63]. Another technique combines magnetic twisting cytometry with traction force microscopy [64]. This technique differs from the other two in that external forces are applied through ECM-coated beads attached to the apical surface of cells. Incorporating fluorescence microscopy techniques such as fluorescence recovery after photobleaching (FRAP) and fluorescence resonance energy transfer (FRET) can add another layer of complexity by allowing one to correlate protein signaling and dynamics with traction forces [65–67]. Perhaps the most ambitious direction is to develop tools to measure cellular forces in a 3-D setting, which would be a more physiologically relevant environment for most cell types. The ultimate goal would be to translate the knowledge gained from studying traction forces into a useful application, such as engineering tissue scaffolds with precise mechanical properties to induce differentiation of progenitor cells into the correct cell type or developing treatments for diseases that arise from abnormal mechanotransduction.

## 4.5 Summary Points

- Proper design of the mPADs arrays is critical as mPADs substrates are cast repeatedly from a master template. As the spread area and contractility of cell types of interest are often not known a priori, the best strategy for optimizing the geometric parameters for the post array is first to decide on the center-to-center post spacing and use a post diameter that is half this spacing. This will influence which type of fabrication process is required. From a technical standpoint, posts with diameters less than 1 $\mu$m are extremely difficult to see under high magnification. The post height

**Troubleshooting Table**

| Problem | Explanation | Potential Solution |
|---|---|---|
| mPADs substrates have significant post collapse after peeling the mold off. | Posts stick to each other during peeling. This is particularly problematic for taller posts. | Sonicate mPADs in ethanol for 1 minute. Then, dry substrates in a critical point drier. |
| mPADs negative molds are difficult to peel off the mPADs substrates or numerous posts are missing upon peeling. | The molds are insufficiently silanized or have been used too many times. After several uses, posts will gradually be ripped off and remain in the molds. | Plasma-clean the mold for 1.5 minutes to activate the surface prior to silanization. UV-ozone treatment is not sufficient for activation. Discard mold after a maximum of four uses. |
| Stamping the mPADs results in significant post collapse. | Excessive pressure was applied. | Apply gentle pressure (i.e., less than the weight of a pair of tweezers). |
| | If the posts are very soft, the act of peeling off the stamp will impart forces that crush the posts. | Before peeling off the stamp, submerge the mPADs with stamp in ethanol. Let the ethanol diffuse in between the posts. Then, peel off the stamp in ethanol. |
| Cells do not attach or spread well on microposts. | Cells were trypsinized too long or otherwise damaged. | Minimize trypsinization time. Do not excessively shear cells when breaking up clusters. |
| | Poor contact between the PDMS stamp and mPADs can result in incomplete fibronectin transfer. | Apply gentle pressure when microcontact printing. |
| mPADs substrates have significant post collapse after transfer to DI water and PBS during the washing steps. | In the steps before Pluronics treatment, the mPADs are hydrophobic, causing water to dewet on the surface. The surface tension of water droplets causes posts to collapse, particularly on the perimeter of the array. | During the washing steps, transfer the mPADs quickly (i.e., in less than 2 seconds) between successive dishes of DI water or PBS. mPADs tend to float if they are simply dropped on the surface of the liquid. It helps to use deep petri dishes with a large volume of liquid so that mPADs can be submerged easily. |
| Cells have crawled down between the posts. Cells that do this appear to be conforming to the lattice, forming thin protrusions that bend at 60°, 90°, or 120° angles, depending on whether the lattice is rectangular or hexagonal. | If cells are on the mPADs for a long-term experiment, the Pluronics blocking may have degraded or desorbed. | Pluronics adsorption on the mPADs for long experiments has not been investigated. Designing arrays with closer post spacing may prevent cells from crawling down the posts. |
| DiI has accumulated inside of the cell, causing extensive interference when taking images of post deflections. | DiI is normally sequestered in the PDMS, but it can incorporate into the cell membrane at the cell-matrix interface and, from there, be trafficked to vesicles inside the cell. DiI membrane staining is especially severe for long-term experiments (i.e., culturing cells on mPADs for several days to weeks). | DiI concentration can be reduced to 1 $\mu$g/mL and still be visible by fluorescence. Washing the mPADs in DI water or PBS after the Pluronics step for several hours or overnight has been observed to reduce cell-membrane staining.<br>Alternatively, fluorescently labeled bovine serum albumin (0.5 $\mu$g/mL) can be used to label the posts. The concentration is sufficiently low to ensure Pluronics adsorption. |

is therefore the only parameter that must be varied to find the optimal stiffness for a particular cell type.

- mPADs substrate and stamp flatness are key to measuring traction forces accurately. When casting mPADs, applying sufficient pressure to squeeze out excess PDMS between the glass and the mPADs mold will result in a level substrate. This will ensure that all posts are imaged in the correct focal plane at high magnification. Using ultraflat PDMS stamps cast on petri dishes or even silanized silicon wafers will ensure excellent contact with the tips of the posts during microcontact printing.

- Optimizing the cell-seeding conditions can dramatically improve the yield of "usable" cells on an mPADs substrate. For single-cell studies, low seeding densities (500 to 2,000 cells/$cm^2$) can reduce the formation of cell aggregates. In contrast, higher seeding densities may be desired for studies of cell aggregates and monolayers. For cell types that adhere very quickly, it is helpful to seed the cells on mPADs in petri dishes rather than irradiated tissue-culture dishes. This prevents too many cells from adhering to the dish before they can adhere to the mPADs.

## References


[1] Glotzer, M., "Animal cell cytokinesis," *Annu. Rev. Cell. Dev. Biol.*, Vol. 17, No., 2001, pp. 351–386.

[2] Krieg, M., et al., "Tensile forces govern germ-layer organization in zebrafish," *Nat. Cell. Biol.*, Vol. 10, No. 4, 2008, pp. 429–436.

[3] Martin, P., "Wound healing—aiming for perfect skin regeneration," *Science*, Vol. 276, No. 5309, 1997, pp. 75–81.

[4] Katsumi, A., et al., "Integrins in mechanotransduction," *J. Biol. Chem.*, Vol. 279, No. 13, 2004, pp. 12001–12004.

[5] Orr, A. W., et al., "Mechanisms of mechanotransduction," *Dev. Cell*, Vol. 10, No. 1, 2006, pp. 11–20.

[6] Davies, P. F., "Flow-mediated endothelial mechanotransduction," *Physiol. Rev.*, Vol. 75, No. 3, 1995, pp. 519–560.

[7] Chen, C. S., Tan, J., and Tien, J., "Mechanotransduction at cell-matrix and cell-cell contacts," *Annu. Rev. Biomed. Eng.*, Vol. 6, No., 2004, pp. 275–302.

[8] Lauffenburger, D. A., and Horwitz, A. F., "Cell migration: A physically integrated molecular process," *Cell*, Vol. 84, No. 3, 1996, pp. 359–369.

[9] Burridge, K., and Chrzanowska-Wodnicka, M., "Focal adhesions, contractility, and signaling," *Annu. Rev. Cell. Dev. Biol.*, Vol. 12, No., 1996, pp. 463–518.

[10] Geiger, B., and Bershadsky, A., "Assembly and mechanosensory function of focal contacts," *Curr. Opin. Cell. Biol.*, Vol. 13, No. 5, 2001, pp. 584–592.

[11] DePasquale, J. A., and Izzard, C. S., "Evidence for an actin-containing cytoplasmic precursor of the focal contact and the timing of incorporation of vinculin at the focal contact," *J. Cell. Biol.*, Vol. 105, No. 6, Pt 1, 1987, pp. 2803–2809.

[12] DePasquale, J. A., and Izzard, C. S., "Accumulation of talin in nodes at the edge of the lamellipodium and separate incorporation into adhesion plaques at focal contacts in fibroblasts," *J. Cell. Biol.*, Vol. 113, No. 6, 1991, pp. 1351–1359.

[13] Pavalko, F. M., and Burridge, K., "Disruption of the actin cytoskeleton after microinjection of proteolytic fragments of alpha-actinin," *J. Cell. Biol.*, Vol. 114, No. 3, 1991, pp. 481–491.

[14] Lo, S. H., et al., "Interactions of tensin with actin and identification of its three distinct actin-binding domains," *J. Cell. Biol.*, Vol. 125, No. 5, 1994, pp. 1067–1075.

[15] Chrzanowska-Wodnicka, M., and Burridge, K., "Rho-stimulated contractility drives the formation of stress fibers and focal adhesions," *J. Cell. Biol.*, Vol. 133, No. 6, 1996, pp. 1403–1415.

[16] Ridley, A. J., and Hall, A., "The small GTP-binding protein rho regulates the assembly of focal adhesions and actin stress fibers in response to growth factors," *Cell*, Vol. 70, No. 3, 1992, pp. 389–399.

[17] Choquet, D., Felsenfeld, D. P., and Sheetz, M. P., "Extracellular matrix rigidity causes strengthening of integrin-cytoskeleton linkages," *Cell*, Vol. 88, No. 1, 1997, pp. 39–48.

[18] Riveline, D., et al., "Focal contacts as mechanosensors: Externally applied local mechanical force induces growth of focal contacts by an mDia1-dependent and ROCK-independent mechanism," *J. Cell. Biol.*, Vol. 153, No. 6, 2001, pp. 1175–1186.

[19] Chen, C. S., et al., "Geometric control of cell life and death," *Science*, Vol. 276, No. 5317, 1997, pp. 1425–1428.

[20] McBeath, R., et al., "Cell shape, cytoskeletal tension, and RhoA regulate stem cell lineage commitment," *Dev. Cell*, Vol. 6, No. 4, 2004, pp. 483–495.

[21] Engler, A. J., et al., "Matrix elasticity directs stem cell lineage specification," *Cell*, Vol. 126, No. 4, 2006, pp. 677–689.

[22] Lee, J., et al., "Traction forces generated by locomoting keratocytes," *J. Cell. Biol.*, Vol. 127, No. 6, Pt 2, 1994, pp. 1957–1964.

[23] Burton, K., and Taylor, D. L., "Traction forces of cytokinesis measured with optically modified elastic substrata," *Nature*, Vol. 385, No. 6615, 1997, pp. 450–454.

[24] Burton, K., Park, J. H., and Taylor, D. L., "Keratocytes generate traction forces in two phases," *Mol. Biol. Cell.*, Vol. 10, No. 11, 1999, pp. 3745–3769.

[25] Harris, A. K., Wild, P., and Stopak, D., "Silicone rubber substrata: A new wrinkle in the study of cell locomotion," *Science*, Vol. 208, No. 4440, 1980, pp. 177–179.

[26] Balaban, N. Q., et al., "Force and focal adhesion assembly: A close relationship studied using elastic micropatterned substrates," *Nat. Cell. Biol.*, Vol. 3, No. 5, 2001, pp. 466–472.

[27] Munevar, S., Wang, Y., and Dembo, M., "Traction force microscopy of migrating normal and H-ras transformed 3T3 fibroblasts," *Biophys. J.*, Vol. 80, No. 4, 2001, pp. 1744–1757.

[28] Galbraith, C. G., and Sheetz, M. P., "A micromachined device provides a new bend on fibroblast traction forces," *Proc. Natl. Acad. Sci. USA*, Vol. 94, No. 17, 1997, pp. 9114–9118.

[29] Tan, J. L., et al., "Cells lying on a bed of microneedles: An approach to isolate mechanical force," *Proc. Natl. Acad. Sci. USA*, Vol. 100, No. 4, 2003, pp. 1484–1489.

[30] Pelham, R. J., Jr., and Wang, Y., "Cell locomotion and focal adhesions are regulated by substrate flexibility," *Proc. Natl. Acad. Sci. USA*, Vol. 94, No. 25, 1997, pp. 13661–13665.

[31] Pelham, R. J., Jr., and Wang, Y., "High resolution detection of mechanical forces exerted by locomoting fibroblasts on the substrate," *Mol. Biol. Cell.*, Vol. 10, No. 4, 1999, pp. 935–945.

[32] Dembo, M., and Wang, Y. L., "Stresses at the cell-to-substrate interface during locomotion of fibroblasts," *Biophys. J.*, Vol. 76, No. 4, 1999, pp. 2307–2316.

[33] Butler, J. P., et al., "Traction fields, moments, and strain energy that cells exert on their surroundings," *Am. J. Physiol. Cell. Physiol.*, Vol. 282, No. 3, 2002, pp. C595–C605.

[34] Schwarz, U. S., et al., "Calculation of forces at focal adhesions from elastic substrate data: The effect of localized force and the need for regularization," *Biophys. J.*, Vol. 83, No. 3, 2002, pp. 1380–1394.

[35] Sabass, B., et al., "High resolution traction force microscopy based on experimental and computational advances," *Biophys. J.*, Vol. 94, No. 1, 2008, pp. 207–220.

[36] Willert, C. E., and Gharib, M., "Digital particle image velocimetry," *Exper. Fluids*, Vol. 10, No. 4, 1991, pp. 181–193.

[37] Lemmon, C. A., et al., "Shear force at the cell-matrix interface: Enhanced analysis for microfabricated post array detectors," *Mech. Chem. Biosyst.*, Vol. 2, No. 1, 2005, pp. 1–16.

[38] Lorenz, H., et al., "SU-8: A low-cost negative resist for MEMS," *J. Micromech. Microeng.*, Vol. 7, No. 3, 1997, pp. 121–124.

[39] Sniadecki, N. J., and Chen, C. S., "Microfabricated silicone elastomeric post arrays for measuring traction forces of adherent cells," *Methods Cell. Biol.*, Vol. 83, No., 2007, pp. 313–328.

[40] Addae-Mensah, K. A., et al., "A flexible, quantum dot-labeled cantilever post array for studying cellular microforces," *Sens. Actuators A*, Vol. 136, No. 1, 2007, pp. 385–397.

[41] Kovacs, G. T. A., Maluf, N. I., and Petersen, K. E., "Bulk micromachining of silicon," *Proc. IEEE*, Vol. 86, No. 8, 1998, pp. 1536–1551.

[42] Klaassen, E. H., et al., "Silicon fusion bonding and deep reactive ion etching: A new technology for microstructures," *Sens. Actuators A*, Vol. 52, Nos. 1–3, 1996, pp. 132–139.

[43] Laermer, F., and Schilp, A. "Method of anisotropically etching silicon," Robert Bosch GmbH, U.S. Patent No. 5,501,893, filed Aug. 5, 1994, and issued Mar. 26, 1996.

[44] Zhao, Y., and Zhang, X., "Cellular mechanics study in cardiac myocytes using PDMS pillars array," *Sens. Actuators A*, Vol. 125, No. 2, 2006, pp. 398–404.

[45] Gottscho, R. A., Jurgensen, C. W., and Vitkavage, D. J., "Microscopic uniformity in plasma-etching," *J. Vac. Sci. Technol. B*, Vol. 10, No. 5, 1992, pp. 2133–2147.

[46] du Roure, O., et al., "Force mapping in epithelial cell migration," *Proc. Natl. Acad. Sci. USA*, Vol. 102, No. 7, 2005, pp. 2390–2395.

[47] Zhao, Y., et al., "Cellular force measurements using single-spaced polymeric microstructures: Isolating cells from base substrate," *J. Micromech. Microeng.*, Vol. 15, No. 9, 2005, pp. 1649–1656.

[48] Bacher, W., Menz, W., and Mohr, J., "The LIGA technique and its potential for microsystems—a survey," *IEEE Trans. Ind. Electron.*, Vol. 42, No. 5, 1995, pp. 431–441.

[49] Campbell, S. A., *The Science and Engineering of Microelectronic Fabrication*, 2nd ed., New York: Oxford University Press, 2001, p. 624.

[50] Jaeger, R. C., *Introduction to Microelectronic Fabrication*, 2nd ed., Upper Saddle River, NJ: Prentice Hall, 2002, p. 316.

[51] Madou, M. J., *Fundamentals of Microfabrication*, 2nd ed., Boca Raton, FL: CRC Press, 2002, p. 752.

[52] Xia, Y. N., and Whitesides, G. M., "Soft lithography," *Angew. Chem. Int. Ed.*, Vol. 37, No. 5, 1998, pp. 551–575.

[53] Freshney, R. I., *Culture of Animal Cells: A Manual of Basic Technique*, 5th ed., Hoboken, NJ: Wiley-Liss, 2005, p. 642.

[54] Jiang, X., et al., "Directing cell migration with asymmetric micropatterns," *Proc. Natl. Acad. Sci. USA*, Vol. 102, No. 4, 2005, pp. 975–978.

[55] Thery, M., and Bornens, M., "Cell shape and cell division," *Curr. Opin. Cell. Biol.*, Vol. 18, No. 6, 2006, pp. 648–657.

[56] Pirone, D. M., et al., "An inhibitory role for FAK in regulating proliferation: A link between limited adhesion and RhoA-ROCK signaling," *J. Cell. Biol.*, Vol. 174, No. 2, 2006, pp. 277–288.

[57] Nelson, C. M., et al., "Emergent patterns of growth controlled by multicellular form and mechanics," *Proc. Natl. Acad. Sci. USA*, Vol. 102, No. 33, 2005, pp. 11594–11599.

[58] Sniadecki, N. J., et al., "Magnetic microposts as an approach to apply forces to living cells," *Proc. Natl. Acad. Sci. USA*, Vol. 104, No. 37, 2007, pp. 14553–14558.

[59] Yang, M. T., Sniadecki, N. J., and Chen, C. S., "Geometric considerations of micro- to nanoscale elastomeric post arrays to study cellular traction forces," *Adv. Mater.*, Vol. 19, No. 20, 2007, pp. 3119–3123.

[60] Saez, A., et al., "Is the mechanical activity of epithelial cells controlled by deformations or forces?" *Biophys. J.*, Vol. 89, No. 6, 2005, pp. L52–L54.

[61] Cai, Y., et al., "Nonmuscle myosin IIA-dependent force inhibits cell spreading and drives F-actin flow," *Biophys. J.*, Vol. 91, No. 10, 2006, pp. 3907–3920.

[62] Deshpande, V. S., McMeeking, R. M., and Evans, A. G., "A bio-chemo-mechanical model for cell contractility," *Proc. Natl. Acad. Sci. USA*, Vol. 103, No. 38, 2006, pp. 14015–14020.

[63] Gavara, N., et al., "Mapping cell-matrix stresses during stretch reveals inelastic reorganization of the cytoskeleton," *Biophys. J.*, Vol. 95, No. 1, 2008, pp. 464–471.

[64] Wang, N., et al., "Mechanical behavior in living cells consistent with the tensegrity model," *Proc. Natl. Acad. Sci. USA*, Vol. 98, No. 14, 2001, pp. 7765–7770.

[65] Wang, Y., et al., "Visualizing the mechanical activation of Src," *Nature*, Vol. 434, No. 7036, 2005, pp. 1040–1045.

[66] Kumar, S., et al., "Viscoelastic retraction of single living stress fibers and its impact on cell shape, cytoskeletal organization, and extracellular matrix mechanics," *Biophys. J.*, Vol. 90, No. 10, 2006, pp. 3762–3773.

[67] Pertz, O., et al., "Spatiotemporal dynamics of RhoA activity in migrating cells," *Nature*, Vol. 440, No. 7087, 2006, pp. 1069–1072.

CHAPTER

# 5

# A Microfluidic Tool for Immobilizing *C. elegans*


S. Elizabeth Hulme,[1] Christopher V. Gabel,[2] and Sergey S. Shevkoplyas[3]

[1]Department of Chemistry and Chemical Biology, Harvard University, Cambridge, MA 02138
[2]Department of Physics, Harvard University, Cambridge, MA 02138
[3]Department of Biomedical Engineering, Tulane University, New Orleans, LA 70118



## Abstract

In this chapter, we present a microfluidic device, the worm clamp, for physically immobilizing *C. elegans* for microscopy and laser ablation. We describe the design of the device (including a discussion of basic principles for the design of microfluidic tools for *C. elegans*), its fabrication with soft lithography, the preparation of worms for loading into the device, and the operation of the device. The operation of the device is simple: a suspension of worms is placed at the inlet of the device, vacuum is applied to the outlet, and the device automatically distributes the worms into the ordered array of clamps for immobilization. The version of the device presented here is capable of immobilizing over 100 worms in less than 15 minutes. The online supplementary information for this book includes an electronic file containing the design for this device (http://www.methodsinbioengineering.com/library.asp).



Key terms: *C. elegans*
worm
soft lithography
PDMS
microfluidics
lab-on-a-chip
immobilization
microscopy
laser ablation

## 5.1 Introduction

The worm *C. elegans*, a soil-dwelling nematode, is a very popular model organism used in nearly all areas of modern biology, including development, aging, learning and memory, behavior, genetics, genomics, neurobiology, and oncology. Two of the principal advantages of using *C. elegans* as a model organism are that the body of the worm is small (adults are approximately 50 $\mu$m wide and 1 mm long) and transparent. As a consequence, it is possible to perform a broad range of in vivo experiments using *C. elegans*, including: (1) the examination of the internal organs and tissues of the worm with subcellular resolution using light microscopy, (2) the monitoring of the biochemical state of the worm using fluorescence microscopy to measure the expression of fluorescently labeled proteins, and (3) the performance of laser-mediated microsurgery to ablate neurons or other cells.

*C. elegans* is typically cultured on nematode growth media (NGM) agar plates seeded with *E. coli* (typically strain OP50), which serves as a food source. A healthy worm crawls quite rapidly on the plate: approximately 200 $\mu$m—roughly 20% of the length of its body—per second (see "Supplemental Movie 5.1.avi"). Observation of the rapidly moving worm requires either real-time tracking microscopy or temporary immobilization of the animal. Conventional methods used by *C. elegans* biologists for immobilizing the animals include: (1) gluing worms to an agarose pad using cyanoacrylate glue, (2) treating the worms with paralytic drugs, and (3) cooling the worms to 4°C. Gluing immobilizes the worm but is experimentally tedious (it takes about 2 minutes per animal); experiments on many animals—a statistically significant population—become prohibitively time-consuming. In addition, gluing is irreversible; it is impossible to recover worms after the experiment. Treating worms with paralytic drugs and cooling worms to 4°C both have the advantage that many worms may be immobilized at once. These approaches, however, unavoidably change the internal biochemical state of the worms. Studies involving the immobilization and examination of very large populations—hundreds or even thousands—of worms in a noninvasive manner are therefore practically impossible using conventional techniques.

This chapter describes the design, fabrication, and operation of a microfluidic device for rapidly immobilizing many live *C. elegans* (Figure 5.1). The device consists of an array of *microfluidic worm clamps* [1, 2]—wedge-shaped microchannels—that physically prevent the worms from moving. Using the device, it is possible to immobilize more than 100 worms in less than 15 minutes. The operational procedure is very simple: a suspension of worms is placed at the inlet of the device, vacuum is applied to the outlet, and the device automatically distributes the worms into the ordered array of clamps for immobilization and subsequent imaging. This process is reversible: following immobilization, worms can be released from the device for further examination.

The device is fabricated via soft lithography [3] out of polydimethylsiloxane (PDMS), which is a particularly useful material for the construction of tools for handling living organisms because it is: (1) compliant, (2) nontoxic, (3) permeable to gases, and (4) optically transparent to wavelengths longer than 230 nm. Because of the optical transparency of PDMS, the device is compatible with conventional microscopy (fluorescence, bright field, DIC, phase contrast) as well as laser-mediated microsurgery.

In a microfluidic device, fluid acts as an extension of the experimenter's hands: precisely controlled flows can gently manipulate the animals. The small size of *C. elegans* (the adult is approximately 50 $\mu$m in width) and its ability to live in liquid make it ideal

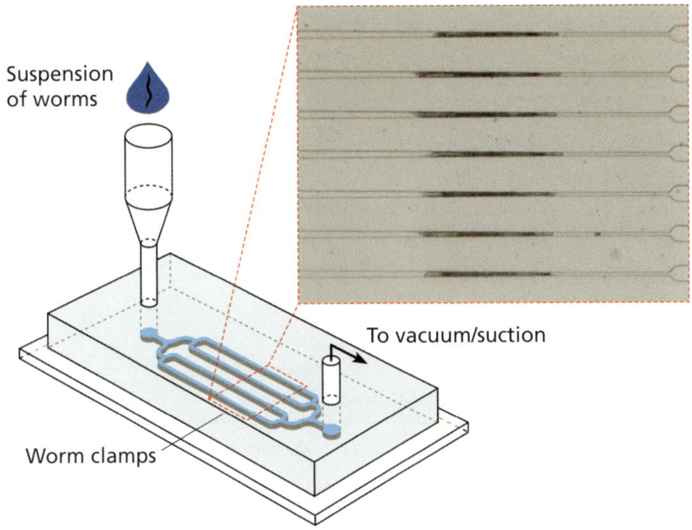


**Figure 5.1** The microfluidic worm clamp device for immobilizing *C. elegans*. The device automatically distributes a suspension of worms into an array of clamps, or tapered microchannels that physically restrict the motion of the worms.

for conventional chip-sized microfluidic devices. A major advantage of using a microfluidic approach to immobilize *C. elegans* is that it is possible to expose animals to suspensions or solutions of stimuli while they are within a device.

The microfluidic worm clamp device represents a fast, simple, reversible, and noninvasive method for immobilizing many live worms simultaneously. This tool is valuable for any study in which worms must be temporarily immobilized. Because so many worms can be immobilized at once, the device facilitates the rapid collection of data for a statistically significant number of animals.

This chapter begins with a description of the design of the device (including a discussion of basic design principles for microfluidic devices for *C. elegans*) and proceeds with a presentation of the procedures for fabricating (obtaining a photomask, producing a silicon master, and replica-molding the master in PDMS) and assembling the device, preparing *C. elegans* for loading into the device, and using the device to immobilize worms.

## 5.2 Materials

Table 5.1 lists the materials and equipment necessary for fabricating the worm clamp device and immobilizing *C. elegans*.

## 5.3 Methods

### 5.3.1 Overview and timeline

Table 5.2 presents a general timeline of the sequence of steps one needs to perform to make the device and to use it for immobilizing worms successfully. These steps are described in detail below.

**Table 5.1** Materials and Equipment

| Reagents/Materials | Facilities/Equipment |
|---|---|
| **Worm preparation** | |
| 25 gravid adult worms | Incubator, 20°C |
| 1 nematode growth media (NGM) agar plate | Nutating mixer |
| 10 NGM agar plates seeded with bacteria | Tabletop centrifuge |
| bleach, 500 $\mu$L | Stereomicroscope |
| 1 M NaOH, 500 $\mu$L | Inverted microscope |
| M9 Buffer, 1L | |
| 2 mL microcentrifuge tubes | |
| 15 mL centrifuge tubes | |
| **Master fabrication** | |
| 1 silicon wafer, 3" diameter | Photoresist spinner |
| SU-8 50 photoresist (MicroChem Corp., Newton, MA) | Hot plate |
| Propylene glycol monomethyl ether acetate | UV exposure tool (mask aligner) |
| Isopropanol | Vacuum desiccator |
| Tridecafluoro(1,1,2,2 tetrahydrooctyl)- trichlorosilane | Access to Class 1000 clean room |
| **Replica molding** | |
| Sylgard 184 poly(dimethyl siloxane) (Dow Corning) | Vacuum desiccator |
| Plastic cups | Oven, 65°C |
| Petri dishes, 3" diameter | |
| Scalpel with No. 11 blades | |
| Tweezers | |
| Cylindrical biopsy punches, 1.5 mm diameter | |
| Transparent adhesive tape | |
| **Device assembly and use** | |
| 1 large glass coverslip (48 × 65 mm, No. 1 thickness) or 1 large glass slide (50 × 75 mm, 1 mm thickness) | Barrel plasma cleaner |
| Polyethylene tubing with 1.5 mm outer diameter (e.g., PE-160 tubing) | Glass bottle with rubber septum cap (for liquid trap) |
| 5 mL plastic syringes | Source of vacuum |
| Syringe needles (18G) | Rubber vacuum tubing |
| 0.2 $\mu$m syringe filters | 1 barbed tube adapter for male Luer-Lok |
| 1 glass Pasteur pipet | |

### 5.3.2 Designing the device and ordering the photomask

For simplicity, we illustrate the design of the device using a four-clamp version [Figure 5.2(a)]. The device consists of five elements: (1) the inlet, (2) a branching network of channels leading from the inlet to the clamps, (3) the array of clamps, (4) a branching network of channels leading from the clamps to the outlet, and (5) the outlet. The channels leading to and from the clamps are 100 $\mu$m in width; the bifurcations of the network are blunted [lower inset in Figure 5.2(a)]. Each clamp is a wedge-shaped channel that narrows from 100 $\mu$m to 10 $\mu$m in width over the span of 5 mm [upper inset of Figure 5.2(a)]. A difference in pressure between the entrance and the exit of the clamp drives the flow of liquid through the clamp and holds the worm in place [Figure 5.2(b)].

The functional significance of the branching network is to enable the device to distribute worms into the clamps automatically [Figure 5.2(c)]. At each branching point in the network of inlet channels, the path of the worm is determined by the ratio of the flow rates into the two downstream branches. The first worm to enter the device has an equal probability of entering each of the four clamps [Figure 5.2(c)(i, ii)]. Once the worm enters a clamp, however, the fluidic resistance of that clamp increases, and the balance of flow rates at the branching points changes. The next worm to enter the device is more

**Table 5.2** Suggested Timeline for Experiments

| Preparing the Device | Time Allotment | Preparing the Worms |
|---|---|---|
| Design the device* | As needed | |
| Order the photomask | 24 to 48 hours from order placement to delivery | |
| Fabricate the master | | |
| Fabricate in clean room | 3 hours | |
| Perform silanization | 3 hours | |
| | | Prepare worms for loading |
| | 1 hour | Release eggs from adults with alkaline bleach |
| | 24 hours | Allow eggs to hatch in absence of food |
| | 48 hours | Return hatchlings to food; grow to young adults |
| Replica mold master in PDMS | 12 hours | |
| Assemble microfluidic device | 30 minutes | |
| | 1 hour | Collect synchronous population of adult worms in M9 buffer |
| Prepare device for loading | 30 minutes | |
| | 15 minutes | Load worms into device |
| | Up to 14 hours | Perform experiments on immobilized worms |
| | 10 minutes | Unload worms from device |

*The design file for the worm clamp device with 128 clamps is available on the Web site for this book.

likely to be diverted to a branch that does not contain a worm [Figure 5.2(c)(iii)]. This process continues until each of the clamps contains a single worm [Figure 5.2(c)(iv, v)].

To ensure that the distribution process proceeds in this way, the fluidic resistance of every branch of the device—the pathway from the inlet, through a clamp, to the outlet—must be equal in the absence of a worm. To achieve paths of equal fluidic resistance, the network of distribution channels contains only bifurcated branching points (and not trifurcations or higher-order branching geometries).

The main type of damage that the immobilization device may inflict on the worm is mechanical damage to the cuticle. A large pressure gradient along the body of the worm could rupture the pressurized cuticle. For this reason, the device is designed to operate under a constant pressure difference between its inlet and outlet. This mode of operation automatically places an upper boundary on the pressure gradient experienced by the worm. Indeed, consider what happens to the pressure difference along the worm in the clamp [Figure 5.2(b)] if the pressure difference along the clamp remains constant [(5.1)].

$$\lim_{R_{12}\to\infty}(P_1 - P_2) = \lim_{R_{12}\to\infty}\left(\frac{1}{1+\dfrac{R_{01}+R_{23}}{R_{12}}}(P_0 - P_3)\right) = (P_0 - P_3) \tag{5.1}$$

In (5.1) [and in Figure 5.2(b)], $P_0$ is the pressure at the inlet of the clamp, $P_1$ is the pressure at the tail of the worm, $P_2$ is the pressure at the mouth of the worm, $P_3$ is the pressure at the outlet of the clamp, $R_{01}$ is the fluidic resistance of the inlet part of the clamp, $R_{12}$ is the resistance of the part of the clamp containing the immobilized worm, and $R_{23}$ is the

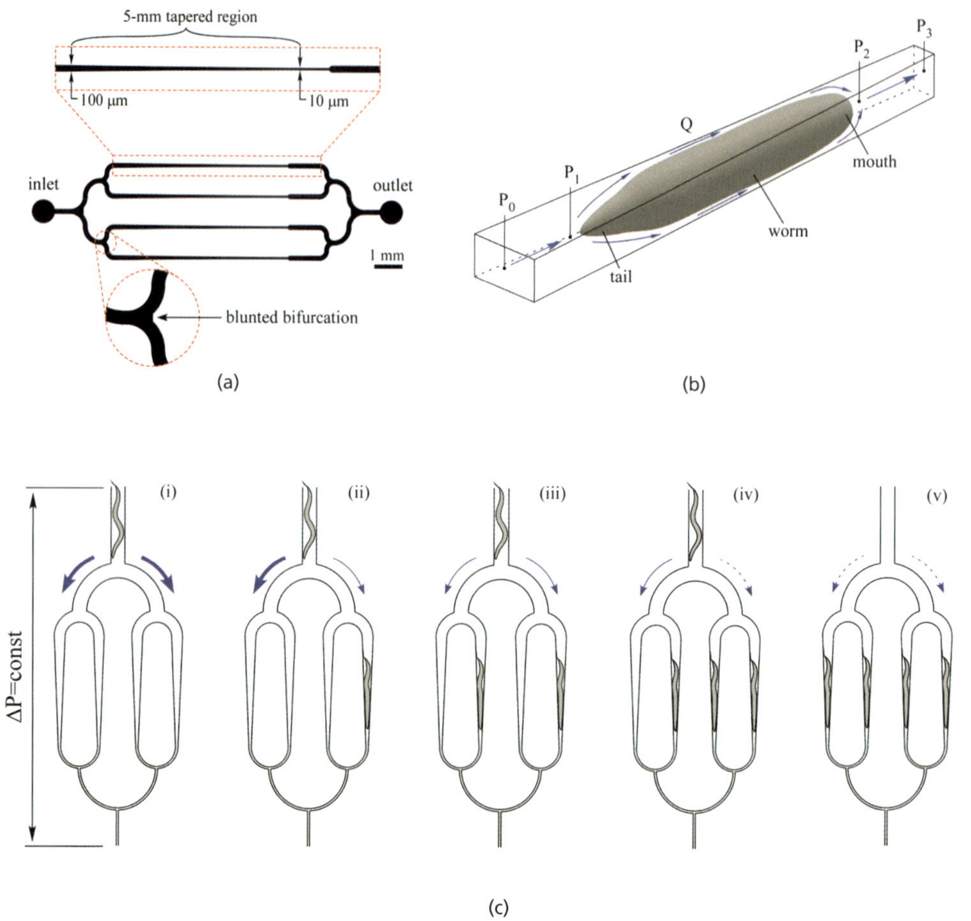


**Figure 5.2** The design of the microfluidic worm clamp device. (a) The design of a four-clamp device. The device consists of an array of four clamps and a network of branching channels that lead from the inlet to the clamps and from the clamps to the outlet. (b) The distribution of pressure ($P_i$) along a clamp containing a worm. The volumetric flow rate through the clamp is indicated by Q. (c) The automatic distribution of worms into the device. The presence of a worm within a clamp increases the fluidic resistance of that clamp and thus diverts subsequent worms to other branches of the device.

resistance of the outlet part of the clamp. Even as the worm fits more tightly into the clamp ($R_{12} \rightarrow \infty$), the pressure gradient along the body of the worm ($P_1 - P_2$) never exceeds the pressure difference applied to the system ($P_0 - P_3$).

If, instead, the device were operated with constant flow—as would be the case if a syringe pump were used to drive the device—the pressure gradient along the body of the worm could increase unbounded, as shown by (5.2), where $Q$ is the volumetric flow rate of liquid through the clamp.

$$\lim_{R_{12} \rightarrow \infty} (P_1 - P_2) = \lim_{R_{12} \rightarrow \infty} (R_{12} Q) = \infty \tag{5.2}$$

In addition, to avoid another type of mechanical damage—puncture of the cuticle of the worm by sharp features—all of the bifurcations in the device are blunted [lower inset in Figure 5.2(a)].

As discussed above, for the automatic distribution of *C. elegans* into the device to proceed properly, the branching points of the network of distribution channels must be bifurcations. Because of this geometrical design constraint, the number of clamps in a device is equal to $2^N$, where $N$ is the number of bifurcations. For example, the device in Figure 5.2(a) has two bifurcations and four clamps; the larger version of the device (the design for which is available in the online supplementary information for this book) has seven bifurcations and therefore has $2^7 = 128$ clamps. It is possible to scale the number of clamps in the device as needed, provided that the number of clamps follows the $2^N$ rule. In practice, however, the footprint of the device should fit onto a standard-sized silicon wafer (2", 3", or 4" diameter), with a generous border (at least 1 cm) between the edge of the wafer and the design. The footprint of the design with 128 clamps is the largest that can fit comfortably within the borders of a 3" silicon wafer.

To generate a photomask for the device, one would need: (1) either to draw an electronic version of the design in accordance with the principles discussed above using polygon editor/CAD software (e.g., CleWin, available from www.wieweb.com) or to download "Supplemental File 5.1 wormclamps.dxf," which contains the design for the array of 128 clamps, then (2) to send the design, in *.dxf format, to a company that specializes in producing transparency photomasks from electronic designs for a fee (e.g., CAD/Art Services Inc., www.outputcity.com).

A transparency mask consists of black ink printed onto a thin, flexible film of plastic (similar to a transparency for an overhead projector) using a high-resolution laser printer. This type of mask is inexpensive (typically less than $100 per mask) but relatively limited in resolution (the minimum feature size is typically 10 $\mu$m). Because of the size of *C. elegans*, the resolution of transparency masks should be sufficient for most applications. The photomask should have clear features on an opaque background. (It is possible to fabricate photomasks in the laboratory; this process, however, requires specialized equipment and will not be described here.)

### 5.3.3 Fabricating the master for the device

A master is a flat substrate, a Si wafer, containing the design of the channels in positive relief. It is possible to mold (replicate) devices from a single master repeatedly. This section describes the fabrication of a master for the 128-clamp device with a feature height of 50 $\mu$m using a transparency mask. This height of 50 $\mu$m is appropriate for immobilizing young adult worms.

Because of the microscale size of the features of a master, it is a common practice to fabricate masters for soft lithography in a Class 1000 clean room—a laboratory environment with a low count of particle contaminants in the air—with yellow/amber filters over the lights to avoid unintended exposure of the photoresist. Figure 5.3 shows an overview of the procedure for creating a master.

1. Start with a clean (dust-free), new Si wafer, 3" in diameter (test grade). Contamination of the wafers with dust can be reduced by opening a new package of wafers only inside the clean room.
2. Pour approximately 3 mL of SU-8 50 photoresist onto the center of the wafer. This resist is extremely viscous and must be poured slowly to avoid introducing bubbles into the resist.

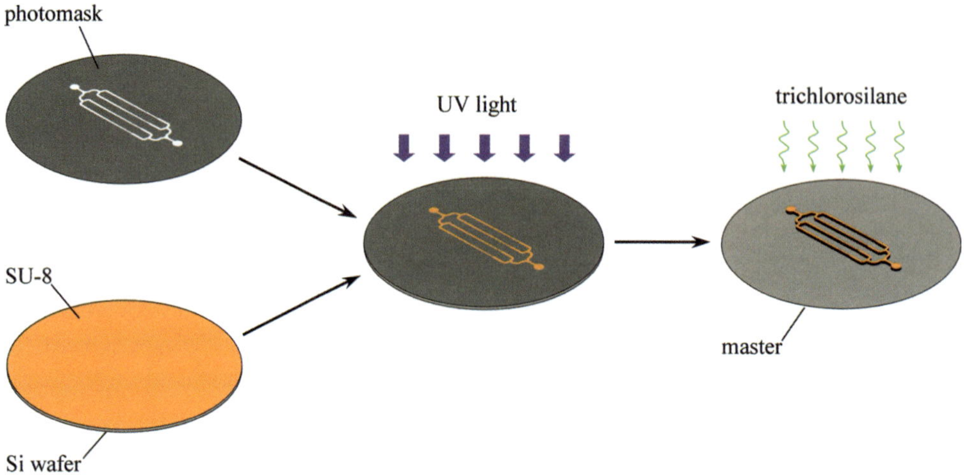


**Figure 5.3** Fabricating a silicon master. A photomask is used to selectively expose regions of a photoresist (SU-8)–coated silicon wafer to UV light. This process crosslinks the exposed regions of photoresist; the resulting structure is a silicon wafer patterned with raised features in SU-8. This silicon master is treated with tridecafluoro(1,1,2,2 tetrahydrooctyl) trichlorosilane, which prevents elastomer from sticking to the master in subsequent molding steps.

3. Center the wafer with resist onto the chuck of the spinner. The speed of spinning determines the thickness of the resist layer, thus the height of the features of the master. For a film of photoresist that is approximately 50 $\mu$m in thickness, program the spinner to ramp from 0 to 500 rpm over 5 seconds, hold at 500 rpm for 5 seconds, ramp from 500 to 2,000 rpm over 5 seconds, and hold at 2,000 rpm for 30 seconds. The necessary spin speed may vary depending on factors such as the temperature of the room and the age of the photoresist. To achieve features that are precisely 50 $\mu$m in height, it will be necessary to adjust the spinning recipe through trial and error.
4. Place the resist-coated wafer onto a hot plate at 65°C for 8 minutes. Leaving the wafer on the hot plate, ramp the temperature from 65°C to 90°C and hold at 90°C for 25 minutes. Allow the hot plate to return to room temperature.

   This step removes solvent from the resist. The wafer can be maintained at 90°C for longer than 25 minutes to ensure that all solvent is removed from the layer of resist. It is important that the hot plate is leveled horizontally for this step in order to produce features of uniform height in the master. After the wafer is cooled, the film of photoresist should appear smooth and should not be tacky.

   We base our description of steps 5 to 13 on the ABM mask aligner (ABM Inc., www.abmfg.com), a typical UV-exposure tool for photolithography. The details of this procedure may vary depending on the particular exposure tool that is used.
5. Place the resist-coated wafer onto the stage of the exposure tool. The tool may have vacuum suction to hold the wafer in place; if it does, engage the vacuum suction.
6. Place the transparency photomask directly onto the resist-coated wafer with the printed side of the mask facing the wafer (Figure 5.3).
7. Place a square piece of clear (unprinted) quartz glass in the frame of the exposure tool where a chrome photomask would normally be. The tool may have vacuum suction to hold the glass in place; if it does, engage the vacuum suction.

8. Raise the stage so that the wafer and transparency mask are pressed against the clear quartz glass. When Newton's rings—an interference pattern of alternating dark and light concentric rings—are visible, the wafer, mask, and blank glass are in good contact. Good contact between the resist-coated wafer and the mask is important for producing a high-quality master.
9. Open the shutter to expose the wafer with near UV light (365 to 400 nm) (Figure 5.3). The exposure time depends on the intensity of the light source of the tool [see (5.3)]; the total exposure dose should be 200 to 400 mJ/cm$^2$.

$$\text{exposure time}\,[s] = \frac{\text{exposure dose}\left[\dfrac{\text{mJ}}{\text{cm}^2}\right]}{\text{intensity}\left[\dfrac{\text{mW}}{\text{cm}^2}\right]} \tag{5.3}$$

   Exposure to UV light produces crosslinkages in SU-8 resist; the resist becomes crosslinked only in those regions that are below the clear regions of the photomask.
10. Lower the stage, and remove the exposed wafer.
11. Place the wafer onto a hot plate at 65°C for 1 minute. Leaving the wafer on the hot plate, ramp the temperature from 65°C to 90°C, and hold at 90°C for 4 minutes. Allow the hot plate, with the wafer in place, to return to room temperature. The postexposure bake completes the UV-initiated crosslinking reaction. Avoid rapidly changing the temperature of the exposed wafer; the crosslinked resist may crack under thermal stress.
12. Place the wafer into a shallow bath of propylene glycol monomethyl ether acetate (PGMEA, sold as "Baker BTS-220" from Mallinckrodt Baker Inc. or as "SU-8 Developer" from MicroChem Corp.) for approximately 5 minutes, with occasional manual agitation in order to dissolve the uncrosslinked resist. It is important to use PGMEA inside of a chemical fume hood to avoid inhalation. Once the uncrosslinked resist appears to be removed from the wafer, rinse the wafer once in fresh PGMEA, followed by isopropanol (IPA) to rinse away the PGMEA. Dry the IPA from the wafer using a stream of nitrogen. If purple or white streaks are visible during the drying step, the master still contains uncrosslinked SU-8 and should be returned to the PGMEA bath for an additional 1 to 2 minutes for further development. Once the master appears free of uncrosslinked SU-8, the photolithographic portion of the fabrication process is complete, and the master may be removed from the clean room. The master should be inspected under a microscope to ensure that the features have transferred from the mask to the wafer properly. Slanted sidewalls or unresolved features may indicate overexposure. Lift-off of the SU-8 features from the silicon wafer may indicate underexposure, poor wafer-to-mask contact during exposure, or poor adhesion of SU-8 to the wafer.
13. Place the master in a vacuum chamber along with a small glass vial containing a drop of tridecafluoro(1,1,2,2 tetrahydrooctyl) trichlorosilane. Evacuate the chamber, and incubate the master with the trichlorosilane for 3 hours (Figure 5.3). The trichlorosilane chemically adsorbs to the surface of the master and coats the master with fluorocarbon chains. This coating will prevent PDMS from sticking to the master in the later molding steps.

### 5.3.4 Replica-molding the master in PDMS

Once the fabrication of the master is complete, the master may be replica-molded many times to produce microfluidic devices in polydimethylsiloxane (PDMS). Provided that precautions are taken to avoid damaging the master during replica molding, it is possible to use a single master to produce hundreds of devices. Figure 5.4 illustrates the major steps involved in replica molding.

1. Sylgard 184 (Dow Corning) is a formulation of PDMS that comes in two parts: a PDMS prepolymer and a crosslinking agent. Combine the two parts with 10g of prepolymer for every 1g of crosslinking agent (a 10:1 ratio). Approximately 30g of PDMS is required to cast a single device. A disposable plastic cup is a convenient container for mixing PDMS. The container used to mix PDMS should be large enough so that the desired amount of PDMS only fills about one-third of the container. Mix the two components together—either by hand or with an electric mixer—for at least 5 minutes, or until the components appear well mixed. Mixture will appear foamy as it is full of air bubbles.
2. To remove the air bubbles from the PDMS mixture, place the container of PDMS under vacuum until the mixture appears transparent and free of bubbles—approximately 30 minutes. The degassing process will initially enlarge the bubbles in the PDMS and thus cause the total volume of the material to increase. To prevent the overflow of PDMS from its container, PDMS should occupy only one-third of the volume of the container prior to degassing. The crosslinking reaction is initiated as soon as the two components of PDMS are mixed. Therefore, the PDMS should be used for replica molding within approximately 2 hours of degassing.
3. Place the silicon master in the center of a 3″-diameter petri dish. Pour degassed PDMS over the master so that the master is covered by a layer of PDMS that is at least 5 mm deep (if the PDMS layer is too thin, it will later be difficult to connect tubing to the inlet and outlet of the device) (Figure 5.4). Place the master on a leveled surface in an oven at 65°C overnight to cure the PDMS.

   If a master is being molded for the first time, ensure that there are no air bubbles between the bottom of the silicon wafer and the surface of the petri dish. Pockets of air underneath the master may cause it to tilt during curing, and the resulting PDMS device will not be level.

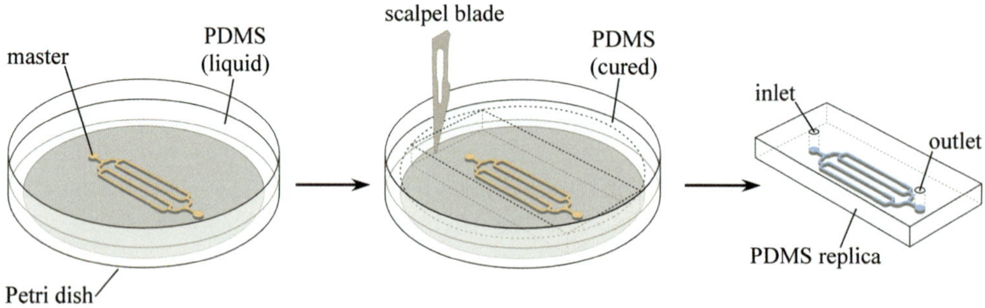


**Figure 5.4** Replica molding. The pattern on the master is imprinted in PDMS by pouring PDMS prepolymer over the master and allowing it to cure. The PDMS replica is removed from the master with a scalpel, and inlet and outlet ports are punched.

Faster rates of curing will occur at temperatures higher than 65°C; at 150°C, PDMS will cure in approximately 10 minutes. Because polystyrene petri dishes melt between 70°C and 80°C, an alternative material, such as glass, should be used to contain the master at temperatures higher than 65°C.

A PDMS-filled master should not be stored in the oven for extended periods (several weeks); after prolonged storage, it may be difficult to remove the PDMS from the master. Instead, it is preferable to store the masters at room temperature with the PDMS removed.

4. Use a scalpel with a fresh blade (No. 11) to remove the PDMS from the master. Insert the blade into the PDMS and start cutting, keeping the tip of the blade in contact with the wafer. Cutting the PDMS near the edge of the wafer (~5 mm from the edge) can reduce the risk of damage to the master (Figure 5.4). The master can easily crack if excessive force is applied: do not push vertically; instead, push in the direction of cutting only. Using a sharp scalpel blade minimizes the force that has to be applied to the wafer during cutting and helps to avoid cracking.
5. Using tweezers, slowly peel the cut PDMS replica from the master. Place the PDMS replica onto a clean surface (a clean glass slide) with the molded features facing up, and use a scalpel to trim excess PDMS. Leave at least a 5 mm gap between the molded features and the edge of the trimmed PDMS.
6. Locate the inlet and outlet of the channels molded in the PDMS. Using a sharp, cylindrical punch (disposable biopsy punches work well), create access holes to the inlet and outlet by punching through the PDMS replica at these locations. The locations of the inlet and outlet are more visible on the molded side of the replica. For the 128-clamp device, use a punch that is 1.5 mm in diameter.
7. Clean dust and debris from the PDMS surface by first covering the molded surface of the PDMS with transparent adhesive tape (Scotch Tape), then peeling the tape off. This process may be repeated several times. If it is not necessary to use the device immediately, the PDMS may be stored for several months with the molded features covered by tape. The tape protects the features from dust contamination.

### 5.3.5 Preparing *C. elegans* for loading

The 128-clamp device works best when a synchronous population of *C. elegans* is used. The dimensions of the device are optimized for young adult worms. The following procedure for generating a synchronous population of young adult worms is adapted from [4] and is a common protocol in *C. elegans* biology. Note that the procedure takes approximately 3 days and should thus be started 3 days prior to assembling the microfluidic device for loading (see Section 5.3.6).

1. Begin with a nematode growth media (NGM) agar plate seeded with a lawn of *E. coli* (strain OP50) and containing an unsynchronized population of *C. elegans*. The plate should contain many (25+) gravid (egg-containing) adults, but the population should not be starved (i.e., there should still be *E. coli* visible on the plate).
2. On a fresh NGM agar plate containing no bacteria, place a drop of freshly mixed 1:1 bleach:1 M NaOH. For best results, the bleach should be stored at 4°C and brought to room temperature just prior to use.

3. Using a worm pick, transfer 25 gravid adults from the unsynchronized plate into the drop of alkaline bleach. Maintain the worms within the droplet for approximately 10 minutes, with occasional agitation. The solution of alkaline bleach will dissolve the bodies of the adult worms, thereby releasing the eggs held within the adults. The eggs are surrounded with chitinous shells, which protect them from dissolution. The progress of the egg release should be monitored using a stereomicroscope.
4. Add 0.5 mL of M9 buffer to the NGM agar plate containing the released eggs. Swirl the plate for about 30 seconds to move the eggs from the surface of the agar into the M9 buffer. Using a pipet, transfer the M9 buffer, now containing a suspension of eggs, to a microcentrifuge tube. Repeat this step two more times to ensure that all eggs are collected from the plate.
5. Spin the suspension of eggs in a centrifuge at 1,300 × g for 30 seconds, aspirate to 0.1 mL, resuspend the pelleted sample by agitating the microcentrifuge tube with a vortex mixer, and add 1 mL of M9 buffer. Repeat this step two additional times.
6. Transfer the suspension of eggs to a 15 mL centrifuge tube. Add M9 buffer for a total volume of 5 mL. Incubate the suspension for 24 hours at 20°C on a nutating mixer. In the absence of a food source, the eggs will hatch but arrest growth in the first larval stage (L1).
7. Obtain five NGM agar plates seeded with *E. coli*. Transfer 1 mL of the suspension of arrested L1 worms to each plate. Incubate the plates for 48 hours at 20°C. The NGM agar plate will absorb the liquid from the suspension within 1 to 3 hours. In the presence of a food source, the L1s will resume growth and develop into young adults in 48 hours at 20°C.
8. Add 0.5 mL of M9 buffer to each of the five NGM plates. Agitate the plates for about 30 seconds to transfer the adult worms from the surface of the plate into the buffer, and collect the suspensions of worms from all five plates into a single 15 mL centrifuge tube. Repeat this step two additional times.
9. Spin the suspension of adult worms in a centrifuge at 750 × g for 60 seconds.
10. Aspirate the suspension to 1 mL. Resuspend the worms, and add 4 mL of fresh M9 buffer.
11. Repeat steps 9 and 10 two additional times.
12. Estimate the concentration of worms in the suspension by pipeting 100 $\mu$L of the suspension onto an NGM agar plate and counting the number of worms. Adjust the volume of the suspension so that the concentration of worms is approximately 100 worms per microliter. The suspension of worms should be loaded in the microfluidic device within 1 hour.

### 5.3.6 Assembling the microfluidic device

The molded PDMS contains the ceiling and sidewalls of a set of microchannels; the PDMS must be sealed to a substrate to form a complete microfluidic device. The device may be sealed to either a glass slide or a thin glass coverslip. A glass slide is less fragile than a glass coverslip; however, if the device is to be used for high-magnification microscopy of worms within the device, it is necessary to use a glass coverslip to accommodate the short working distance of a high-magnification objective. Figure 5.5 illustrates the sealing process and shows a photograph of an assembled device.

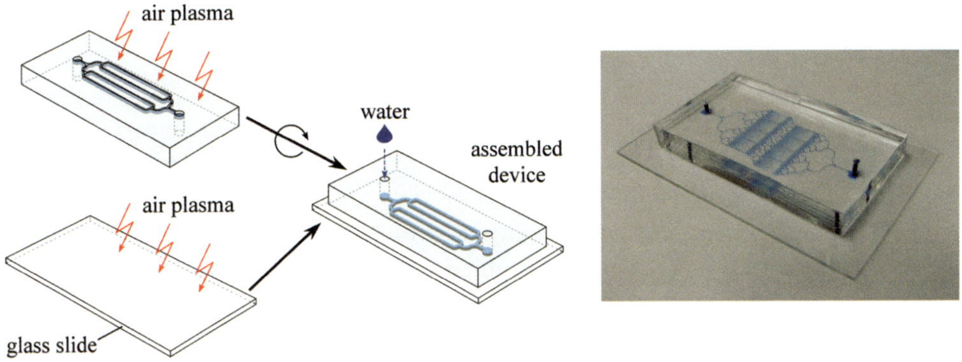


**Figure 5.5** Assembling the microfluidic device. Oxidation of the PDMS replica and a glass slide with an air plasma makes it possible to bond the two materials together covalently. The newly formed microchannels are then filled with water. The photograph shows the actual 128-clamp device. The channels contain blue ink to increase their visibility.

1. Place a clean glass slide (or coverslip) in a plasma cleaner (a barrel plasma etcher may also be used). A plasma cleaner exposes samples to an air plasma, which oxidizes the surface of a sample. Treat the surface of the glass with an air plasma at 1 Torr for 4 minutes. It is possible to bond PDMS to glass (and also PDMS to PDMS) covalently following exposure to an air plasma. The required exposure time for these materials depends on multiple factors, including the pressure in the plasma cleaner during oxidation, the composition of the atmosphere inside the plasma cleaner, and the power of the plasma.
2. Remove the tape from the molded PDMS replica. Open the plasma cleaner and insert the replica, molded side facing up. The PDMS and the glass slide should not be in contact. Expose both the PDMS replica and the glass to an air plasma at 1 Torr for 1 minute.
3. Immediately following oxidation, place the glass slide on a clean surface with the oxidized side facing up, and place the PDMS replica onto the glass with the molded features facing the glass. A conformal seal between the PDMS and the glass substrate should be visible. It may be helpful to roll the rounded edge of a petri dish over the PDMS gently to ensure that the two surfaces are in contact.
4. Place the newly sealed device in an oven at 65°C for 15 minutes to complete the covalent bonding of the glass and PDMS.
5. Remove the device from the oven, and allow it to cool to room temperature.
6. Fill a 5 mL plastic syringe with filtered water. Attach the syringe to the inlet of the cooled device using a syringe needle and polyethylene (PE) tubing. The outer diameter of the PE tubing should be slightly larger than 1.5 mm to ensure a snug fit between the inlet of the PDMS device and the tubing. Fill the microchannels of the device completely with water. The oxidation process renders the surfaces of the PDMS and glass hydrophilic; over time, however, dry PDMS will revert back to its normal hydrophobic state. Therefore, the channels should be filled with water immediately following the 15-minute baking step and subsequent cooling. Once filled with water, the device may be stored for several days at 4°C, with the inlet and outlet covered with tape; it is best, however, to use the device immediately.

### 5.3.7 Preparing the device for loading

Figure 5.6 shows photographs of the experimental setup for the device.

1. To prepare the device for loading, connect a syringe containing M9 buffer to the outlet of the device, and depress the plunger of the syringe to replace the water in the device with M9 buffer.
2. Attach a 5 cm piece of PE tubing (1.5 mm in diameter) to the end of a glass Pasteur pipet. Together, the pipet and tubing form the inlet reservoir of the device. Insert the tubing into the access hole of the inlet.
3. Place the device with the inlet reservoir onto the stage of an inverted microscope. If necessary, use a clamp or tape to maintain the mouth of the inlet reservoir facing upward (Figure 5.6).
4. Fill the inlet reservoir with M9 buffer by connecting a buffer-filled syringe to the outlet of the device and depressing the plunger of the syringe. Filling the inlet reservoir in this manner (through the outlet of the device) prevents the introduction of bubbles into the device.
5. Once the inlet reservoir is filled with buffer, remove the syringe and tubing from the outlet.
6. Connect the outlet of the device to a source of vacuum through a liquid trap. The liquid trap prevents liquid from the device from being sucked directly into the vacuum source. A simple liquid trap consists of a glass bottle with a rubber cap. PE tubing connects the outlet of the device to a syringe needle that passes through the rubber cap of the trap. A second needle passes through the rubber cap and connects to a length of rubber vacuum tubing, which in turn connects to the source of vacuum. A barbed adaptor is necessary to connect the vacuum tubing to the syringe needle. Initially, the source of vacuum should be off. The operation of the 128-clamp device requires a vacuum gauge pressure of approximately –95 kPa, or –14 psi,

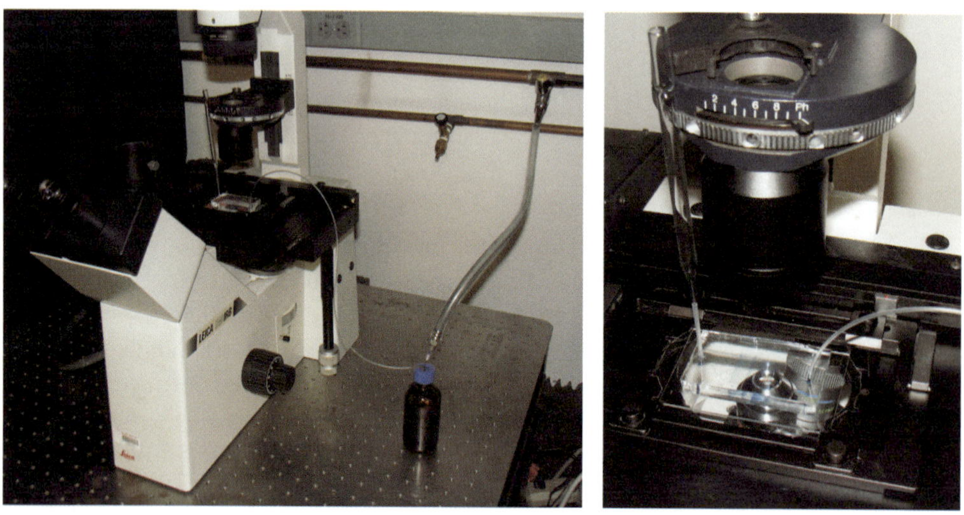

**Figure 5.6** Photographs of the experimental setup. The image on the left shows the device mounted on the stage of a microscope. The inlet reservoir (a glass Pasteur pipet) is attached to the inlet of the device. Polyethylene tubing connects the outlet of the device to the liquid trap (a brown glass bottle sealed with a rubber cap). Rubber vacuum tubing connects the liquid trap to the nozzle of the house-supplied vacuum. The image on the right shows a closer view of the device mounted on the microscope stage.

relative to the atmosphere. This pressure difference can be obtained using a house-supplied vacuum (if available) or a vacuum pump.

### 5.3.8 Loading worms into the device

The operational sequence—or "life cycle"—of the device is shown in Figure 5.7.

1. To initiate loading, add 0.5 mL of the suspension of worms (preparation described above) to the inlet reservoir of the device. Within a few minutes, worms will sediment—due to gravity—to the bottom of the inlet reservoir.
2. Turn the vacuum on. Worms will fill the clamps in the device rapidly. The progress of the loading procedure should be monitored using the microscope. Because the

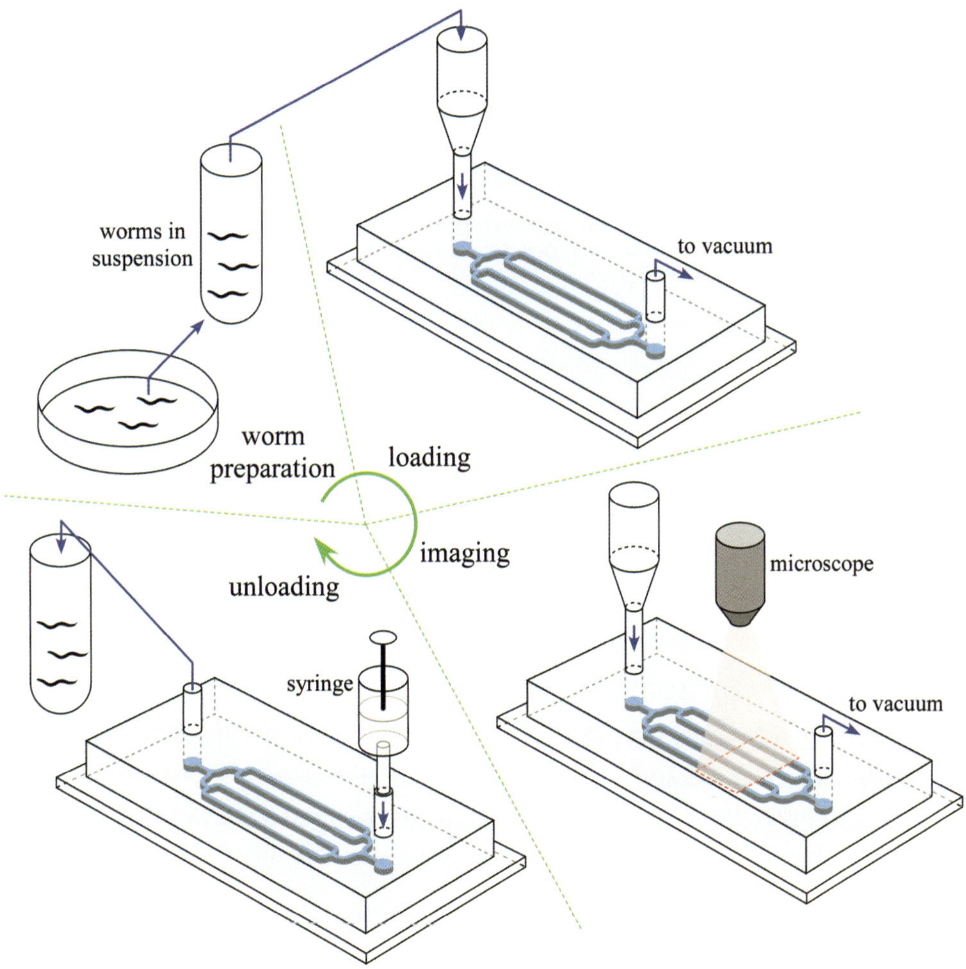


**Figure 5.7** The operational sequence for the device. Worm preparation: Transfer a population of synchronous worms from an NGM plate to a suspension in M9 buffer. Loading: Load the suspension of worms into the inlet reservoir of the assembled device. Turn on the source of vacuum to drive the flow of liquid and worms through the device. Imaging: The device is compatible with conventional microscopy. It is possible to maintain the worms within the device for at least 14 hours. Unloading: Reverse the direction of flow of liquid through the device by attaching a buffer-filled syringe to the outlet of the device. Manually depress the plunger of the syringe to push the worms out of the device. If desired, the worms may be returned to NGM plates seeded with bacteria for further culture.

device operates in the regime of constant pressure, clamped animals will not be damaged (see Section 5.3.2).

3. When it appears that worms are no longer entering the device, add an additional 0.5 mL aliquot of the suspension of worms to the inlet reservoir. Repeat this step until approximately 90% of the clamps of the device contain worms.
4. Leave the source of vacuum connected to the device to maintain the worms in the immobilized state. It is possible to move the device around as long as it remains attached to the vacuum.
5. Because the vacuum continuously drives liquid through the device, it is important to monitor the level of liquid in the inlet reservoir and add fresh buffer into the reservoir as needed.
6. Perform imaging or laser microsurgery. Keep the source of vacuum on to maintain immobilization of the worms. Immobilized worms may be maintained in the device for at least 14 hours. For long-term immobilization, it may be desirable to use an inlet reservoir with a greater volume than the glass Pasteur pipet.
7. (optional) To treat the immobilized worms with a solution of a soluble chemical compound or a suspension of particles, add the solution or suspension to the inlet reservoir.

### 5.3.9 Unloading worms from the device

It is possible to recover the worms from the device following immobilization. To accomplish this,

1. Remove the tubing that connects the source of vacuum to the device from the access hole of the outlet.
2. Remove the inlet reservoir from the device. To prevent the contents of the inlet reservoir from draining onto the device and microscope stage, place a gloved thumb over the mouth of the inlet reservoir.
3. Attach a piece of PE tubing, approximately 15 cm in length, to the inlet of the device.
4. Place the free end of the PE tubing into a 15 mL centrifuge tube.
5. Connect a 5 mL syringe containing M9 buffer to the outlet of the device and depress the plunger of the syringe to push the worms out of the clamps through the PE tubing at the inlet and into the 15 mL centrifuge tube.
6. Spin the suspension of adult worms in a centrifuge at 750 × g for 60 seconds.
7. Aspirate the suspension to 1 mL.
8. Transfer the 1 mL suspension to an NGM agar plate seeded with *E. coli*. This protocol enables successful recovery of at least 99% of the clamped worms from the device.

## 5.4 Data Acquisition, Anticipated Results, and Interpretation

Ideally, after loading worms in the array of clamps, each clamp should contain a single worm. Figure 5.8 shows a photomicrograph of 118 worms immobilized within the array of clamps. Typically, the array of 128 clamps successfully immobilizes a single worm in approximately 90% of the clamps (Figure 5.8). Three factors can reduce this success

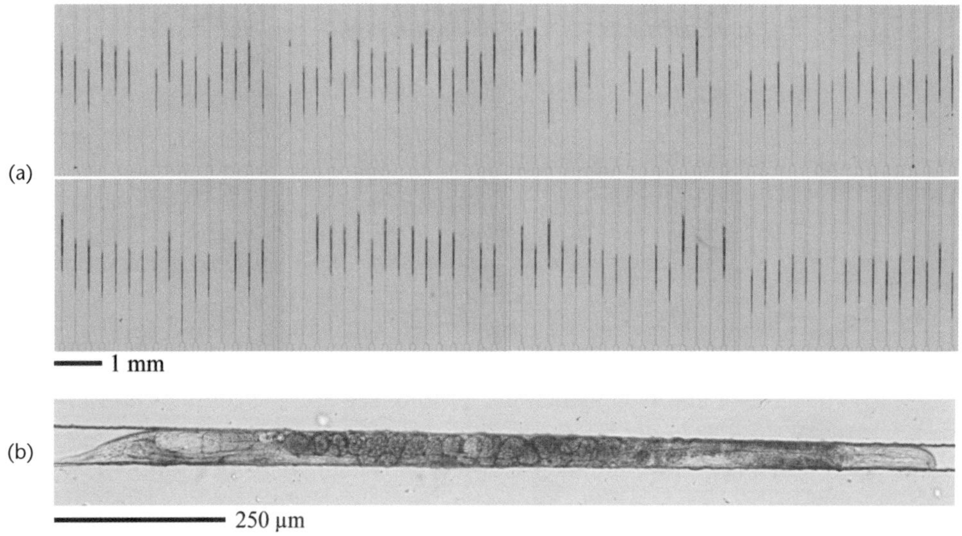


**Figure 5.8** (a) A composite image of over 100 worms immobilized within the 128-clamp device. Worms reach different points in the clamps because of the variation in the size of the worms. (b) A magnified view of a single worm within a clamp. Body structures such as the head, tail, pharynx, intestine, oocytes, and fertilized eggs are visible.

rate. First, microscale debris can clog the channels and prevent worms from entering the clamps. It is possible to avoid this outcome by washing the suspension of synchronous worms thoroughly prior to loading in order to ensure that debris from the NGM plate (e.g., bacteria, eggs, shed cuticles) is removed from the suspension. Second, two or more worms may enter a single clamp. Adjustment of (1) the concentration of worms in the suspension, (2) the rate at which aliquots of the suspension are added to the inlet reservoir, and (3) the pressure of the vacuum driving flow through the device can significantly reduce the occurrence of these "multiples." With optimization of these three parameters, it is possible to reduce the frequency of "multiples" to below 1%. Finally, a fraction (<10%) of the clamps in the device may remain empty despite the presence of additional worms in the inlet reservoir. The cause of these empty clamps is a decrease in the overall rate of flow into the device at the inlet as the clamps become populated with worms due to the increase of the total fluidic resistance of the device. A more detailed discussion of this phenomenon can be found in [1]. Worms can survive within the clamps without food for at least 14 hours; longer immobilization times will likely require the addition of a food supply.

The type of data collected with the array of clamps depends on the nature of the experiment being performed. In general terms, one would load a population of worms into the device, score a particular phenotype, and remove the population from the device. In addition, during immobilization, it is possible to apply a physical or chemical perturbation, such as laser ablation, radiation exposure, or the addition of dissolved chemicals or a suspension of particles. A simple example would be the measurement of body length in a population of worms. For this measurement, one would image each worm in the device using bright-field microscopy and determine the length of each worm from the photomicrographs. The mean length (±SD) of the worms in Figure 5.8 is (1.2 ± 0.1) mm.

## 5.5 Discussion and Commentary

Many experiments in *C. elegans* biology require the immobilization of live worms. The microfluidic worm clamp device described here provides a fast, high-throughput, simple, noninvasive, and reversible means for immobilizing worms.

The template for the design of the microfluidic array of 128 clamps can be found in the online supplementary information for this book ("Supplemental File 5.1 wormclamps.dxf"). For specific applications, it may be desirable to alter the design of the device. Certain features of the device are critical and are useful principles for the design of any microfluidic device for *C. elegans*. First, to avoid placing mechanical stress on the worms, the regions of the device intended for them should not contain sharp corners, sharp edges, or abrupt constrictions. Second, the use of a constant pressure difference between the inlet and the outlet to drive the flow of liquid through the device is a useful way of placing an upper limit on the pressure gradients experienced by the worms in the device.

Because the fabrication of devices involves replica-molding a single master many times, it is beneficial to spend extra time making sure that the master is of high quality. Under inspection with a microscope, the features of a developed master should appear smooth and free of cracks. If the features do not appear smooth, it may be beneficial to extend the preexposure bake of SU-8-coated wafer to overnight at 90°C on a level hot plate. This long baking time allows the resist to become completely level; imperfections due to bubbles or uneven evaporation of solvent will be smoothed out. In addition, avoiding any rapid changes in temperature can reduce the formation of cracks in the features of the master.

In general, sources of failure during the assembly and use of microfluidic devices are: (1) the lack of a good seal between the PDMS replica and the underlying glass substrate, (2) the presence of bubbles within the device, and (3) the formation of clogs within the microchannels. To produce a strong bond between the PDMS replica and the glass substrate, it may be necessary to optimize the duration of the plasma oxidation step. If either material is not sufficiently oxidized, the surfaces will not be reactive enough to bond. If the surface of PDMS is overoxidized, it will become glassy and brittle and will not form the necessary conformal contact with the glass substrate. In addition, chemical contaminants in the plasma cleaner, on the glass, or in the PDMS itself may prevent bonding. If the failure of PDMS to seal to glass is a persistent problem, it may be necessary to spin-coat the glass substrate with a thin layer of PDMS. Typically, PDMS seals more reliably to itself than it seals to glass. To produce a glass slide with a 50 $\mu$m layer of PDMS, transfer approximately 3 mL of uncured PDMS onto the glass slide, spin the slide for 30 seconds at 2,000 rpm, and cure the slide overnight in an oven at 65°C.

Air bubbles can form in PDMS devices if the walls of the device become hydrophobic. To avoid this occurrence, use the devices as soon as possible after they are filled with liquid. In addition, if the inlet reservoir of the device contains bubbles, they can enter the device. Before the source of vacuum is attached to the device, the inlet reservoir and the inlet of the device should be visually inspected to check for bubbles. Also, if the pressure drop provided by the vacuum is too large, bubbles may be pulled into the device. It may be necessary to adjust the vacuum to find the ideal pressure drop for the device.

The obstruction of flow due to clogging can be fatal to microfluidic devices. Sources of clogs in the device include debris in the suspension of worms (bacteria, eggs and young larvae, shed cuticles) and dust from the laboratory environment. To minimize the formation of clogs, care should be taken to prevent dust and debris from entering the device: suspension of worms should be washed at least three times, all liquids should be filtered through a 0.2 $\mu$m membrane prior to use, and the PDMS replica should be cleaned with adhesive tape prior to the assembly of the device.

Several problems may arise when loading worms into the device. If multiple worms appear to be stuck at the inlet of the device, the concentration of worms may be too high. Alternatively, it may be that the tubing of the inlet reservoir is inserted too far into the PDMS device and is blocking the interface between the inlet and the microchannels. If this issue persists, it may be necessary to widen the inlet channel in the design of the device.

It is also important for the size of the device to match the size of the worms being used. If worms in the clamps are able to move in and out of the plane of the device, the channels are too tall; a new master should be fabricated with features of lower height. If worms pass completely through the device, it may be that the worms are too small for the device (the device presented here is optimized for young adult worms). In this case, it would be necessary to redesign the device with narrower clamps. Alternatively, if the pressure drop provided by the vacuum is sufficiently large, even adult worms will be pulled completely through the device. In this situation, the pressure drop provided by the vacuum should be reduced. The problems most likely to be encountered during the fabrication and operation of the device, explanations, and suggested solutions are given in the Troubleshooting Table.

A number of alternative microfluidic methods exist in the literature for manipulating and immobilizing worms. A unifying feature of these devices is that single worms are loaded one at a time into an on-chip immobilization chamber. In the simplest case, the immobilization chamber consists of a shallow microchannel, which partially immobilizes worms by trapping them between the ceiling and floor of the channel [5]. In another approach, the wall of the immobilization chamber contains an array of small (compared to the worm) aspiration channels, which employ suction to hold a worm in place [6, 7]. In other devices, the ceiling of the immobilization chamber is a thin, flexible membrane that can be deflected downward to immobilize the worm [7, 8]. In contrast to these physical methods for immobilizing worms, another device uses an integrated cooling system to cool worms to 4°C [9]. In these devices, worms are loaded into, and unloaded from, the immobilization zone either: (1) by manually loading single worms into the device one at a time [5] or (2) by employing a collection of computer-controlled microfluidic pumps, switches, and valves to move worms through the device [6–9]. The inclusion of pumps, switches, and valves enables relatively rapid manipulation of worms within the devices; however, these control elements complicate both the fabrication and operation of these devices. The principle advantages of the microfluidic worm clamp device described here over these alternative methods are *simplicity* and *parallel processing*: hundreds of worms are immobilized automatically and simultaneously without a need for any sophisticated control elements.

## 5.6 Application Notes

The possible applications are limited only by the imagination and ingenuity of researchers. The device is compatible with most types of conventional microscopy,

### Troubleshooting Table

| | Problem | Possible Cause | Solution |
|---|---|---|---|
| 1 | PDMS replica does not peel off the master | PDMS has been stored on master too long; master is insufficiently silanized (relevant only the first time master is used) | Remove PDMS replicas from master within 1 week after curing; repeat silanization step |
| 2 | PDMS doesn't seal or leaks | Oxidation time in plasma cleaner is too long/short for either PDMS or glass; glass is dirty | Decrease or increase oxidation times to optimize sealing; clean glass (etching may be necessary) |
| 3 | There are air bubbles in channels | PDMS channels have reverted to hydrophobic state; vacuum is too high | Fill device with liquid and use for experiment directly after bonding PDMS to glass; reduce the strength of the vacuum |
| 4 | There is debris in channels | PDMS replica was contaminated with dust; solutions/suspensions contain debris | Clean PDMS replica with adhesive tape prior to device assembly; filter all solutions/wash all suspensions prior to adding to device |
| 5 | Worms are stuck in inlet | Concentration of worms is too high; inlet channel is too narrow | Lower the concentration of worms; redesign device with a wider inlet channel |
| 6 | Worms are sucked through device or get in too quickly | Vacuum is too high; fluidic resistance of the device is too low | Reduce the strength of the vacuum |
| 7 | Worms wiggle in or out of the plane of the device | Channels are too tall | Fabricate master with shallower features |
| 8 | Worms zip through the clamps; there is no clamping | Worms are too small for the device | Make clamps more tapered or use older (i.e., larger) worms |

including bright-field, fluorescence, DIC, and phase-contrast imaging. For example, Figure 5.9 shows bright-field [Figure 5.9(a)] and fluorescence [Figure 5.9(b)] images of a *C. elegans* mutant that expresses green fluorescent protein in the pharynx. Potential observables include, but are not limited to, the intensity and distribution of a fluorescent biomarker (e.g., the presence or absence of a fluorescently labeled protein or the fluorescent signal of a FRET-based calcium dye for monitoring the activity of a nerve cell) and the morphology of a body structure within the worm (e.g., the length of a regenerating neuronal axon or the sarcopenic degeneration of muscles in aging worms).

In addition, it is possible to use the device: (1) to immobilize worms for laser-mediated microsurgery [2], (2) to selectively expose regions of the bodies of immobilized worms to radiation, and (3) to treat worms with solutions or suspensions of stimuli during immobilization. Figure 5.10 shows the use of the device for laser-mediated microsur-

(a)

(b)

**Figure 5.9** It is possible to perform both (a) bright-field imaging, and (b) fluorescence imaging of worms within the device. The worm in the images is a mutant that expresses green fluorescent protein in the pharynx. In the bright-field image, the whole body of the worm is visible; in the fluorescence image, only the pharynx of the worm is visible.

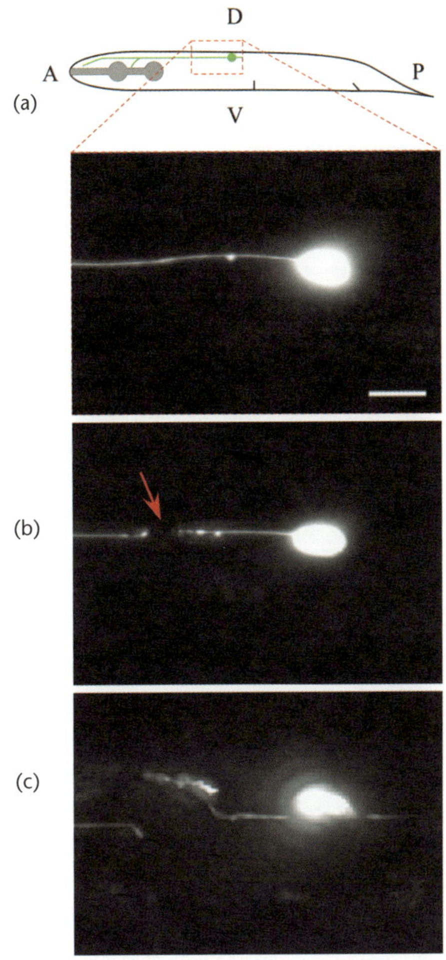


**Figure 5.10** Laser ablation of an individual nerve fiber in *C. elegans* immobilized within the worm clamp device. (a) In vivo fluorescence image of the ALM mechanosensory neuron immediately before laser surgery (scale bar 10 $\mu$m). The schematic diagram shows the location of the ALM neuron in the body of the worm (A, P, D, and V indicate the anterior, posterior, dorsal, and ventral sides of the worm). (b) The same neuron shortly after laser surgery showing a clear break in the nerve fiber; the red arrow indicates the location of the break. (c) Regeneration of the severed nerve fiber 9 hours after laser surgery.

gery: it was possible to immobilize a worm with a fluorescently labeled neuron within a clamp and to ablate a region of the axon of the neuron using a femtosecond laser. The clamp enabled the collection of a time series of fluorescent images that recorded the regeneration of the ablated neuron (see "Supplemental Movie 5.2.avi").

## 5.7 Summary Points

- Immobilization is a necessity in *C. elegans* research.
- An ideal immobilization technique should be reversible, noninvasive, and nonharmful to the animals.

- The small size of *C. elegans* and its ability to live in liquid make it possible to develop lab-on-a-chip microfluidic tools for studying *C. elegans*.
- The microfluidic worm clamp device provides a method for reversibly immobilizing over 100 worms in less than 15 minutes.
- Because the worm clamp device is optically transparent, it is compatible with conventional optical microscopy and laser ablation experiments.

## Acknowledgments

We would like to express our gratitude to our mentors Professor George M. Whitesides and Professor Aravinthan Samuel, as well as to acknowledge the invaluable contribution of Professor Walter Fontana and Dr. Javier Apfeld to the development of the worm clamp.

### Annotated References

[1] Hulme, S. E., et al., "A microfabricated array of clamps for immobilizing and imaging *C. elegans*," *Lab on a Chip*, Vol. 7, No. 11, 2007, pp. 1515–1523.
Reference 1 presents the design, fabrication, and operation of the 128-clamp immobilization device.

[2] Pinan-Lucarre, B., et al., "Apoptotic death caspase CED-3 promotes neuronal regeneration in caenorhabditis elegans," in preparation.
Reference 2 uses the 128-clamp device to perform laser ablation and time-lapse microscopy on *C. elegans*.

[3] Xia, Y., and Whitesides, G. M., "Soft lithography," *Angewandte Chemie*, Vol. 37, No. 5, 1998, pp. 550–575.
Reference 3 provides an overview of the techniques and capabilities of soft lithography.

[4] Chalfie, M., and Mendel, J., (eds.), "WormBook: The online review of *C. elegans* biology," www.wormbook.org (last accessed on September 16, 2008).
WormBook is a comprehensive and up-to-date online review of the biology of *C. elegans*, as well as a compilation of protocols for working with *C. elegans*.

[5] Chronis, N., Zimmer, M., and Bargmann, C. I., "Microfluidics for in vivo imaging of neuronal and behavioral activity in *Caenorhabditis elegans*," *Nature Methods*, Vol. 4, No. 9, 2007, pp. 727–731.

[6] Rohde, C. B., et al., "Microfluidic system for on-chip high-throughput whole-animal sorting and screening at subcellular resolution," *Proc. Natl. Acad. Sci. USA*, Vol. 104, No. 35, 2007, pp. 13891–13895.

[7] Zeng, F., Rohde, C. B., and Yanik, M. F., "Sub-cellular precision on-chip small-animal immobilization, multi-photon imaging and femtosecond-laser manipulation," *Lab on a Chip*, Vol. 8, No. 5, 2008, pp. 653–656.

[8] Guo, S. X., et al., "Femtosecond laser nanoaxotomy lab-on-a-chip for in vivo nerve regeneration studies," *Nature Methods*, Vol. 5, No. 6, 2008, pp. 531–533.

[9] Chung, K., Crane, M. M., and Lu, H., "Automated on-chip rapid microscopy, phenotyping, and sorting of *C. elegans*," *Nature Methods*, Vol. 5, No. 7, 2008, pp. 637–643.
References 5 to 9 are recent examples of microfluidic tools for studying *C. elegans*.

### Supplementary electronic materials and resources

- "Supplemental Movie 5.1.avi": a worm crawling on an NGM plate
- "Supplemental Movie 5.2.avi": regeneration of an ablated neuron in a worm immobilized within the worm clamp
- "Supplemental File 5.1 wormclamps.dxf": design file for the 128-clamp device

These are all available at http://www.methodsinbioengineering.com/library.asp.

CHAPTER

# 6

# Osmolality Control for Microfluidic Embryo Cell Culture Using Hybrid Polydimethylsiloxane (PDMS)–Parylene Membranes

Yun Seok Heo,[1] Andreja Jovic,[1] Lourdes M. Cabrera,[2] Gary D. Smith,[2,3,4] and Shuichi Takayama[1,5]

[1]Department of Biomedical Engineering
[2]Department of Obstetrics and Gynecology
[3]Department of Urology
[4]Department of Molecular and Integrated Physiology
[5]Macromolecular Science and Engineering
University of Michigan, Ann Arbor, MI 48109

## Abstract

An important first step for adapting microfluidics to embryo culture and development is to understand how the microenvironment can be controlled and improved to support these cells. One oft-overlooked aspect of this technology is the detrimental osmolality shifts due to evaporation of culture medium through membranes, which transduce mechanical deformation into flow or valving. Here we present a method for reducing evaporation in polydimethylsiloxane (PDMS)–based platforms, ultimately preventing osmolality shifts to promote successful embryo development, from the single-cell stage into a blastocyst. A hybrid PDMS-Parylene-PDMS membrane was developed that significantly reduced evaporation and was also amenable to real-time imaging and to deformation-based actuation for microfluidic valving and pumping. This technique is well suited for a wide range of microfluidic cell culture systems where greater control of the microenvironment is necessary to reduce evaporation-associated osmolality shifts.

Key terms
embryo culture
osmolality measurement
evaporation in cell culture
microfluidics

## 6.1 Introduction

The advent of microfluidic technology has led to the ability to culture cells with minimal use of reagents, to address cells with exquisite spatial [1] and temporal [2] control, and to create microenvironments that better mimic in vivo fluid-to-cell-volume ratios [3]. Despite these advantages over conventional cell culture techniques, the technology is limited by problems associated with evaporation through the materials commonly used for fabrication of microfluidic systems, such as polydimethylsiloxane (PDMS). PDMS is an attractive material for microfluidics because of its favorable mechanical properties, optical transparency, biocompatibility [4, 5], and straightforward manufacture by rapid prototyping [6]. On the other hand, evaporation through this material poses an immense problem because of the significant increases in osmolality that can result from the loss of even minute volumes of fluid from the system. Ion balance [7], cellular growth rate [8], metabolism, antibody production rate [9], signaling [10], and gene expression [9, 11] are all affected by shifts in osmolality. Particularly sensitive to these shifts are mammalian gametes and embryos, from which stem cells can be derived. Development in these cells is hindered by osmolalities significantly lower or higher than 265 to 285 mmol/kg [12], representing a particularly narrow range compared to other cell lines, which are more robust in this context. It is therefore important to develop methods that address this important aspect of a cell's microenvironment.

Previously, several techniques have been developed to overcome the problems associated with evaporation through PDMS-based devices. This includes placing water-filled reservoirs on-chip [13, 14], submerging the whole chip in water [15], applying PCR-tape to the chip [16, 17], using oil to cover aqueous liquids [18, 19], and changing the curing-agent-to-base ratio [20]. While these methods effectively address the evaporation issue, there are significant drawbacks in terms of adaptability to deformation-based actuators, as well as optical accessibility and limitations in the environment in which the setup can be used. Here we present a method that greatly reduces evaporation-associated osmolality shifts through the development of a hybrid PDMS-Parylene-PDMS membrane with low water permeability that is thin enough to be adapted to deformation-based microfluidic systems and real-time imaging of cells cultured on such devices. We demonstrate how the method can be applied to successfully culture mouse embryos to develop from the one-cell stage into blastocysts in a Braille display-actuated microfluidic setup. The insights and methods described should be broadly useful for advancing microfluidic cell cultures and assays where prolonged microliter cell cultures need to be performed in PDMS devices.

The challenge of maintaining a supportive environment for embryo culture is significant. Despite the humidified environment (typically 85% humidity) of a cell culture incubator, significant evaporation was measured, with thinner PDMS-membrane devices having a more dramatic shift in osmolality and a correspondingly worse development rate for embryos, where a majority of embryos stop development at the two-cell stage, or in approximately 24 hours. In heated, yet nonhumidified, environments, such as portable handheld cultures with real-time video microscopy [21], in-channel evaporation is even faster, leading to cell death in less than 1 hour in the absence of media refreshment. In this chapter, we first describe how to quantify the rate of evaporation and its effect on the composition of cell culture media by measuring the rate of osmolality shifts in PDMS chips with different-thickness membranes over a 4-day period inside a humidified cell culture incubator.

## 6.2 Experimental Design

### 6.2.1 Hypothesis

Long-term embryo culture and development on PDMS-based microfluidic platforms is constrained by evaporation-mediated osmolality shifts or techniques that limit evaporation but inhibit practicality; this can be overcome with the introduction of PDMS-Parylene-PDMS membranes. For experiments involving assessment of embryo development, 14 to 15 embryos were analyzed per experiment, and this was performed at least in triplicate for a total of 42 to 45 embryos per experimental condition. To properly evaluate the osmolality levels in wells with PDMS membranes of varying thicknesses over 96 hours, experiments were performed at least in triplicate. In order to ensure reproducibility of results for these experiments, the osmometer was calibrated daily with ampoules of one-time-use calibration solution.

## 6.3 Materials

### 6.3.1 Reagents

- Six- to eight-week-old B6C3F1 female mice (Charles River)

  *Caution:* All studies must be conducted with an approved animal protocol from your institution, and all animal experiments must comply with national regulations and U.S. National Institutes of Health guidelines for the care and use of experimental animals.

  *Alternative:* Should this prove impractical, mouse embryos are available for purchase through companies like Embryotech, which provide straws of embryos that can be thawed for eventual use in experiments.
- Slide glasses (75 × 25 × 1 mm) (Corning Glass Works, Corning, New York)
- Potassium simplex optimized medium (KSOM; Specialty Media, Phillipsburg, New Jersey)
- Serum substitute supplement (SSS; Irvine Scientific, Santa Ana, California)
- Pregnant mare serum gonadotropin (PMSG; Sigma Chemical Co., St. Louis, Missouri)
- 10 IU human chorionic gonadotropin (hCG; Sigma Chemical Co.)
- Modified human tubal fluid medium (HTF-H; Irvine Scientific)
- HEPES (1×) (Sigma Chemical Co.)
- Hyaluronidase (Sigma Chemical Co.)
- PDMS prepolymer (Sylgard 184, Dow Corning)
- Wescor's Optimol osmolality standards (Wescor, Utah)
- Parylene C

  *Alternative:* Another viable option is to use Parylene coated on wafers through batch processing by a company, such as Specialty Coating Systems.
- SU-8 (MicroChem, Newton, Massachusetts)
- 4" silicon wafer

*Tip:* Any type works, but it is often less expensive and sufficient to purchase "reject-" or "test-grade" silicon wafers.

- Tridecafluoro-1,1,2,2-tetrahydrooctyl-1-trichlorosilane (United Chemical Technologies, Bristol, Pennsylvania).
- 14G blunt needle (BD)
- Oil for embryo culture (light mineral oil) (Irvine Scientific)

### 6.3.2 Equipment

- Wescor's VAPRO Vapor Pressure Osmometer (Wescor, Utah)
- PDS 2010 labcoater (Specialty Coating Systems)
- Wire thermocouple (5TC-TT-J; Newport, Santa Ana, California)
- Plasma oxidizer (SPI PlasmaPrep II, SPI Supplies, West Chester, Pennsylvania)
- Stereomicroscope

## 6.4 Methods

The flow diagram depicted in Figure 6.1 gives a general idea of the setup for this method and approximately how long particular sets of steps take.

### 6.4.1 PDMS-Parylene-PDMS membrane preparation

1. Thoroughly mix a 10:1 solution of PDMS base to curing agent.
2. Remove bubbles from the mixture through vacuum at 200 mm Hg.

   *Tip:* Do not leave more than 1 hour under vacuum to avoid premature curing.
3. React vapor of tridecafluoro-1,1,2,2-tetrahydrooctyl-1-trichlorosilane onto silicon wafers for 15 minutes in vacuum at 200 mm Hg.

   *Caution:* Reagent is highly toxic and can cause nausea, severe eye irritation, and chemical burns. This reagent must be used in a fume hood, and nitrile gloves must be worn at all times during handling. Throw away gloves and any other materials exposed to this solution in a container separate from regular waste.
4. Spin-coat PDMS mixture onto silanized silicon wafers, and place into 120°C oven and cure for 15 minutes.

   *Tip:* Spin-coating settings commonly used are 370 rpm for 4 minutes. This results in a PDMS layer of approximately 100 $\mu$m.
5. Coat a layer 2.5 to 5 $\mu$m thick of Parylene C at 4 mTorr on the backside of cured PDMS membranes using a PDS 2010 labcoater (Specialty Coating Systems).

   *Note:* The furnace of the machine should be set to around 690°C, and the pressure sensor should be heated to around 135°C to prevent the Parylene from sticking to it. The cold trap should be set to about –80°C so that Parylene particles do not reach the vacuum pump.

   *Alternative:* Should this prove impractical, silicon wafers with spin-coated PDMS can be sent to a company, such as Parylene Coating Inc., and a batch-process order can be placed. It may take on the order of 1 month to receive the finished samples, but hundreds of wafers can be coated at a time.

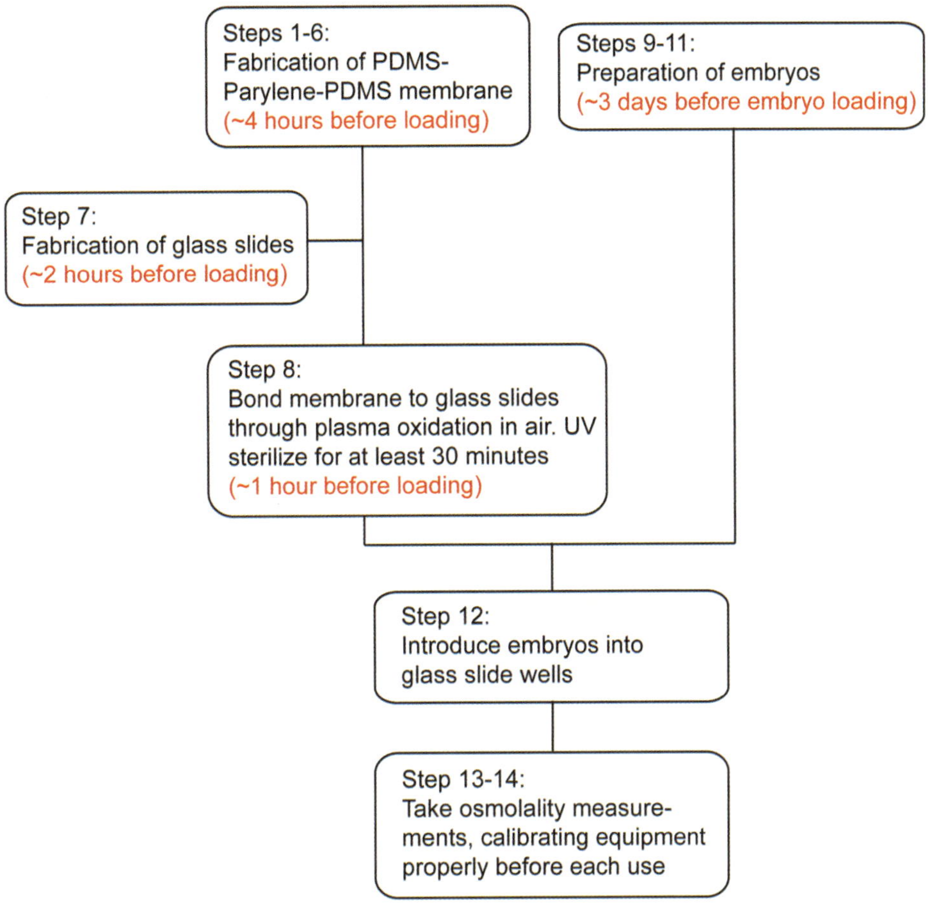


**Figure 6.1** Generalized protocol flow diagram. The steps of the protocol can be divided into the following categories: device membrane fabrication, embryo preparation, glass-slide fabrication, device bonding and sterilization, embryo loading, and osmolality measurement. The red lettering indicates the approximate times when the designated steps should to be carried out.

*Alternative:* As an alternative, we showed previously [22] that it is possible to use plastic tape in place of the PDMS-Parylene-PDMS membrane to reduce evaporation in PDMS-based microfluidic setups. However, this method is ultimately impractical because, at high temperatures, the adhesive presumably becomes cytotoxic to embryos.

6. Spin-coat another PDMS layer on this membrane, and cure for an addition 15 minutes in a 120°C oven.

   *Tip:* Use the same spin-coating settings as step 4.

### 6.4.2 Preparation of glass slides and bonding to hybrid membranes

1. Have four holes drilled (by a glass-blowing shop) into glass slides (75 × 25 × 1 mm) with diameters of 6.5 mm spaced 1.5 cm apart.

*Note:* Glass slides were used to avoid contact of oil with PDMS, which could potentially lead to unwanted side effects of oil adsorption into the PDMS membrane (Figure 6.2).

*Alternative:* Using the proper drill-bit diameter, holes can be drilled into the glass slides directly. Use caution when doing so.

2. Plasma-oxidize the glass slide and the membrane under air for 30 seconds and bond together. UV-sterilize for at least 30 minutes. Dispense 50 $\mu$l of KSOM media and cover with 500 $\mu$l of oil. Media needs to equilibrate overnight in environment of 5% $CO_2$ in air at 37°C.

   *Tip:* In our lab, effective, rapid bonding is achieved by using an SPI PlasmaPrep II plasma oxidizer where samples are exposed to 400 mTorr of pressure during oxidation under air plasma for 30 seconds. (See Troubleshooting Table.)

### 6.4.3 Embryo preparation

1. Superovulate 6- to 8-week-old B6C3F1 female mice by an intraperitoneal administration of 10 IU of pregnant mare serum gonadotropin (PMSG; Sigma Chemical Co.) by injection, followed 44 hours later with an injection of 10 IU human chorionic gonadotropin (hCG; Sigma Chemical Co.).

   *Alternative:* To avoid handling mice, frozen embryo straws can be ordered from companies like Embryotech. Thaw the embryo and continue to step 3.
2. Collect zygotes 18 hours later by dissecting oviducts in HEPES-buffered human tubal fluid medium (HTF-H; Irvine Scientific) supplemented with 0.1% w/v hyaluronidase to remove surrounding cumulus cells, then washed zygotes three times in warmed HTF-H supplemented with 0.3% (w/v) BSA.
3. Place zygotes into 50 $\mu$L of KSOM plus 0.1% SSS overlaid with mineral oil in organ culture dishes (control) and glass-slide wells. (See Troubleshooting Table.)
4. Place zygotes and culture devices into a humidified environment of 5% $CO_2$ in air at 37°C.

### 6.4.4 Osmolality measurements

1. Calibrate the osmometer using osmolality standards before every run.
2. For measurements, extract 10 $\mu$L of sample and place it on the sample-loading area.

**Troubleshooting Table**

| Step Number | Solution |
|---|---|
| Section 6.4.2, step 2 (device bonding) | Should the bonding prove problematic, it is advised to peel off the PDMS-Parylene-PDMS membrane from the silicon wafer and oxidize the bottom PDMS surface and use this layer for bonding to the sample with features. This is helpful because the PDMS-Parylene bond between the initial PDMS layer onto which Parylene was deposited is stronger than the bond between the PDMS layer that is later spin-coated onto Parylene. Also, if bonding problems persist, a corona treater can be used to oxidize the surfaces of the sample with features and the membrane. |
| Section 6.4.3, step 3 (embryo preparation) | During even short intervals of embryo handling while they are being loaded into microdevices and overlaid with mineral oil, osmolality shifts can result due to evaporation. Embryo viability can sometimes be improved, especially when working with very small volumes, by initially handling them in media with a lower osmolality than is normally used for embryos (in the range of 230 to 250 mmol/kg). |

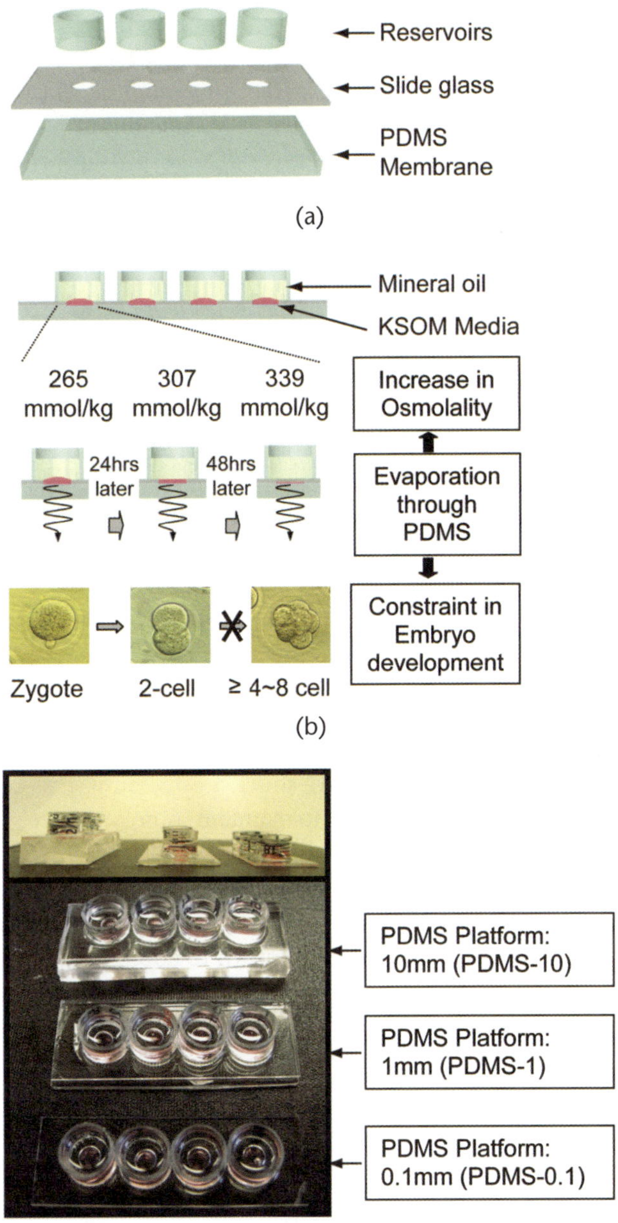


**Figure 6.2** (a) Schematic design of glass slide with 6.5 mm diameter holes attached to PDMS membranes of 10, 1.0, or 0.1 mm thickness to form wells. (b) A cross section showing the glass-PDMS hybrid structure filled with media and overlaid with mineral oil. Evaporation through PDMS increases osmolality in media and constrains embryo development. The displayed values are the measured average osmolality in PDMS-0.1 at 24-hour intervals. At 48 hours the osmolality shifts from 265 mmol/kg to 339 mmol/kg, and embryo development is blocked accordingly at the approximately 4- to 8-cell stage. (c) A photograph of an actual device filled with media and mineral oil. (Figure adapted from [22] and used with permission from Analytical Chemistry.)

## 6.5 Data Acquisition, Anticipated Results, and Interpretation

Using the protocol presented here, we made measurements to quantify the osmolality shifts in wells created in glass slides with PDMS membranes of varying thicknesses over a period of 4 days. We show how these osmolality shifts directly contribute to the inhibition of embryo development from the zygote stage. Furthermore, we repeat these quantitative measurements of evaporation-associated osmolality shifting with wells having hybrid PDMS-Parylene-PDMS membranes. Not only do these membranes vastly reduce evaporation and maintain osmolality over a 4-day period, but it was also found that they were mechanically adaptable to a deformation-based actuation system for microfluidic systems and optically feasible for real-time imaging.

Holes were drilled into glass slides (by a glass-blowing shop) and bonded through plasma oxidation to PDMS membranes of various thicknesses (100 $\mu$m, 200 $\mu$m, 1 mm, 1 cm). The resulting wells were filled with 50 $\mu$L KSOM, which was then covered with mineral oil; this setup ensured that evaporation occurred through the PDMS membrane for the most part. As a control, a microdrop consisting of 50 $\mu$L of KSOM covered with mineral oil in a culture dish was used. The samples were placed into a humidified cell culture incubator (85% humidity, 5% $CO_2$ in air, 37°C), and measurements of osmolality were taken over a 4-day period. For these measurements, 10 $\mu$L of media were extracted for each condition, and the osmolality was measured using an osmometer. It was found that compared to the control condition, PDMS membranes with thicknesses greater than 1 mm had comparable osmolalities over 96 hours (Figure 6.3). By contrast, PDMS membranes at 200 $\mu$m and smaller experienced large shifts in osmolality. This demonstrated that while thicker PDMS membranes were able to maintain osmolality levels over a period of several days, they still posed a problem in terms of adaptability to microfluidic actuation systems since deformation-based setups are unable to properly deform such thick membranes.

In order to gain useful quantitative insights into the material properties of PDMS, Fick's First Law of Diffusion, which is known to approximate transport through PDMS reasonably well, was used to model the rate of evaporation through the PDMS membranes (Figure 6.3). Osmolality refers to the number of dissolved particles per kilogram of water; this includes ions and undissociated molecules, such as glucose or proteins [23]. The total number of dissolved particles is represented by $O^*$; this value is assumed to be constant over time. $O$ is the measured osmolality (millimoles per kilogram), and $m$ is the total mass (kilograms) of the media. This gives us the following relationships:

$$O(mmol/kg) \times m(kg) = O^*,\ m(kg) = \frac{O^*}{O} \tag{6.1}$$

$$\therefore \frac{m_2}{m_1} = \frac{O^*/O_2}{O^*/O_1} = \frac{O_1}{O_2} \tag{6.2}$$

We can subsequently define $J$ to be the mass flux through a unit area per unit time (kilograms per square meter-second). $\Delta m$ is mass change, and $A$ is the area (square meters). For our setup, the area of the well was 6.5 mm, and $t$ is time (seconds), which is 24 hours in our case.

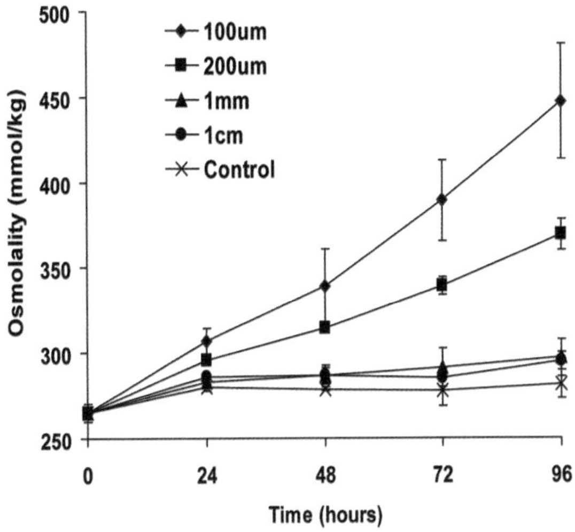

(a)

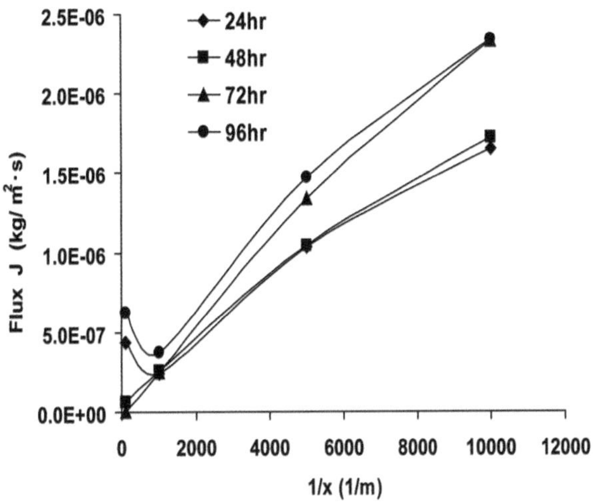

(b)

**Figure 6.3** (a) Osmolality of media (KSOM + SSS) over time when cultured in control (organ culture dish) and glass-slide wells with PDMS membrane bottoms of varying thickness. Osmolality changes over time were significantly different for PDMS 0.1 mm compared to remaining treatment groups $P < 0.01$. (b) Based on (6.3), Flux J was plotted with respect to $1/x$ using the measured osmolalities with varying PDMS thicknesses. (Figure adapted from [22] and used with permission from Analytical Chemistry.)

$$J = -\frac{\Delta m}{A \cdot t} = \frac{1}{A \cdot t} \cdot (m_1 - m_2) = \frac{1}{A \cdot t} \cdot \left( m_1 - \frac{O_1}{O_2} \cdot m_1 \right) = \frac{m_1}{A \cdot t} \cdot \left( 1 - \frac{O_1}{O_2} \right) \tag{6.3}$$

Fick's first law tells us that the flux can be described as follows:

$$J = -D \cdot \frac{dC}{dx} = D \cdot S \cdot \frac{(P_s - P)}{dx} = D \cdot S \cdot P_s \cdot (1 - P/P_s) \cdot \frac{1}{dx} \approx k \cdot \frac{1}{x} \propto \frac{1}{x} \tag{6.4}$$

Figure 6.3(b) depicts a plot of $J$ with respect to $1/x$ as described in (6.4) The slopes, $k$, are between $10^{-10}$ and approximately $2 \times 10^{-10}$ kg/m-s, which yields diffusion coefficients, $D$, of $3 \times 10^{-9}$ to approximately $6 \times 10^{-9}$ m$^2$/s. This range of values is consistent with those previously reported [24–26]. The humidity of our incubator was measured to be approximately 0.85 (85% humidity); the humidity at the media surface is considered to be 1 (100% humidity); $P_s$ represents the saturated vapor pressure; $P$ is the vapor pressure outside of the chip's environment, and thus $P/P_s$ is 0.85; $S$ is the water solubility coefficient in PDMS, which has a value of approximately 1.04 [$(cm^3)_{water}/cmHg\cdot(cm^3)_{PDMS}$ [27]; and the saturated-vapor-specific volume ($V_g$) at 37°C is $V_g \approx 22.94$ (m$^3$/kg), $\rho_g \approx 0.0436$ (kg/m$^3$) [28].

To definitively portray the detrimental effects of the evaporation-associated osmolality shifts in these wells, mouse embryos were cultured in them over a period of 4 days in a humidified incubator. Then, 13 to 15 zygotes were introduced into glass-slide wells with PDMS membranes of thicknesses of 10, 1, and 0.1 mm (Figure 6.2). The same control was used as mentioned previously. Under these conditions, it was found that embryo development had progressed similarly after 1 day to the two-cell stage for all membrane thicknesses tested. However, after 2 days, there was a significant drop in embryo development to the four-cell stage for cells that had been cultured with PDMS membranes 0.1 mm thick; all other membrane thicknesses showed no significant change compared to the control. This demonstrated that for embryo culture, PDMS membranes 1 mm and thicker provided adequate barriers to osmolality shifts due to evaporation. Nevertheless, this result reinforced the need for membranes that could maintain osmolality over several days and were of a reasonable thickness to be adapted to deformation-based microfluidic actuation systems.

To this end, PDMS-Parylene-PDMS membranes were created. Figure 6.4(b) depicts the significant improvement in osmolality maintenance of this hybrid membrane compared to PDMS membranes that were not coated with 2.5 $\mu$m of Parylene. During 96 hours of incubation, the osmolality shifted from 265 to 285 mmol/kg for membranes with Parylene. This presented a range well suited for embryo culture. As portrayed in Figure 6.4(c), utilization of Parylene-coated PDMS membranes substantially improved blastocyst development (87 ± 14%) compared to noncoated PDMS membranes (2 ± 4%; $P < 0.01$). For all these experiments, changes in osmolality were tested for statistical difference by a mixed linear regression model for repeat measures; statistical comparisons of blastocyst development were done with chi-square analysis. $P < 0.05$ was considered statistically significant.

We also tested the viability of applying tape to PDMS membranes to reduce evaporation. While the method did succeed in doing so, its application is not advisable because of its potential cytotoxic effects at physiological temperatures. Ultimately, this demonstrated that hybrid membranes are well-suited and practical barriers to detrimental osmolality shifts; the mechanical feasibility for deformation-based microfluidics was assessed next.

## 6.6 Discussion and Commentary

Microfluidics is becoming a popular research tool for culturing and manipulating cells. With its obvious attractive possibilities for reducing the use of reagents, parallelizing

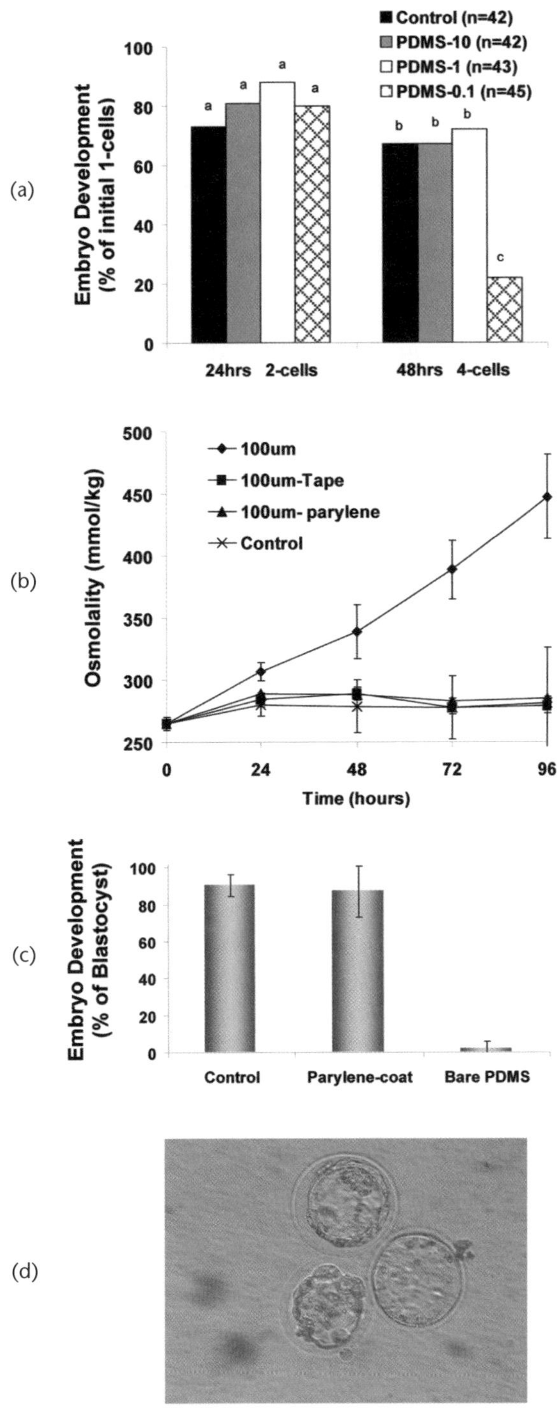


**Figure 6.4** (a) Mouse pronuclei (PN) zygote development in control (organ culture dish) and glass-slide wells with different thickness PDMS membranes. (b) Parylene was coated on the mounted PDMS outside surface with 2.5 $\mu$m thickness, and then osmolality change was measured. Osmolality of KSOM drifted from 265 to 285 mmol/kg during 96 hours of incubation in Parylene-coated PDMS-0.1. (c) Percentage blastocyst development was significantly improved with Parylene coating (87 ± 14 blastocyst development) compared to no Parylene (2 ± 4; $P < 0.01$). (d) Blastocyst images taken from Parylene-coated PDMS-0.1 well. (Figure adapted from [22] and used with permission from Analytical Chemistry.)

experiments, and mimicking physiological flow conditions, many proof-of-principle experiments have been reported. Much of the initial work, however, has been performed on relatively hearty cell lines with a broad range of acceptable culture parameters. As the microfluidic cell culture field matures, however, the type of cells that need to be cultured is expanding. Many of the cells of interest are much more delicate and sensitive to their environments, requiring higher-level control over the culture conditions. Embryos are among the most sensitive cells that exist. Thus, in addition to the importance of culturing embryos for reproductive science and infertility treatments, embryo culture provides a sensitive test of the reliability, stability, and biocompatibility of microfluidic systems. In our example, embryos serve, in a sense, as a sensor. The osmometer requires 10 $\mu$L for analysis. Microchannels only hold nanoliters. High embryo viability provides confidence that the environment in the nanoliters-volume microchannels is stable.

In terms of methods for embryo culture with the purpose of performing reproductive studies and therapies, conventional approaches include microdrop cultures or center-well dishes. Although these methods have been tremendously useful for both research and clinical practice, a large gap still exists between the physiological embryo culture environment and the in vitro one. Microfluidics can bridge this gap by controlling not only the chemical microenvironment, as demonstrated here, but also the mechanical microenvironment. Ciliary action and muscular contractions are vital for embryo transport and development, although the specifics of these mechanisms have not been fully elucidated.

What is required are model experimental systems to test hypotheses about the role of physiological flow conditions on embryo development. Such systems may also eventually have an impact in the clinic by enhancing in vitro embryo development. The methods described in this chapter provide the groundwork upon which such systems and tests may be constructed. They were developed after numerous failed efforts to culture viable embryos in PDMS devices in which many other commonly used cell lines proliferate without problem. Although evaporation was suspected relatively quickly to be a major problem, the degree of osmolality control required for embryo culture necessitated a more quantitative understanding and more stringent control of osmolality than was available previously. Alternative efforts to reduce evaporation that have failed include the use of a humidified culture incubator, presoaking of the PDMS devices in media to make them water laden, absorption of oil to increase the moisture-barrier properties, and the formation of a simple one-layer coating of Parylene without sandwiching the Parylene between another protective layer of PDMS. In this last case, the Parylene coating cracked readily upon deformation of the channels, leading to the loss of its moisture-barrier function. We want to point out that the evaporation problem was particularly acute in the type of devices described here with a thin (hundreds of micrometers), flexible PDMS membrane. This type of thin-membrane floor-channel configuration was required in our particular example because of the need: (1) to have good optical microscopy (which requires transparent thin material), and (2) to use deformation-based fluid actuation with actuators that push into the channels from the chip bottom. Applications where optical access is not required or is less demanding, where deformation-based actuation is not required, or where evaporation is not a problem or is even desired may be better served by other materials or chip designs.

In vivo, the fallopian tube or oviduct plays an essential role in gamete transport, fertilization, and the early development of the embryo. From ovary to uterus, the oviduct commonly consists of five sections: fimbria, ampulla, ampullary-isthmic junction (AIJ), isthmus, isthmus-uterotubal junction (UTJ), as shown in Figure 6.5. Embryos developing from zygotes to blastocysts in vivo travel slowly through the isthmus in the oviduct toward the uterus in a few days under dynamic activities in terms of mechanical and chemical factors. The oviductal musculature and the cilia have been considered as potential effectors of the movement. It has been reported that smooth muscle contraction plays a major role in transporting the early-stage embryo from isthmus to the uterus in a few days [29], while oocytes or zygotes are transported into the oviductal ampulla and rapidly toward the ampullary-isthmic junction by ciliary action [30]. Since fertilized embryos spend most time in the isthmus before settling down in the uterus, smooth muscle contraction plays a major role in the dynamic conditions for early-stage embryo development, which is also strengthened by the fact that the isthmus of the oviduct is generally lined with far fewer ciliated cells than the ampulla [31] and possesses a thicker muscular tunic. Mean contraction frequencies in the isthmus were approximately 8.1 per minute, which is equivalent to 0.135 Hz [32]. Further studies on models of contraction showed that a sequence of contractions in the uterine direction (a pseudoperistalic wave) will lead to positive ovum transport [33]. In addition to embryo transport, the purpose of these segmental muscular contractions may be to stir tubal contents and ensure the mixing of gametes and embryos with tubal secretions [34]. Important consid-

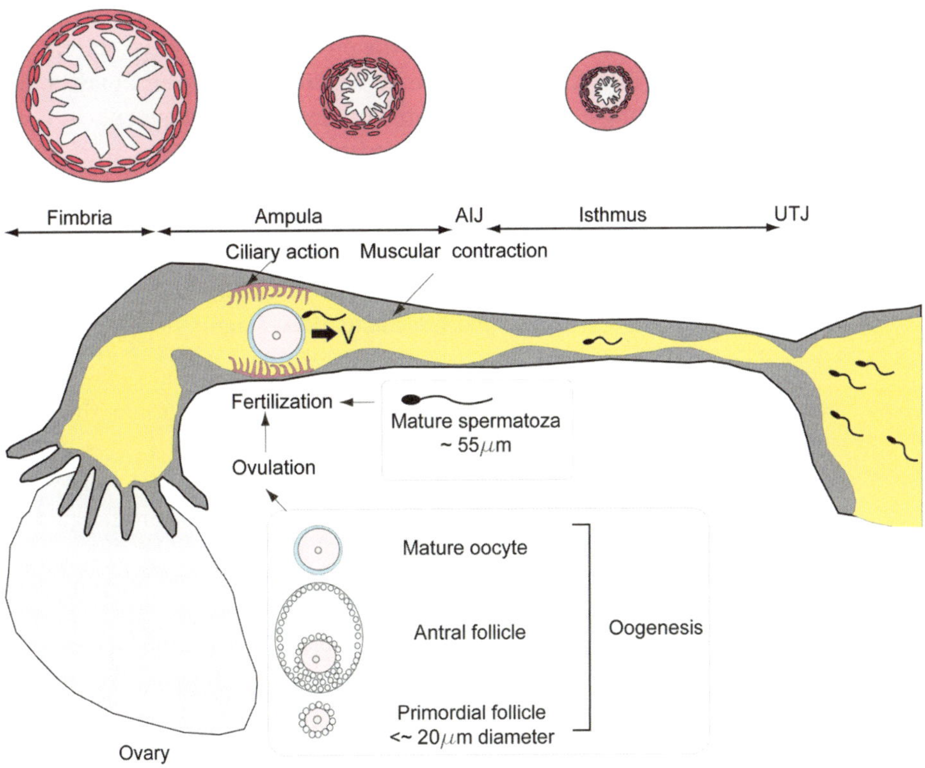


**Figure 6.5** Sperm migration, oogenesis, and microenvironment in human reproduction.

erations for recapitulating the embryo microenvironment include the fluid dynamic properties dictating embryo transport and development and the hormones controlling these mechanical parameters. Quantification of these parameters will facilitate the development of accurate mathematical models for understanding the effects that particular mechanical and chemical cues have on embryo development and allow for the development of enhanced experiments for testing embryo development under more dynamic, realistic conditions. To this end, great efforts have been made to quantify the ciliary beating frequency (CBF) of the cilia that line the oviduct; various animal, and even human, models have been used to quantify this parameter under different conditions. Using a laser Doppler flowmeter system to measure CBF in rabbit cilia of the oviduct, Holloway et al. [35] found that the mean frequency was approximately 8.5 Hz at ambient temperature and 25 Hz at 36°C; furthermore, it was found that there was a linear relationship in this range between CBF and temperature. Despite the paucity of data regarding muscular contraction in humans and CBF at physiological temperatures, the available information has helped to develop useful mathematical models for comprehending the influence of mechanical cues on embryo development and transport. With the straightforward method presented here, it is believed that microfluidics can play an immense role in furthering understanding and quantification of embryo microenvironment parameters, as well as provide a well-suited platform to recapitulate the vital mechanical and chemical cues needed for optimal embryo development.

## 6.7 Application Notes

As described briefly in the previous section, the physiological embryo-culture condition is dynamic and pulsatile. Braille display-based microfluidic systems may be useful for exploring the effect of various flow patterns on embryo development. In order to be appropriately interfaced to a Braille display actuator (Navigator, Telesensory, Sunnyvale, California), the hybrid PDMS-Parylene-PDMS membrane needed to be reversibly deformable by the Braille pins. Figure 6.6 demonstrates that this did indeed occur.

After fabrication of the hybrid membrane (following completion of step 6 in Section 6.4.1), the following protocol can be used to create the microfluidic device that will be bonded to it and to setup the Braille display actuation system:

1. Spin-coat SU-8 to a thickness of 30 $\mu$m onto a thin (~200 $\mu$m) glass wafer or cover glass and apply backside diffused-light photolithography using the mask with channel features for Braille deformation-based actuation.
2. Spin-coat SU-8 to a thickness of 200 $\mu$m onto the same glass wafer, and apply conventional front-side photolithography using the mask for rectangular channel features.
3. Silanize the resulting wafer with tridecafluoro-1,1,2,2-tetrahydrooctyl-1-trichlorosilane using a similar procedure to that described in Section 6.4.1, step 3.
4. Prepare a thick (~1 cm) PDMS slab with channel features by casting prepolymer at a 1:10 curing-agent-to-base ratio against positive relief features composed of SU-8.
5. Cure the PDMS slab at 60°C for 120 minutes, and punch holes with a blunt needle or biopsy punch to create reservoirs and inlets and outlets.

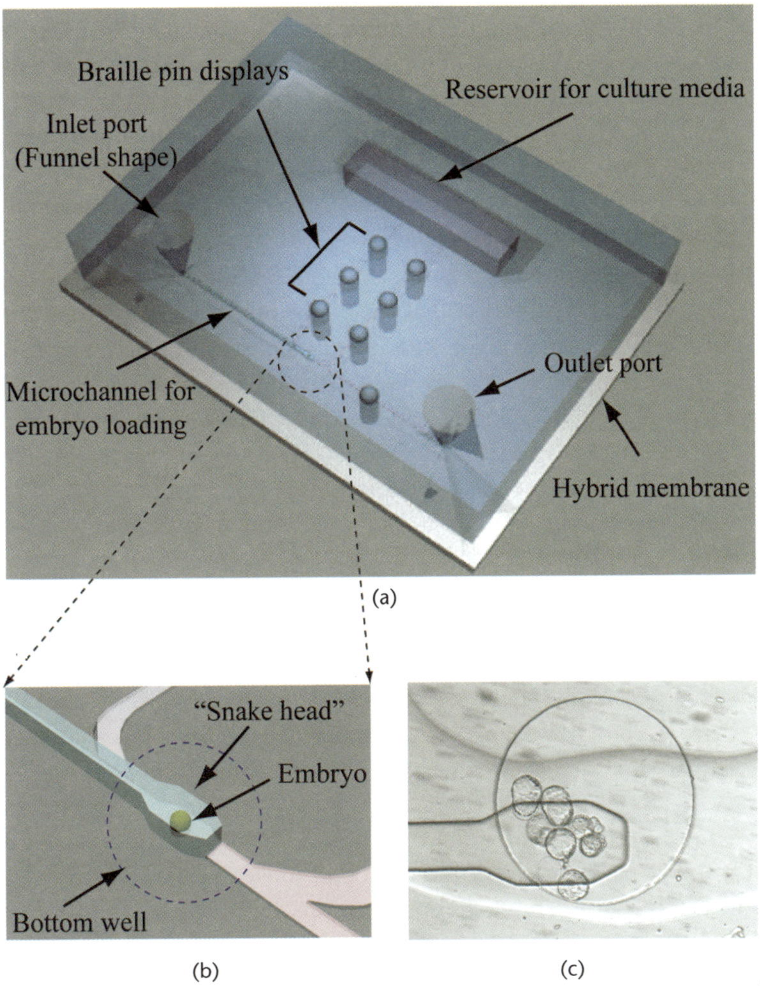


**Figure 6.6** (a) Schematic design of PDMS embryo culture device. (b) Enlarged culture area. (c) It consists of three layers: upper channel (200 $\mu$m height) for introducing embryo, middle channel for Braille system (30 $\mu$m height), and PDMS-Parylene-PDMS hybrid membrane with 80-$\mu$m-deep well to keep the introduced embryos in the culture chamber [i.e., "Snake head" (300 $\mu$m wide × 200 $\mu$m deep)].

*Tip:* Visually inspect the sample to make sure that flatness has been maintained and channel features have not been distorted in any way. Use packaging tape to remove all debris from the surface of the sample. This is absolutely vital to ensure proper bonding. Small particulates can create large voids in, or completely prevent, bonding.

6. Plasma-oxidize the PDMS slab with features and the membrane, in air, for 30 seconds, and bond together. Immediately introduce liquid into the microchannels, and UV-sterilize for at least 30 minutes.

   *Tip:* In our lab, effective, rapid bonding is achieved by using an SPI PlasmaPrep II plasma oxidizer where samples are exposed to 500 mTorr of pressure during oxidation under air plasma for 30 seconds.

*Troubleshooting:* Should the bonding prove problematic, it is advised to peel off the PDMS-Parylene-PDMS membrane from the silicon wafer, oxidize the bottom PDMS surface, and use this layer for bonding to the sample with features.

*Troubleshooting:* If bonding problems persist, a handheld corona treater BD-20AC from Electro-Technic Products Inc. (www.electrotechnicproduct.com) can be used to oxidize the surfaces of the sample with features and the membrane. This device is also much less expensive and is good for labs working on low budgets.

7. Plug in Braille display to laptop through a USB cable. Before using the setup with microfluidic chips, ensure that the Braille pins are responsive to commands.

   *Note:* Braille displays come with operating instructions. Implementing microfluidic actuation requires translating desired pin movements for microfluidic actuation into a series of letters. Pin configurations for each $2 \times 4$ Braille cell are compared to a large lookup table for all possible Braille characters (eight pins per cell equal 256 possible characters to cover all configurations). The resulting equivalent sequence of characters is then passed to standard commercial Braille screen-reader software, which controls the actual Braille pins. In our lab, Visual Basic was used to create a custom program to easily develop desired pumping sequences instead of relying on prearranged text that induced desired Braille pins for actuation.
8. Align the appropriate channel features to the pins of the Braille display by eye. This can be accomplished more rigorously through the use of a stereomicroscope.
9. Once it is properly aligned on the Braille display, make sure that the chip stays firmly in place by placing a weight on top (a few pounds should suffice).

To demonstrate that embryo culture was achievable on a Braille display–actuated microfluidic platform, the device described above was used. Figures 6.6 and 6.7 demonstrate this practical setup for embryo culture.

Embryos were loaded into the upper channel and conveyed to the well area of the device, as depicted in Figure 6.6(b). Portrayed in Figure 6.6(c) are zygotes that were successfully cultured utilizing this microfluidic setup into blastocysts over 96 hours without changing the media. When the same setup was used without the hybrid PDMS-Parylene-PDMS membrane, zygotes did not develop into blastocysts. This ultimately demonstrated that the hybrid membrane was necessary for embryo culture and that it could be successfully adapted to a Braille-actuated microfluidic setup for dynamic embryo culture. This method should broaden the applicability of microfluidic embryo cell cultures and thereby vastly improve assisted reproductive technologies in the context of biomedical research (transgenic mice), genetic gain and domestic-animal production, and human infertility.

## 6.8 Summary Points

- Evaporation is a substantial problem in PDMS-based microfluidic cell culture because of the small volumes of liquid involved, resulting in large shifts in osmolality.
- Shifts in osmolality are detrimental to embryo development.
- The PDMS-Parylene-PDMS hybrid membranes provide a moisture barrier, are readily incorporated into commonly used PDMS fabrication protocols, and are amenable to effective interfacing with deformation-based microfluidic actuation.

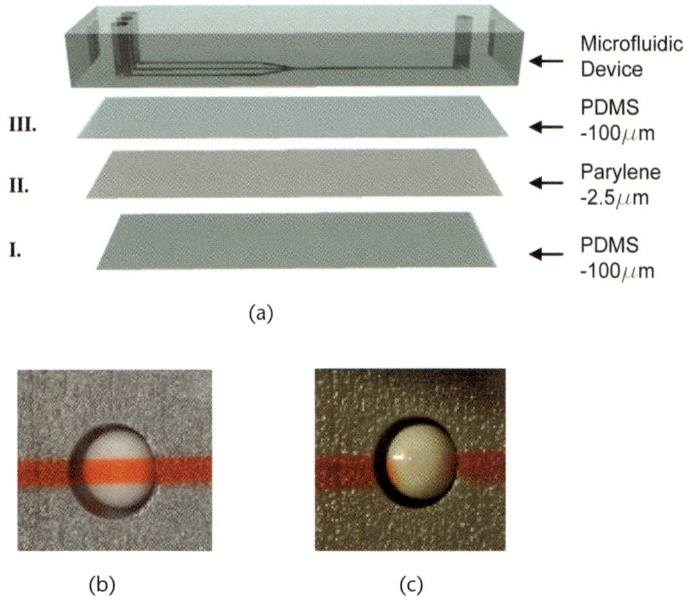


**Figure 6.7** Schematic representation of Braille display–based microfluidics. (a) A typical design for Braille display–based microfluidics is composed of two layers: upper bulk PDMS with microchannel and bottom membrane. To test the suitability of the Parylene-coated PDMS with Braille displays, bottom membrane consists of three layers: 100 $\mu$m PDMS, 2.5 $\mu$m Parylene, and 100 $\mu$m PDMS. (b) When Braille pin pushed against the PDMS-Parylene-PDMS hybrid membrane, the channel was fully closed. (c) When the pin was released, the membrane was restored, and the channel was opened. (Figure adapted from [22] and used with permission from Analytical Chemistry.)

- Application of a PDMS-Parylene-PDMS membrane can drastically reduce evaporation and resultant osmolality shifts. We show that it is well suited for embryo development from the one-cell stage to blastocyst.
- This hybrid membrane is also useful for high numerical aperture, short-working-distance optical microscopy, and real-time imaging of cell culture because it is thin and optically transparent.
- Mouse embryo development is commonly used to evaluate the biocompatibility and lack of toxicity of biomedical materials and devices. Similarly, mouse-embryo-development assays are useful to validate the stability and reliability of microfluidic cell culture systems. Here we specifically apply this principle to osmolality maintenance.
- Since the physiological embryo culture environment inside the oviduct is dynamic and pulsatile, microfluidic embryo culture with pulsatile deformation-based microfluidic actuation may enhance embryo development.

## Acknowledgments

This work is supported by the National Institutes of Health (NIH), the U.S. Department of Agriculture (USDA), the Michigan Economic Development Corporation (MEDC), the Wallace H. Coulter Foundation, and the U.S. Army Research Office.

## References


[1] Takayama, S., et al., "Selective chemical treatment of cellular microdomains using multiple laminar streams," *Chemistry & Biology*, Vol. 10, No. 2, 2003, pp. 123–130.

[2] Hersen, P., et al., "Signal processing by the HOG MAP kinase pathway," *Proc. Natl. Acad. Sci. USA*, Vol. 105, No. 20, 2008, pp. 7165–7170.

[3] Walker, G. M., Zeringue, H. C., and Beebe, D. J., "Microenvironment design considerations for cellular scale studies," *Lab on a Chip*, Vol. 4, No. 2, 2004, pp. 91–97.

[4] Quake, S. R., and Scherer, A., "From micro- to nanofabrication with soft materials," *Science*, Vol. 290, No. 5496, 2000, pp. 1536–1540.

[5] Johnson, T. J., et al., "Laser modification of preformed polymer microchannels: Application to reduce band broadening around turns subject to electrokinetic flow," *Analytical Chemistry*, Vol. 73, No. 15, 2001, pp. 3656–3661.

[6] Duffy, D. C., et al., "Rapid prototyping of microfluidic systems in poly(dimethylsiloxane)," *Analytical Chemistry*, Vol. 70, No. 23, 1998, pp. 4974–4984.

[7] Moor, A. N., Murtazina, R., and Fliegel, L., "Calcium and osmotic regulation of the Na+/H+ exchanger in neonatal ventricular myocytes," *J. Molecular and Cellular Cardiology*, Vol. 32, No. 6, 2000, pp. 925–936.

[8] Zhou, W. C., et al., "Fed-batch culture of recombinant NS0 myeloma cells with high monoclonal antibody production," *Biotechnology and Bioengineering*, Vol. 55, No. 5, 1997, pp. 783–792.

[9] Lin, J. Q., et al., "Enhanced monoclonal antibody production by gradual increase of osmotic pressure," *Cytotechnology*, Vol. 29, No. 1, 1999, pp. 27–33.

[10] Lezama, R., et al., "Epidermal growth factor receptor is a common element in the signaling pathways activated by cell volume changes in isosmotic, hyposmotic or hyperosmotic conditions," *Neurochemical Research*, Vol. 30, No. 12, 2005, pp. 1589–1597.

[11] Wu, M. H., et al., "The effect of hyperosmotic pressure on antibody production and gene expression in the GS-NS0 cell line," *Biotechnology and Applied Biochemistry*, Vol. 40, No. 2004, pp. 41–46.

[12] Brinster, R. L., "Studies of development of mouse embryos in-vitro. I. Effect of osmolarity and hydrogen ion concentration," *J. Experimental Zoology*, Vol. 158, No. 1, 1965, pp. 49–57.

[13] Urbanski, J. P., et al., "Digital microfluidics using soft lithography," *Lab on a Chip*, Vol. 6, No. 1, 2006, pp. 96–104.

[14] Song, J. W., et al., "Computer-controlled microcirculatory support system for endothelial cell culture and shearing," *Analytical Chemistry*, Vol. 77, No. 13, 2005, pp. 3993–3999.

[15] Zheng, B., Roach, L. S., and Ismagilov, R. F., "Screening of protein crystallization conditions on a microfluidic chip using nanoliter-size droplets," *J. American Chemical Society*, Vol. 125, No. 37, 2003, pp. 11170–11171.

[16] Zhao, Z., et al., "Monolithically integrated PCR biochip for DNA amplification," *Sens. Actuators A: Physical*, Vol. 108, No. 1–3, 2003, pp. 162–167.

[17] Koh, C. G., et al., "Integrating polymerase chain reaction, valving, and electrophoresis in a plastic device for bacterial detection," *Analytical Chemistry*, Vol. 75, No. 17, 2003, pp. 4591–4598.

[18] Lee, D. S., et al., "Bulk-micromachined submicroliter-volume PCR chip with very rapid thermal response and low power consumption," *Lab on a Chip*, Vol. 4, No. 4, 2004, pp. 401–407.

[19] Khandurina, J., et al., "Integrated system for rapid PCR-based DNA analysis in microfluidic devices," *Analytical Chemistry*, Vol. 72, No. 13, 2000, pp. 2995–3000.

[20] Chang, W. J., et al., "Poly(dimethylsiloxane) (PDMS) and silicon hybrid biochip for bacterial culture," *Biomedical Microdevices*, Vol. 5, No. 4, 2003, pp. 281–290.

[21] Futai, N., et al., "Handheld recirculation system and customized media for microfluidic cell culture," *Lab on a Chip*, Vol. 6, No. 1, 2006, pp. 149–154.

[22] Heo, Y. S., et al., "Characterization and resolution of evaporation-mediated osmolality shifts that constrain microfluidic cell culture in poly(dimethylsiloxane) devices," *Analytical Chemistry*, Vol. 79, No. 3, 2007, pp. 1126–1134 [the main reference that describes the osmolality measures, hybrid membrane development, and application to embryo and endothelial cell culture in microfluidic devices].

[23] Kaplan, A., et al., *Clinical Chemistry: Interpretation and Techniques*, Malvern, PA: Williams & Wilkins, 1995, p. 151.

[24] Watson, J. M., and Baron, M. G., "The behaviour of water in poly(dimethylsiloxane)," *J. Membrane Science*, Vol. 110, No. 1, 1996, pp. 47–57 [discusses basic and experimental analysis of permeation of water through polydimethylsiloxane].

[25] Tamai, Y., Tanaka, H., and Nakanishi, K., "Molecular simulation of permeation of small penetrants through membranes. 2. Solubilities," *Macromolecules*, Vol. 28, No. 7, 1995, pp. 2544–2554.

[26] Favre, E., et al., "Sorption, diffusion and vapor permeation of various penetrants through dense poly(dimethylsiloxane) membranes—a transport analysis," *J. Membrane Science*, Vol. 92, No. 2, 1994, pp. 169–184.

[27] Barrie, J. A., and Machin, D., "Sorption and diffusion of water in silicone rubbers. 1. Unfilled rubbers," *J. Macromolecular Science—Physics*, Vol. B 3, No. 4, 1969, pp. 645–672.

[28] Annamalai, K., and Puri, K. I.,, *Advanced Thermodynamics Engineering*, Washington, D.C.: CRC Press, 2002, p. 673.

[29] Tsafriri, A., *The Physiology of Reproduction*, New York: Raven, 1994, pp. 817–860.

[30] Halbert, S. A., Tam, P. Y., and Blandau, R. J., "Egg transport in rabbit oviduct—roles of cilia and muscle," *Science*, Vol. 191, No. 4231, 1976, pp. 1052–1053.

[31] Gaddumrosse, P., and Blandau, R. J., "Comparative observations of ciliary currents in mammalian oviducts," *Biology of Reproduction*, Vol. 14, No. 5, 1976, pp. 605–609.

[32] Bourdage, R. J., and Halbert, S. A., "In-vivo recording of oviductal contractions in rabbits during the periovulatory period," *American J. Physiology*, Vol. 239, No. 3, 1980, pp. R332–R336.

[33] Blake, J. R., Vann, P. G., and Winet, H., "A model of ovum transport," *J. Theor. Biol.*, Vol. 102, No. 1, 1983, pp. 145–166.

[34] Lyons, R. A., Saridogan, E., and Djahanbakhch, O., "The reproductive significance of human fallopian tube cilia," *Human Reproduction Update*, Vol. 12, No. 4, 2006, pp. 363–372.

[35] Holloway, G. A., Halbert, S. A., and Lee, W. I., "Fibre-optic laser instrument for measuring ciliary activity of oviducts in-vitro," *Medical & Biological Engineering & Computing*, Vol. 26, No. 6, 1988, pp. 655–658.

CHAPTER

# 7

# Image-Based Cell Sorting Using Microscale Electrical and Optical Actuation

Joseph R. Kovac,[1,2,3] Brian M. Taff,[1,2,3] and Joel Voldman[1,2,3]

[1]Research Laboratory of Electronics
[2]Microsystems Technology Laboratory
[3]Department of Electrical Engineering and Computer Science
Massachusetts Institute of Technology, Cambridge, MA

## Abstract

Imaging cells, a common technique in both basic biology and applied biotechnology, is used for everything from routine imaging of cell cultures to multidimensional fluorescence imaging of subcellular dynamics. Though sorting cells following imaging is extremely challenging, it would be enabling for applications such as screening cells based upon complex phenotypes and purifying populations. This chapter presents two complementary methods for sorting cells following imaging. One is based upon dielectrophoresis, using electric fields to create localized, switchable traps that can manipulate cells for sorting. The second method uses optical scattering forces to eject cells from silicone wells and sort them into a passing fluid flow. In this chapter, we describe the methods for building, packaging, and using these two methods for cell sorting, highlighting the similarities and differences between the two approaches.

Key terms
cell sorting
microscopy
dielectrophoresis
cell trap
optofluidics
optical tweezers
microfluidics

## 7.1 Introduction

Cell sorting is an integral step in both basic biology and applied biotechnology. The ability to separate cell subpopulations from each other allows one to perform genetic screens—correlating genotype to phenotype, or purifying populations—creating pure populations for bulk assays, among other uses. Even though cell sorting is central to many cell biology investigations, the methods available to sort cells from each other are limited. Although cells exhibit complex subcellular and morphological behavior and carry out functions over time, the ability to sort based on these types of processes is limited. These complex phenotypes are often only apparent when using microscopy. They might not be recognized, let alone sorted, by techniques based on physical parameters, such as cell size and density, proliferation, biomolecule secretion, or other properties recognizable by physical or probe-based detection and sorting techniques.

A methodological gap thus exists in our ability to observe cells and then isolate them based upon different imaged phenotypic markers. Observation and sorting are inextricably linked, and the functionality available between them directly affects the types of assays that one can perform. For example, many assays involve visual analysis and thus require optical observation techniques. The premiere optical observation technique, microscopy, is severely limited, however, in its ability to isolate cells, especially mammalian cells that adhere to substrates. The premiere isolation technique, fluorescence-activated cell sorting (FACS), only obtains whole-cell information and is thus unable to image cells. A technological gap therefore exists between observation and isolation.

Within this gap lies a rich set of potential phenotypes upon which one cannot sort (Figure 7.1). Morphology is the most basic phenotype of cells, yet conveys rich information, from general cell health to indications of cell type. In addition, eukaryotic cells organize work and information in organelles; thus, knowing where a protein resides gives information as to its function. Finally, cells are dynamic systems whose processes occur over time. Yet, sorting based upon morphology, localization, or dynamics is currently prohibitive. For instance, while trivial to observe via microscopy, morphology is fundamentally incompatible with flow sorting, which manipulates cells that are in suspension. Dynamic information cannot be used with FACS because cells are observed at

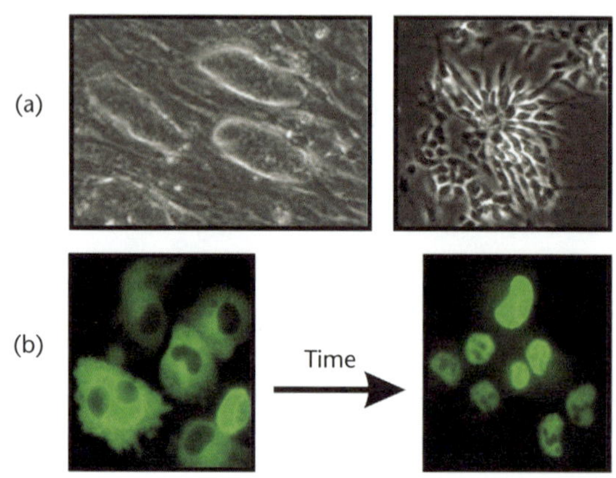


**Figure 7.1** Applications of image-based cell sorting. (a) Different cells exhibit different morphologies, such as these embryonic stem cells, which appear different in their undifferentiated state (left) versus when differentiating into neurons (right). (b) Localization and dynamics are also important components of phenotype. Here cells contain an NFAT-GFP reporter, which translocates upon exposure to ionophore, demonstrating both a dynamic and localized response. (Images courtesy F. McKeon, HMS.)

only one time point, and because FACS doesn't image, it cannot distinguish localization-based phenotypes.

We have developed two approaches in our lab to enable image-based cell sorting. Both approaches offer ways to manipulate micron-sized cells in order to separate them from one another. Although a number of different force fields can be used to exact forces on cells, we have chosen to use electrical forces—in the form of dielectrophoresis, or DEP—and optical forces—primarily the optical scattering force—to enact cell manipulation. The two methods are complementary and provide useful case studies on general approaches for manipulating cells in microfluidic environments.

In this chapter, we describe how we build, package, and use these two systems for cell sorting, highlighting both the similarities and differences between the two approaches. The devices and assays described here have been reported elsewhere [1, 2], so here we focus more on the methods for sorting cells using these devices than on the results of the cell sorts.

### 7.1.1 Electrical and optical microscale cell manipulation

We use two distinct manipulation strategies in our cytometry systems (Figure 7.2). In choosing methods for sorting cells, one seeks forces that can be spatially localized and readily turned on and off, that minimally impact cell health, and that apply forces sufficient for manipulation. In addition, we wish to have actuation methods that can scale over the millimeter to centimeter areas needed for large arrays of cells. Electrical and optical forces meet these criteria.

The electrical forces that we use take the form of dielectrophoresis (DEP) [3]. DEP refers to the force on a cell (or any polarizable body) in a spatially nonuniform electric field. When a cell is placed in any electric field, an electric dipole is induced in the cell. When the field is spatially nonuniform (which it commonly is), the electrical forces pull-

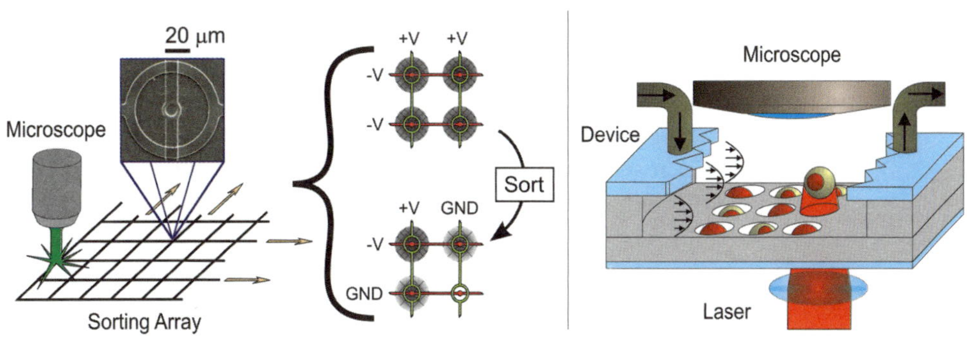


**Figure 7.2** Two approaches to cell manipulation for cell sorting. (a) Electrical sorting uses an array of dielectrophoretic traps formed by the intersection of two metal electrodes (grid and ring-dot image at left). Each ring and associated central dot forms a p-DEP trap. By arraying these traps across a substrate, we can create an array cytometer (left). The traps are arranged in rectangular arrays in a row/column trap-addressing scheme (right). When all electrodes are energized, all traps are in the "on" state (dark gray highlights). Grounding the row and column electrodes associated with a particular trap sets it to the "off" state (no gray highlights). Traps on the same row and column are "on" but activated in a "quarter-strength" configuration (mid-tone gray highlights), which is sufficient to hold cells in place. (b) The optical-sorting approach uses a transparent glass and silicone device that contains arrays of cell-confining microwells. A laser input from the bottom selectively levitates cells in desired locations; these cells are then swept downstream by the fluid flow.

ing on each half of the cell are unbalanced, resulting in a net force that propels the cell to either the maximum electric field intensity (positive DEP, or p-DEP) or the minimum field intensity (negative DEP, or n-DEP). This DEP force for a spherical particle in an electric field $E$ is given by

$$\mathbf{F}_{d\epsilon p}(\mathbf{r}) = 2\pi e_m R^3 \operatorname{Re}[K(\omega)] \cdot \nabla |\mathbf{E}(\mathbf{r})|^2$$

where $\varepsilon_m$ is the permittivity of the medium, $R$ is the particle radius, $\omega$ is the radian frequency, and $\underline{K}$ is the complex Clausius-Mossotti (CM) factor. The CM factor describes the relative polarizability of the cell and medium, and the sign of the real part of the CM factor determines whether p-DEP or n-DEP occurs. For a uniform particle, the CM factor is given by

$$\underline{K}(\omega) = \frac{\underline{\varepsilon}_p - \underline{\varepsilon}_m}{\underline{\varepsilon}_p + 2\underline{\varepsilon}_m}$$

where $\underline{\varepsilon}_m$ and $\underline{\varepsilon}_p$ are the complex permittivities of the medium and the particle, respectively, and are each given by $\underline{\varepsilon} = \varepsilon + \sigma / j\omega$, where $\varepsilon$ is the permittivity of the medium or particle, $\sigma$ is the conductivity of the medium or particle, and $j$ is $\sqrt{-1}$. If the relative polarizability of the cell is greater than that of the medium, then $\operatorname{Re}[\underline{K}(\omega)]$ will be positive, while $\operatorname{Re}[\underline{K}(\omega)]$ will be negative if the medium is more polarizable than the cell.

A mammalian cell, which contains a relatively insulating plasma membrane surrounding a conducting cytoplasmic compartment, is more complicated to model than the simple expression for $\underline{K}(\omega)$ described above. Nonetheless, the fundamental physics of the CM factor is constant in that it describes the interplay between the particle and the medium at a particular frequency.

The discussion above shows that the direction of the force depends on the properties of the cell, the medium, and the applied electric field. DEP traps can be made that use both p-DEP and n-DEP. Importantly, cells in conductive media such as saline or DMEM *only* experience n-DEP because the medium has both higher conductivity and higher permittivity than the cells and will thus have higher polarizability at all frequencies. Conversely, p-DEP traps require low-conductivity buffers in order for the cell to be more polarizable (at least at some frequencies) than the medium. Detailed descriptions of the physics of DEP are found in several reviews [3–6] and monographs [7–9].

DEP-based cell actuation meets the needs described above. First, the forces are readily localized within defined spatial regions determined by the size and spacing of the electrodes comprising an individual cell trap. Second, because DEP uses electric fields, the voltages that create those fields can be turned on and off, as well as individually addressed by having distinct electrodes routed to each trap. Third, DEP has been shown to be compatible with cell health, especially for acute exposure, as long as any solution heating and electric fields are minimized [10, 11]. Fourth, DEP can apply approximately pN forces to cells, which is sufficient to overcome the weight of cells in media (~1/2 pN). Finally, it is straightforward to create large arrays of traps.

Optical forces can take the form of either scattering forces or gradient forces [12]. Both types of forces result from momentum conservation when light is reflected and refracted as it interacts with a particle. Scattering forces push a particle along the axis of a light beam, while gradient forces pull a particle into the region of highest intensity, in the case of a cell suspended in physiological media. The most common optical manipulation technique, optical tweezers, uses the high divergence created by sending light through a high numerical-aperture objective to balance gradient and scattering forces along the beam axis [13]. In contrast, we use a low numerical-aperture objective to create a light beam where scattering forces dominate, creating essentially an "optical fire hose."

Quantitatively modeling optical forces is straightforward in two regimes. In the Rayleigh regime, where the wavelength is much larger than the particle size, the scattering and gradient forces reduce to closed-form equations [12]. In the ray-optics regime, where particle size greatly exceeds the wavelength, Ashkin showed that one can divide the incident beam into rays that independently contribute to the total scattering and gradient force [14]. By ray-tracing each ray, performing a momentum balance for each ray, and summing the total momentum change imparted to the particle due to all rays, the predicted total force can be determined.

Unfortunately, in the context of cell manipulation, the particle radius (~1–10 $\mu$m) is often on the order of the wavelength (~1 $\mu$m), making neither model extremely accurate. Nahmias et al. performed numerical simulations using generalized Lorenz-Mie theory (GLMT) to predict optical forces and showed that GLMT accurately describes forces in the Rayleigh, ray-optics, and intermediate regimes [15]. They found that knowledge of a nondimensional term in a laser/particle system,

$$\beta = \frac{a^4}{\lambda \omega_0^3}$$

where $a$ is particle radius, $\lambda$ is light wavelength, and $\omega_0$ is the beam waist in the focal point, along with knowledge of the particle and surrounding medium refractive index and incident power, allows prediction of average axial force by using curves plotted in their manuscript. Nahmias et al. also presented similar nondimensional analysis for use in radial force prediction.

Our optical actuation scheme, like that of DEP, can also be used for cell manipulation. The forces are localized to the diameter of a laser beam, which can be focused to cell size (or smaller). By either modulating the laser or a shutter in the optical path, one can turn the force on and off. Optical forces, as long as they are at the right wavelengths and of reasonable power, can be used to manipulate cells with minimal impact on cell health [16–18], and optical scattering forces can provide the pN forces necessary to levitate cells. Finally, by moving a beam relative to a chip containing cells, the optical forces can be addressed over large areas.

While DEP and optical manipulation both meet our cytometry system requirements, each offers different relative benefits and trade-offs. DEP places most of the system complexity on-chip in the form of the electrodes that create the traps, while optical manipulation requires only very simple chips and an external optical system. Optical manipulation scales better than DEP manipulation; extra sites for DEP manipulation require extra electrodes and wiring to address them, while for optical manipulation, one only needs a bigger chip. DEP can be used to manipulate cells before assay (e.g., to load prescribed populations) and then sort, while optical manipulation can only be used to sort. Thus, the specific implementation one uses depends on the particulars of the application.

## 7.2 Materials

### 7.2.1 Materials for microfabrication

Fabrication for the two devices used SU-8 processing, PDMS processing, and silicon processing. For SU-8, we used SU-8 2050 and SU-8 2035, both from MicroChem (Newton, Massachusetts), while PDMS was Sylgard 184 (Dow Corning, Midland, Michigan) mixed in a 10:1 base-to-curing-agent ratio and degassed for about 1 hour in a dessicator. Silane for vapor-phase wafer silanization prior to PDMS molding was tridecafluoro-1,2,2-tetrahydrooctyl)-1-tricholorosilane (T2492-KG, United Chemical Technologies, Bristol, Pennsylvania). For silicon microfabrication, we used approximately 7 to 25 -Ωcm phosphorus-doped 150 mm *n*-type wafers (WaferNet Inc., San Jose, California) as the starting material. All other materials used in silicon microfabrication are standard clean room chemicals (i.e., metals, gases). Finally, we used diamond-tipped drill bits to drill through glass slides (Tripple Ripple product line, CR Laurence, Los Angeles, California).

### 7.2.2 Cell lines and culture

We maintained HL-60 cells (ATCC CCL-240, Manassas, Virginia) at 37°C under a humidified 7.5% $CO_2$ atmosphere. The culture medium was RPMI 1640 (Gibco, Grand Island, New York) supplemented with 10% v/v bovine calf serum (Hyclone, Logan, Utah), 1% v/v l-glutamine taken from 200 mM stock (Gibco), 100 U/mL of penicillin (Gibco), and 100 $\mu$g/mL of streptomycin (Gibco).

We cultured BA/F3 pro B cells and WeHi-3B myelomonocytic leukemia cells (a kind gift from Susan Lindquist, Whitehead Institute, Cambridge, Massachusetts) in the same incubated atmosphere. B cell culture medium was RPMI 1640 (21870, Gibco), supplemented with 10% v/v fetal bovine serum (FBS) (SH30088.03, Hyclone), 2% v/v L-glutamine (25030, Gibco), 1% v/v penicillin-streptomycin (15140, Gibco), and 10% v/v WeHi-3B conditioned medium. Leukemia cell medium was Iscove's modified Dulbecco's medium (IMDM) (12440, Gibco) supplemented with 10% v/v FBS, 1% v/v penicillin-streptomycin, and 25 $\mu$M 2-mercaptoethanol (21985, Gibco). We prepared WeHi-3B conditioned media by collecting media from WeHi-3B cells grown in T75 flasks (3 days after seeding), spinning media down at 1,000 rpm for 7 minutes, and collecting the supernatant. After collection, we filtered the media through a 0.2 $\mu$m vacuum filter bottle, aliquotted the media, and stored it at –20°C for future use in B cell culture.

Additionally, we cultured two lines of MCF7 epithelial breast cancer cells, one of which was transfected with a construct expressing the red fluorescent protein mCherry [19] fused to the mouse ornithine decarboxylase PEST sequence and three copies of the SV40 large T-antigene nuclear localization sequence (NLS) under the control of the p21 promoter. Selection with blasticidin established a stable, clonal cell line. A second line was not transfected. Both lines were a kind gift from Galit Lahav (Harvard Medical School, Boston, Massachusetts). MCF7 culture medium for the nontransfected line was RPMI 1640 supplemented with 10% v/v FBS, 1% v/v L-glutamine, and 1% v/v penicillin-streptomycin. Culture medium for the transfected cell line additionally contained 5 $\mu$g/mL blasticidin (ant-bl-1, InvivoGen, San Diego, California).

### 7.2.3 Buffers and reagents

Low-conductivity operating buffer consists of a 10.25% w/v sucrose/deionized water solution. We use semiconductor-grade (18.2 MΩ-cm) water for deionized water. Bovine serum albumin (BSA) was from Invitrogen (BSA fraction V, 15260, Carlsband, California).

### 7.2.4 Staining

We stained cells using either CellTracker Green CMFDA (C7025, Invitrogen) or CellTracker Orange CMTMR (C34551, Invitrogen) prepared in 10 mM working solutions in dimethyl sulphoxide (DMSO) (Sigma-Aldrich, St. Louis, Missouri) that we diluted to final concentrations of 5 to 10 $\mu$M using serum-free culture media. Immediately preceding introduction onto the chips, we rinsed the HL-60s one to three times and suspended them at concentrations ranging from $6.5 \times 10^5$ to approximately $1 \times 10^7$ cells/mL in 10.25% w/v sucrose/deionized water media with electrical conductivity of approximately 0.01 S/m.

### 7.2.5 Equipment

We performed all imaging on an upright Axioplan 2-MOT microscope (Zeiss, Thornwood, New York) equipped with a computer-controllable motorized stage (999000, Ludl, Hawthorne, New York). We used an EXFO X-Cite 120 source (EXFO Photonic Solutions Inc., Richardson, Texas) and Chroma fluorescence filter sets (41001 FITC, 41007a Cy3, 31004 Texas Red, Chroma Technology Corp., Rockingham, Vermont). We captured images using a LaVision Imager 3 QE CCD digital camera (LaVision GmbH, Goettingen, Germany) for all image recording. For optical-sorting experiments, we used a 980 nm fiber-coupled diode laser (3CN00283AL, Avanex, Fremont, California) seated in a butterfly package holder (LM14S2, Thorlabs, Newton, New Jersey) and controlled by a laser diode/thermoelectric cooler (LDTC 2/2, Wavelength Electronics, Bozeman, Montana), a A/D and D/A converter (USB-1408FS, Measurement Computing, Norton, Massachusetts), and the MATLAB Data Acquisition Toolbox (Mathworks, Natick, Massachusetts). We used KG5 filter glass in the filter cube, emission, and light source paths to protect the microscope and camera from laser damage during irradiation. We focused the laser into the specimen plane using a pair of 0.15 NA aspheric lenses (C280TM-B, Thorlabs). For electrical stimulation, we passed waveforms from an Agilent 33220A waveform generator (Agilent, Santa Clara, California) to sets of AD132 op amps (Analog Devices Inc., Norwood, Massachusetts) in single-ended input/differential output mode, outputting phase-matched inverted and noninverted signals that we routed to independently selectable analog switches (ADG333), controllable by a MATLAB PCI interface.

## 7.3 Experimental Design

The particulars of experimental design for image-based cell sorting are essentially identical to what they would be for any cell-imaging assay. The techniques described here simply provide an enabling platform for performing an image-based cell sort. Therefore, the principles of experimental design and controls can be taken from the original assay that is attempting to be ported to the microscale. For fluorescence imaging, it is important to have controls for staining, such as cells stained without primary antibodies (for immunofluorescent assays). Additionally, for multicolor fluorescence, one should check for bleed-through of fluorescence from one channel into the other(s) and adjust filter sets and/or fluorophores to minimize such effects (or account for them).

For assaying sorting itself, it is important to characterize the system in terms of recovery and purity. Recovery is the ability to recover the target sorted population, while purity refers to the presence of unselected cells in the sorted population. These can be assessed by loading, sorting, and then recovering premixed ratios of distinctly labeled cells. We recommend performing an assay loading two prestained cell populations into the chip at ratios approximating those expected to be obtained in the eventual assay, then sorting and measuring the purity and recovery of cells at the outlet.

## 7.4 Methods

Here we present the detailed methods needed to perform cell-sorting experiments using DEP and optical manipulation approaches. In this section we describe the methods surrounding device fabrication, packaging, and creating the experimental setup. We list the methods for each technique separately, drawing attention to the similarities and differences between the two.

### 7.4.1 Material choices and fabrication

#### 7.4.1.1 DEP cell sorting

Our DEP cytometers are fabricated on silicon substrates because of silicon's: (1) ease of processing, (2) high thermal conductivity, (3) negligible autofluorescence, and (4) biocompatibility. First, processing silicon leverages a historical abundance of IC-related expertise for device manufacturing. Our devices require multiple, distinct orthogonally routed on-chip electrodes to trap and retain individual cells in prescribed array-affiliated locations. As a result, we require multiple levels of metal interconnects whose fabrication is most easily performed on silicon substrates. Second, silicon's thermal conductivity is higher than that of the glass substrates commonly used in BioMEMS devices. This high thermal conductivity allows efficient dissipation of the joule heating that occurs when electric fields are introduced into conductive liquid media, minimizing temperature rises and any associated impact on cell physiology. While not critical to the devices described here (which operate in low-conductivity solutions), silicon's high thermal conductivity allows one to operate in high-conductivity cell culture media. Third, silicon and related inorganic materials (silicon dioxide and silicon nitride) have negligible autofluorescence. In assays centered on manipulating and tracking individual cells, it is critical that the autofluorescence from the chip not compete with fluorescence signals from the cells being imaged. Finally, silicon and related materials are known to be compatible with cell growth, which is critical for a cell-based assay platform.

With our substrate chosen, we use established fabrication processes to create the two-level metal electrodes that form the traps and the interconnects [Figure 7.3(a)].

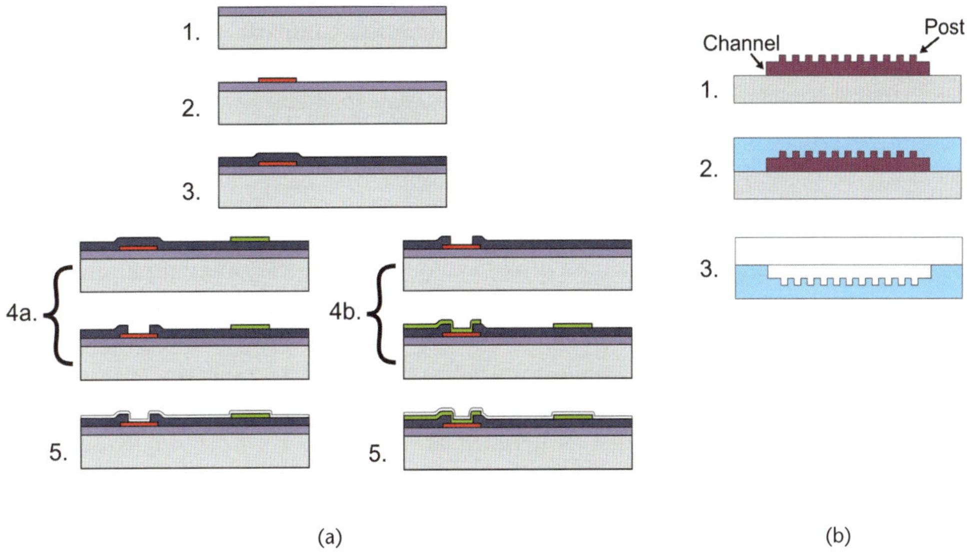


**Figure 7.3** Fabrication steps for the sorting cytometers. (a) The basic processing steps used to fabricate our DEP cytometer devices. (b) Fabrication of PDMS molds for flow chambers/microwell arrays. Individual steps are described in the text.

1. We first thermally grow an insulating layer of silicon dioxide (1.5 $\mu$m thick) to electrically isolate our first metal level from the semiconductor substrate.
2. To deposit our first metal layer (M1), we sputter deposit a 5,000-Å-thick aluminum film. In some cases, it is beneficial also to deposit an additional 500Å of titanium to provide a better etch stop for protecting the M1 level when patterning overlying layers later in the sequence. With standard positive photoresists, contact alignment with chrome-plated glass masks, and a subsequent $BCl_3/Cl_2$ plasma-etch chemistry, we then pattern M1, forming the first level of electrodes.
3. We next deposit a 1.5$\mu$-thick film of PECVD silicon dioxide as an intermetal dielectric (IMD).

4a. In some devices we progress onward by then processing our second metal level (M2) in a manner identical to the first (though with a different mask) and later return to pattern our IMD using contact lithography and plasma-based dry etching.

4b. Alternatively, in some designs, we dry-etch the IMD first and then deposit and pattern the M2 level.

5. Next, we deposit a thin silicon nitride layer (~250Å) using a PECVD process. Similar to reports with chip-based neuronal studies [20], this film reduces nonspecific cell adhesion to the substrate during assays and reduces the degradation of the electrodes in media.
6. Finally (not shown), we section our wafers by protecting their front surfaces with resist and sending them through a die saw.

The two approaches to creating the two-level metal electrodes in steps 4a and 4b arise from the differing needs of different chip designs. If we want orthogonal electrode crossings but do not need via connections between the layers, then there is no need to create vias, and the order in which we pattern the IMD and M2 becomes independent of their deposition sequence (step 4a, which is less complex, would prove appropriate). Alternatively, if we wish to include direct connections between the two levels to provide additional flexibility for routing electrodes, step 4b enables low-resistance vias that span the IMD.

#### 7.4.1.2 Optical cell sorting

Similar and unique concerns drive material and fabrication-method selection in the optical-sorting approach. As with the DEP-based approach, we require materials that are easy to process and thermally compatible, have negligible autofluorescence, and are compatible with cells. Unlike the DEP-based approach, however, we fundamentally require an optically transparent substrate, and because flushing a cell-loaded microwell trap array for reuse would be impractical, we need a low-cost single-use method for device fabrication. These features, together with the lack of a need for electrodes in the device, led us to choose standard SU-8/PDMS replica molding [Figure 7.3(b)] for device fabrication. Importantly, this molding process allows economical production of single-use devices. After assembling the molded PDMS with a glass slide to form a complete device, the resulting architecture enables straightforward imaging with negligible autofluorescence and cell-compatible materials. As with the DEP-based cytometer, one must avoid significant heating, here due to optical absorption of the laser light by the media. Thermal temperature rises of approximately 1K to 10K per 100 mW optical power have been reported in optical tweezer contexts [17, 21, 22], with the lower end of that range reported for cell manipulation; simple thermal models of a glass/PDMS device at expected optical power levels confirmed this prediction.

To fabricate the devices,

1. We first spin-coat a silicon wafer with a 105-$\mu$m-thick SU-8 (SU-8 2050, MicroChem) layer and expose the layer with a first mask to define flow channels. We then spin-coat a 35-$\mu$m-thick SU-8 layer (SU-8 2035, MicroChem) and expose the substrate with a second mask, defining a post array to form the microwells after molding. Next, we develop the two-level SU-8 stack in a single step using PM acetate. We then silanize the wafers for 24 hours with (tridecafluoro-1,2,2-tetrahydrooctyl)-1-tricholorosilane (T2492-KG, United Chemical Technologies) by placing the wafers and a small boat containing approximately seven drops of silane in a dessicator. The silanization step decreases adhesion between the master and the PDMS during the subsequent molding step and prolongs SU-8 master lifetime considerably.
2. We mix PDMS (Sylgard 184, Dow Corning) in the standard 10:1 base-to-curing-agent ratio and degas the PDMS in a desiccator for 1 hour. Next, we pour a 2-mm-thick PDMS layer on the master, cure the PDMS in a convection oven for 2 hours, and carefully peel the PDMS mold from the master for bonding in subsequent packaging steps.
3. The final step in fabrication is to bond the patterned PDMS microwell array to a glass slide, which is described in more detail below.

The specific sorting application dictates the dimensions of the SU-8 layers. The thickness of the first layer defines the volume of cell suspension to sediment into the microwell array, which is important in cases of low cell concentration. Microwell diameter and depth relative to cell size influence microwell single-cell trapping efficiency [23] and trapping stability. The second SU-8 layer thickness and post diameters set these microwell dimensions. We optimized well dimensions by fabricating test devices with a range of well diameters and by fabricating separate wafers with different well depths; this is a straightforward way to optimize well dimensions for a given cell type or application. The PDMS thickness (2 mm in our case) must be minimal to reduce aberrations of the laser beam as it passes through the PDMS layer.

### 7.4.2 Packaging and experimental setup

Both the packaging and testing of devices typically require introduction of fluids and cells onto a chip, imaging to assess phenotype, and application of sorting forces. Both DEP and optical cell sorters have similar fluid interfacing requirements, incorporating off-chip syringe pumps and valves. The primary difference between the two approaches is that the different material sets used for each device lead to different chip-to-world fluidic interfaces. Because both technologies require bright-field and fluorescence imaging, the two technologies are compatible with standard microscopes. In contrast, the two approaches have very different ways of introducing the sorting forces. For the DEP-based approach, we require high-density electrical connections to the device, while the optical sorter moves that complexity to an off-chip laser system.

#### 7.4.2.1 DEP cell sorting

We make fluid connections to our DEP cytometers by inserting tubing into the bottom side of the chip. By forming through holes in the substrate using a handheld drill and diamond-tipped drill bits (Tripple Ripple product line, CR Laurence), we form ports for pumping liquids across the front surface of the chip. With the through holes in place, we then mount each chip onto a printed circuit board (PCB) packaging board using double-sided tape [Figure 7.4(a)]. This rigid packaging board offers stable tubing connections via press-fittings aligned to the underside of the chip's through holes. To encapsulate the chips, we form front-side flow chambers by plasma-bonding coverslips to laser-cut PDMS gaskets. We finally seal the system by applying epoxy to the edges surrounding the flow chambers and on underside press-fit tubing junctions.

As the number of traps increases in large-format arrays, the device wiring complexity increases such that it becomes beneficial to incorporate a secondary adapter PCB. The packaging board, capable of holding any chip variation that we create, interfaces with a series of different adapter boards to properly translate those electrode layouts to standard connectors. The packaging boards used gold-coated traces, allowing direct wire bonding between the chip and the board. As such, if a DEP cytometer chip fails, it is relatively straightforward to replace the packaging board assembly and prepare for another round of experimentation. Though it demands planning from the onset, this modular strategy has continually brought savings in both time and effort.

The PCBs also serve as an interface to an underlying aluminum support frame that we affix to a steel stage insert on our upright microscope via adhesive-attached magnetic matting. This support frame stably holds the entire DEP cytometer system in place during experiments, minimizing imaging artifacts from unwanted device movement. The package also contains aluminum strain reliefs to further isolate the chip from external perturbations.

#### 7.4.2.2 Optical cell sorting

Packaging the optical cell-sorting devices is primarily an issue of fluidic interfacing. After drilling holes through a 1-mm-thick glass slide using a similar process as is used for the DEP devices, we glue PEEK tubing into the holes using epoxy to form fluidic connections to the device, which yields robust, one-time-use chip interfacing. We plasma-bond the PDMS mold to the glass slide, forming a flow chamber with a microwell array in the chamber floor [Figures 7.3(b), 7.4(b), and 7.5(c)].

We have implemented a high-purity cell-retrieval scheme in our optical sorter that impacts packaging, flow channel layout, and device operation. Our device uses two inputs (a cell-loading input and a sheathing buffer input) and two outputs (a waste path and a path to an integrated reservoir for sorted-cell collection and retrieval) [Figure 7.5(b, c)]. We highlight the motivation of using two inputs later. Using two separate collection paths increases sorted-cell purity by avoiding sorted-cell transit through the waste path, which could contain contaminating cells originating from the loading

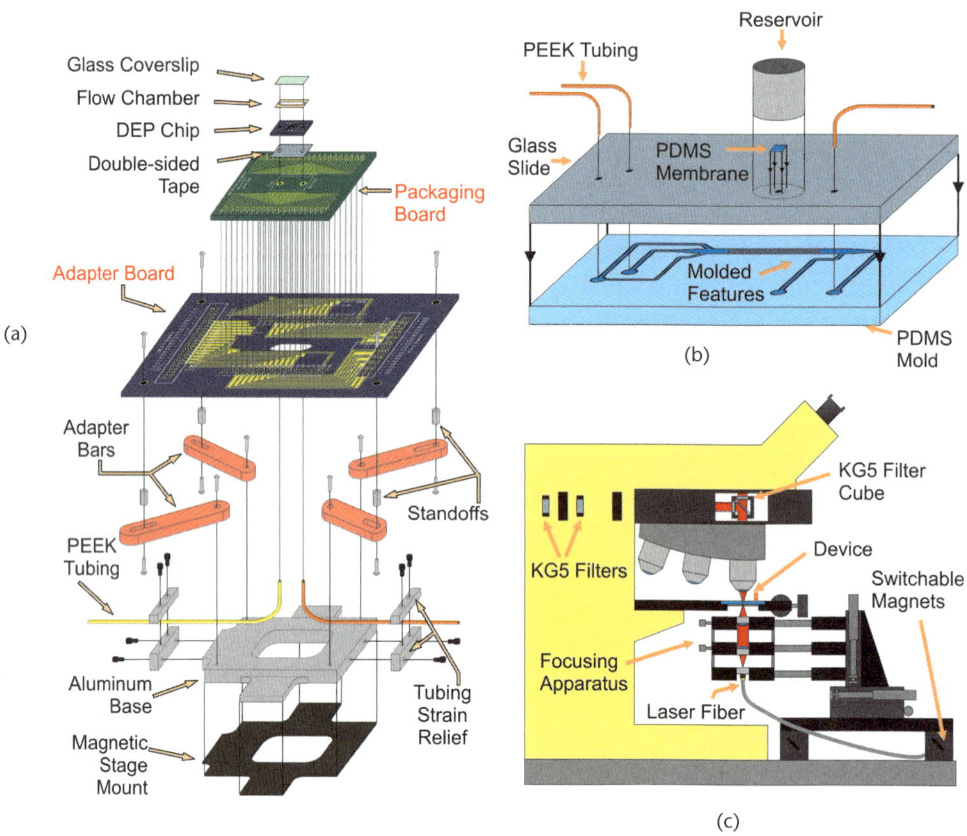


**Figure 7.4** Packaging and testing platforms. (a) An exploded diagram of the DEP cytometer packaging scheme. The setup integrates electrical connections, fluidic plumbing, and imaging means into a single stable platform that easily rests on an upright microscope stage. (b) Our optical cell sorter device consists of a two-layer sandwich of glass and PDMS microfluidic flow chambers. The glass provides a rigid structure for connecting external tubing and readily fits into a standard microscope stage slide holder. (c) Optical subsystem for the optical cell sorter. The chip sits on an automated stage, while a simple optomechanics assembly holds the laser fiber for collimation and focusing and rests on a three-axis stage with a magnetic base.

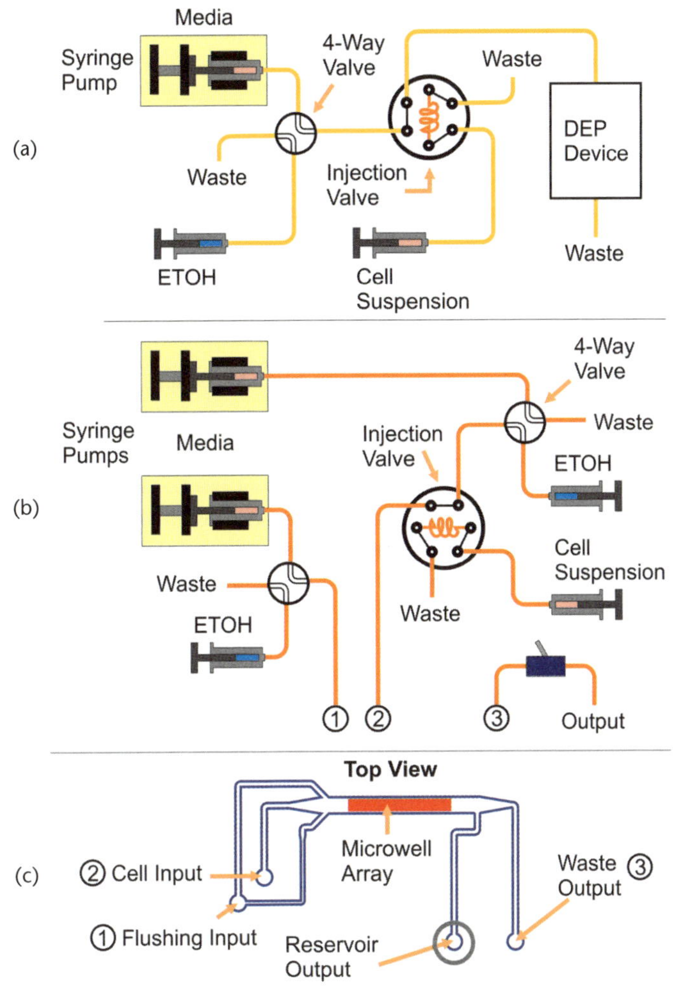


**Figure 7.5** Fluid handling. (a) Our DEP cytometer fluids setup uses a single injection valve, a syringe pump, and a four-way valve to deliver cells and liquids to the device. (b) The optical sorter uses a similar flow setup, with additional valves and input pumps to increase sort purity. (c) Layout of the optical-sorting device detailing fluid routing paths for cell loading and collection.

phase. We leave one drilled hole in the glass fluidic interfacing layer free of tubing (the hole that will interface with the "sorted-cell reservoir" output channel in the PDMS layer) and seal it by plasma-bonding a thin membrane of PDMS over it. We then attach the cylindrical portion of a microcentrifuge tube around the membrane-covered hole using PDMS as an epoxy, forming a reservoir that interfaces with the device when the membrane is punctured by a needle. Using PDMS as an epoxy to attach the tube lowers autofluorescence levels relative to traditional epoxies, facilitating imaging of sorted cells. Before puncture, most of the path leading to the membrane will not fill with fluid, preventing cells from contaminating the path. We use this "one-time valve" as a simple alternative to microfabricated valves during prototyping as a way to divert sorted cells for collection after sorting. After packaging is complete, we interface with the device as shown in Figure 7.5(b).

To mitigate cell adhesion in the optical cell sorter, we first bake the device in a convection oven (65°C, 12 hours) after bonding, before a subsequent bovine serum albumin (BSA) treatment, described later. This bake accelerates PDMS hydrophobicity recovery after bonding. Lee et al. suggested that fibronectin adsorbs more efficiently to hydrophobic surfaces than hydrophilic ones [24]; the same may be true for BSA. We have found that placing the device back into the oven and following the BSA coating steps described later is critical for lowering cell adhesion to surfaces enough to permit optical cell release.

In contrast with the DEP cytometer, where the sorting demands on-chip complexity in the form of electrodes, the optical approach moves that system complexity off-chip into an optical subsystem that interfaces with the microscope [Figure 7.4(c)]. In this subsystem, we attach a fiber-coupled semiconductor diode laser (3CN00283AL, Avanex) to an assembly containing a collimating/focusing pair of 0.15 NA aspheric lenses (C280TM-B, Thorlabs). We use these small aspheric lenses because of their short focal lengths, which allow them to be arranged into a compact assembly easily incorporated into the microscope. We mount the subsystem assembly to a three-axis base resting on a bottom plate with switchable magnets, enabling rapid incorporation/removal and focusing of the laser system in our setup [Figure 7.4(c)]. We use KG5 filter glass (Chroma Technology Corp.) throughout the microscope to protect the camera from damaging laser intensities and never use eyepieces during laser irradiation. Further optical setup detail is available elsewhere [1]. After support optics and fluidics are in place, we position the microfluidic device onto the microscope stage using a commercially available glass-slide stage insert. We construct a strain relief for the device tubing by connecting the rigid PEEK tubing from the device to more flexible tubing running to the rest of the fluidic system with a union. We then tape the union to the microscope stage to complete the strain relief.

## 7.5 Data Acquisition, Anticipated Results, and Interpretation

With the device fabricated, packaged, and placed into the experimental setup, the next steps are to load cells into the device in order to perform an assay, acquire data via microscopy, determine the sort parameters, and then enact the sort itself.

### 7.5.1 Cell culture and assay

Most of our cell culture and assay protocols are as they would be for any other microscopy-based assay. What changes in using our sorting cytometers are the steps after removing cells from the incubator, as we describe below.

#### 7.5.1.1 DEP cell sorting

Just prior to running an on-chip assay, we typically remove our cultures from the incubator and stain them using standard procedures. While labeling the cells, we prepare a low-conductivity operating buffer that consists of a 10.25% w/v sucrose/deionized water solution, which has an electrical conductivity of approximately 0.01 S/m. If desired, one can tune the buffer conductivity upward by adding bovine calf serum. When cell staining is complete, we rinse and resuspend the cells in the low-conductivity operating buffer and load them into the cytometer to proceed with the assay.

Because cell health can be an issue when using low-conductivity operating buffers, we find it best to set up the chip and experimental system well before introducing cells onto the chip. To maintain cells at 37°C, we often use an external heat gun and thermocouple measurements to heat the device controllably.

Establishing a sterile and bubble-free fluidic system is critical to any microfluidic device for use with cells. To prime and sterilize the fluid paths in our system [Figure 7.5(a)], we flush them with 80% ethanol and then incubate for approximately 15 minutes. Then, using off-chip valves, we replace the ethanol with cell-free culture media delivered via glass syringe. To ensure complete removal of the ethanol, we flush out multiple chip volumes using hand-driven flow until several media-tinted drops appear at the device outlet. Finally, we place the syringe into a syringe pump.

With the device readied, we inject our stained-cell suspensions into the fluid delivery system. For most sorting assays, we try to avoid saturating the chip surface with unnecessarily high cell concentrations in an effort to avoid nonspecific binding and chip fouling. We typically use cell concentrations ranging between $6 \times 10^4$ and $10^6$ cells/mL. We inject a cell plug into the system using a 50 $\mu$L glass syringe and an off-chip six-port injection valve with a 15 to 50 $\mu$L sample loop. We have found it easiest to inject cells in a bubble-free manner if we initially flush the sample loop with ethanol, then cell-free operating buffer, and finally with cell-containing operating buffer, before injecting the cells into the device. Immediately following this three-step fluid-handling routine, we turn the injection valve from its initial "load" setting to an "inject" position.

With the cell suspension in the system, we start the syringe pump to introduce the cells into the DEP cytometer chips; we typically set injection flow rates to approximately 100 $\mu$L/min (on-chip flow chambers are typically 4 mm wide and 230 to 250 $\mu$m high). We maintain this flow setting until cells arrive in the device fluid chamber. Once the population begins entering the chip, we stop the syringe pump, flush the connections to the sample loop on the injection valve with ethanol, and then return the valve to the "load" setting. This process limits the number of cells introduced onto the device while maintaining a bubble-free on-chip fluid environment.

To load cells into the DEP sorting arrays, we restart the syringe pump flow at a reduced rate of 10 to 20 $\mu$L/min and activate the trapping electrodes. We have observed effective cell retention at a number of different frequencies and voltage amplitudes in our devices, though we most commonly use a 1 MHz signal delivered at 2 $V_{pp}$. We wait for all trap sites in our devices to fill with single cells and check array loading status using intermittent imaging (to minimize photodamage). Once the array is full, we maintain the voltages and flush remaining untrapped cells out of the system. The cells are now ready for imaging and sorting.

#### 7.5.1.2 Optical cell sorting

Cell loading and assays with the optical method are similar to those using the DEP approach but with a few important differences. As with the DEP architecture, we first culture, stain, or otherwise prepare cells in an appropriate manner for our assay using established protocols. After cell preparation is complete, we resuspend cells in standard culture medium; cells remain in standard medium with the optical-sorting approach throughout an assay unless variation of the medium is required for the assay itself. At this point, cells are ready for loading onto the device. As with the DEP approach, the optical-sorting system should be readied before any time-critical steps occur in preparing the cells. We first flush 80% ethanol through the fluid paths and device for easy filling and sterilization. Using off-chip valves, we purge the ethanol with phosphate-buffered saline (PBS) and flow 75 mg/mL BSA (BSA fraction V, 15260, Invitrogen) into the device to reduce nonspecific cell sticking. We allow the BSA to adsorb for 1 hour, then purge the system with PBS alone. Last, we place syringes in their respective syringe pumps.

We inject cells at concentrations of 1 to 10 $\times$ $10^6$ cells/mL using an injection valve/sample loop similar to that described for the DEP sorter. We load cells by opening the waste output and driving both syringes at flow rates such that the buffer input sheathes the cell-loading input, entirely isolating cells from the entrance to the path leading to the collection reservoir [Figure 7.5(b, c)]. This sheathing action is important and highlights a reason that we use a dual-input device, as allowing cells to enter and possibly become stuck in the channel leading to the collection reservoir could later contaminate the sorted population during retrieval, as discussed earlier. Importantly, using a sample loop to quickly inject the cells reduces cell sedimentation in the tubing, thus facilitates more uniform cell loading in the device.

We inject at a high flow rate (~100 $\mu$L/min in the cell stream input, 15 $\mu$L/min in each sheathing input; our channel is approximately 100 μm high and 3 mm wide) until cells appear in the device. After cells reach the microwell region, we stop flow for approximately 5 minutes, allowing cells to settle into the microwells. We then reinstate cell-free media flow in both inputs (~40 $\mu$L/min, standard cell culture media), then in the sheathing flow input only—avoiding introduction of residual cells in the cell-loading path—and then briefly from the sheathing flow input back through the cell-loading input, pushing residual untrapped cells in the cell-loading input further from the main chamber. Without the use of valves, these steps remove untrapped and marginally trapped cells from the microwell region as well as all peripheral regions where untrapped cells may reside. We found these measures to be critical for high-purity cell recovery during the sorting release phase.

### 7.5.2 Imaging and sorting

#### 7.5.2.1 DEP cell sorting

The DEP-based cytometer uses a coverslip for its flow chamber ceiling, allowing high-resolution imaging with minimal optical aberrations. In addition, because fluid connections are made from the underside of the chips, no protrusions interfere with the objective turret on an upright microscope.

To image cells in our devices, we use standard fluorescence microscopy or DIC imaging. The overall cell-imaging and cell-sorting software runs in MATLAB. Before acquiring images, we run an array registration routine. This software routine takes as inputs specifications of the DEP device layout as well as microscope settings to register and index all trap locations. This protocol then enables the tracking of individual cells in the array over time.

To effect sorting, we leverage the DEP cytometer's built-in row/column-based site addressing scheme [Figure 7.2(a)]. We first determine the row and column indices associated with all cells that display the phenotype of interest. For each cell we wish to sort, we deactivate its row and column electrodes and set the fluid flow to approximately 10 $\mu$L/min using a syringe pump. We visually track the targeted cell until it progresses out of the trapping array, then reactivate the row and column electrodes associated with the now sorted cell. We then recover the cells at the device outlet via fluid droplet deposition in a collection vial. Examples of sorts performed with the device are shown in Figure 7.6.

#### 7.5.2.2 Optical cell sorting

As implemented, the optical approach requires imaging through a 1-mm-thick glass slide. This constraint is not fundamental as the transparency of the device could easily accommodate design variations using an inverted microscope and alternative methods of fabricating microwells on standard coverslips to enable high-resolution imaging. Fluidic connections are presently routed through the top of the device, necessitating consideration of clearances between the objective turret and the tubing; using an inverted microscope would circumvent this issue as well.

In the optical approach, we image the device using MATLAB-based software to automatically scan the stage and record multiwavelength images of the loaded array across the entire device and register image locations to array sites. In a method similar to the DEP cytometer, this registration feature requires the input of the location of the corners

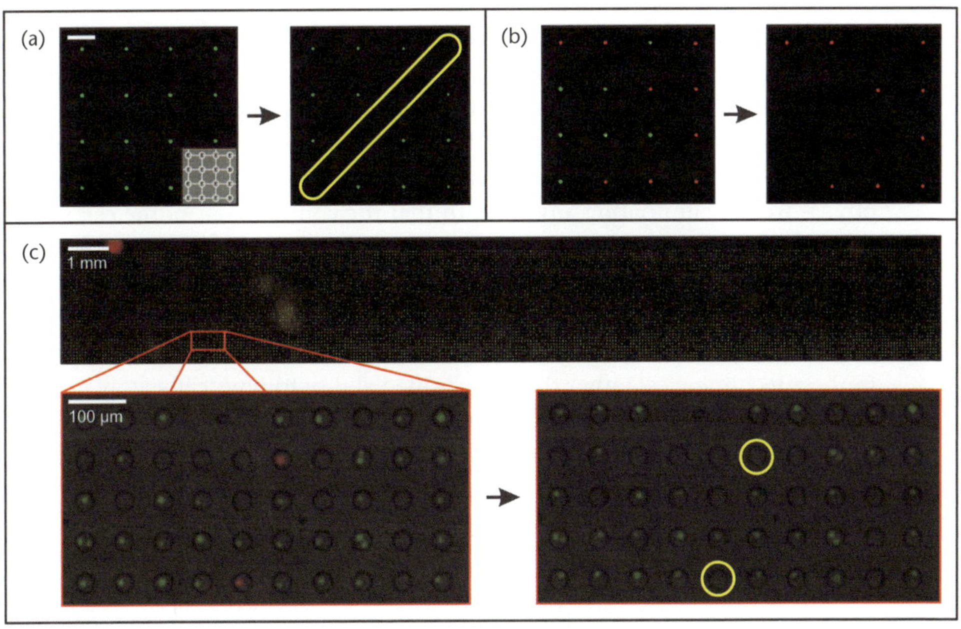


**Figure 7.6** Image-based cell sorting with DEP (a, b) and optical (c) cell sorters. (a) Images of a 4 × 4 trap array (bright-field image in inset) filled with HL-60 cells before and after sorting. Cells were removed diagonally from the array, one at a time, until 12 cells remained. (b) Images of a 4 × 4 array randomly loaded with CellTracker Orange- and Green-labeled HL-60 cells. After imaging, we sorted the green cells from the array. (c) Image of a 10,000-site microwell array (top). The bottom two images show the array before and after removing CellTracker Orange-labeled cells from a background of CellTracker Green-labeled cells.

of the array as well as information about the microwell layout. A center-to-center well spacing of 65 $\mu$m is sufficient for unambiguous site registration and is robust to slight device movements during scanning. A tighter interwell spacing would most likely be sufficient, allowing higher trap densities.

After we complete imaging, we record the locations of cells of interest using our software. To start sorting, we first pierce the PDMS membrane connecting to the collection reservoir, shut the waste path, and reinstate buffer flow solely through the sheathing/buffer input at flow rates of approximately 5 $\mu$L/min (again highlighting the importance of having a path to introduce cell-free buffer that bypasses the cell-loading path where contaminating cells reside). We scan the microscope stage to locations containing cells of interest and activate the laser (100 to 150 mW, spot diameter approximately 8.5 $\mu$m—set by collimated beam width and focusing lens NA), serially focusing it onto target cells. Cells levitate axially along the beam and are then released into the flow stream, where they then flow into the collection reservoir. Release takes approximately 3 to 30 seconds per site and is successful about 80% to 90% of the time. After sorting, we can remove cells from the collection reservoir using a pipettor. Typically, fluid volumes of approximately 300 $\mu$L accumulate in the reservoir over an hour when a flow rate of approximately 5 $\mu$L/min is used during release; the total amount of liquid accumulation in the reservoir must be considered given the reservoir size, experiment duration, and release flow rate to prevent overflow. We show examples of sorts with the device in Figure 7.6.

## 7.6 Discussion and Commentary

In this chapter we have described the detailed methods needed to create and operate electrical and optical microfluidic cell sorters. Both approaches have their relative strengths that make them ideally suited for particular applications. For instance, the electrical approach is reusable and utilizes a uniformly flat substrate, potentially offering better imaging opportunities than the optical approach presents with its topography, as well as eliminating any concern that the confinement of the cell might affect the screened phenotype. On the other hand, the fabrication of the optical sorter is significantly simpler, and array sizes are more easily scaled owing to the fact that trapping at array sites is passive and requires no interconnect routing to individual sites.

Ultimately, imaging constraints often prove to be the fundamental limiting factors in these technologies, rather than the limited ability to physically array and sort cells. For this reason, it is important to appreciate that ultimate scalability is not necessarily the most important metric. For instance, if the phenotype investigated must be imaged once per minute at 40× resolution, it likely does not matter if an array can be scaled to 10,000 sites—there is no possible way to screen all sites quickly enough. Instead, it is often more important to consider the context into which the sorter fits. Does the sorter immediately precede an analysis chamber? Does the fabrication of stages preceding or following the sorter dictate that a particular sorting approach (electrical or optical) be taken? Is there already electrical complexity in the device architecture, or is the device simple and transparent, made using PDMS, and able to accept a laser? These are likely more pertinent details to focus on, rather than being swayed by ultimate scalability.

**Troubleshooting Table**

| Problem | Explanation | Potential Solution |
|---|---|---|
| Cells do not appear to move when DEP field is applied | Electric field is not entering the liquid. This is likely caused by either a poor electrical connection between signal generator and electrodes or a residual film on the electrodes. | Check electrical connections between signal generator and chip using oscilloscope probe.<br>Check processing parameters to ensure that all residual films (photoresist, $SiO_2$, and so forth) have been removed after processing. |
| Cells stick to surfaces in device | There is insufficient surface passivation and/or the cells are very adhesive. | Nonadherent cells stick less than adherent cells. Make sure to preadsorb BSA onto chip surfaces before introducing cells, and ensure that all buffer solutions contain BSA. |
| Optical cell release fails | There is insufficient BSA adsorption to microwell surfaces. | Ensure that devices are cured after plasma bonding in a 65°C convection oven overnight before the experiment, enhancing BSA adsorption during the coating step. |
| Cells randomly release from microwells | Flow rate is too high; microwell diameter/depth is improperly designed. | Use minimum release flow rate that still allows removal of sorted cells. Fabricate device with a range of microwell diameters to test for proper depth. Construct masters with different post heights to vary well depth before deciding on an optimal geometry. |
| SU-8 delaminates during PDMS molding or posts break off | Insufficient/incorrect silane is used, or it is applied incorrectly. | Use described silane; possibly use more silane; only pull vacuum in dessicator jar initially, then shut off dessicator jar valve and allow to sit rather than continuing to pump. |

## 7.7 Summary Points

- Optical and electrical manipulation can be used to trap and/or sort cells following imaging.
- Determining whether to use optical or electrical manipulation depends on the particular assay; optical manipulation scales best, while DEP can be used to manipulate cells before assay.
- Imaging throughput is often the bottleneck in these assays; carefully estimate imaging requirements before settling on an array size or a scheme (electrical or optical) to use.
- The electrical approach is potentially more self-sufficient and easily integrated into multistage devices utilizing traditional MEMS/semiconductor processing, whereas the optical approach offers simple integration into traditional, transparent microfluidics by moving complexity into an off-chip optical subsystem.

## Acknowledgments

The authors acknowledge research support from the NIH (RR19652) and the Singapore-MIT Alliance. In addition, J. R. K. was supported by a fellowship from the American Society for Engineering Education, Department of Defense.

### References

[1] Kovac, J. R., and Voldman, J., "Intuitive, image-based cell sorting using optofluidic cell sorting," *Analytical Chemistry*, Vol. 79, Dec. 15, 2007, pp. 9321–9330.

[2] Taff, B. M., and Voldman, J., "A scalable addressable positive-dielectrophoretic cell-sorting array," *Analytical Chemistry*, Vol. 77, 2005, pp. 7976–7983.

[3] Voldman, J., "Electrical forces for microscale cell manipulation," *Annu. Rev. Biomed. Eng.*, Vol. 8, 2006, pp. 425–454.

[4] Pethig, R., "Dielectrophoresis: Using inhomogeneous ac electrical fields to separate and manipulate cells," *Critical Reviews in Biotechnology*, Vol. 16, 1996, pp. 331–348.

[5] Green, N. G., Ramos, A., and Morgan, H., "Ac electrokinetics: A survey of sub-micrometre particle dynamics," *J. Physics D–Applied Physics*, Vol. 33, 2000, pp. 632–641.

[6] Gascoyne, P. R. C., and Vykoukal, J., "Particle separation by dielectrophoresis," *Electrophoresis*, Vol. 23, 2002, pp. 1973–1983.

[7] Morgan, H., and Green, N. G., *AC Electrokinetics: Colloids and Nanoparticles*, Baldock, Hertfordshire, England: Research Studies Press, 2003.

[8] Hughes, M. P., *Nanoelectromechanics in Engineering and Biology*, Boca Raton, FL: CRC Press, 2003.

[9] Jones, T. B., *Electromechanics of Particles,* Cambridge: Cambridge University Press, 1995.

[10] Docoslis, A., Kalogerakis, N., and Behie, L. A., "Dielectrophoretic forces can be safely used to retain viable cells in perfusion cultures of animal cells," *Cytotechnology*, Vol. 30, 1999, pp. 133–142.

[11] Glasser, H., and Fuhr, G., "Cultivation of cells under strong ac-electric field—differentiation between heating and trans-membrane potential effects," *Bioelectrochemistry and Bioenergetics*, Vol. 47, 1998, pp. 301–310.

[12] Svoboda, K., and Block, S. M., "Biological applications of optical forces," *Annual Review of Biophysics and Biomolecular Structure*, Vol. 23, 1994, pp. 247–285.

[13] Ashkin, A., "Optical trapping and manipulation of neutral particles using lasers," *Proc. Natl. Acad. Sci. USA*, Vol. 94, 1997, pp. 4853–4860.

[14] Ashkin, A., "Forces of a single-beam gradient laser trap on a dielectric sphere in the ray optics regime," *Biophys. J.*, Vol. 61, Feb. 1992, pp. 569–582.

[15] Nahmias, Y. K., Gao, B. Z., and Odde, D. J., "Dimensionless parameters for the design of optical traps and laser guidance systems," *Applied Optics*, Vol. 43, July 2004, pp. 3999–4006.

[16] Liang, H., et al., "Wavelength dependence of cell cloning efficiency after optical trapping," *Biophys. J.*, Vol. 70, Mar. 1996, pp. 1529–33.

[17] Liu, Y., et al., "Physiological monitoring of optically trapped cells: Assessing the effects of confinement by 1064-nm laser tweezers using microfluorometry," *Biophys. J.*, Vol. 71, Oct. 1996, pp. 2158–2167.

[18] Mohanty, S. K., et al., "Comet assay measurements of DNA damage in cells by laser microbeams and trapping beams with wavelengths spanning a range of 308 nm to 1064 nm," *Radiation Research*, Vol. 157, Apr. 2002, pp. 378–385.

[19] Shaner, N. C., et al., "Improved monomeric red, orange and yellow fluorescent proteins derived from Discosoma sp red fluorescent protein," *Nature Biotechnology*, Vol. 22, Dec. 2004, pp. 1567–1572.

[20] Stenger, D. A., and McKenna, T. M., *Enabling Technologies for Cultured Neural Networks*, San Diego, CA: Academic Press, 1994, p. 355.

[21] Liu, Y., et al., "Evidence for localized cell heating induced by infrared optical tweezers," *Biophys. J.*, Vol. 68, May 1995, pp. 2137–2144.

[22] Peterman, E. J., Gittes, F., and Schmidt, C. F., "Laser-induced heating in optical traps," *Biophys. J.*, Vol. 84, Feb. 2003, pp. 1308–1316.

[23] Rettig, J. R., and Folch, A., "Large-scale single-cell trapping and imaging using microwell arrays," *Analytical Chemistry*, Vol. 77, Sept. 1, 2005, pp. 5628–5634.

[24] Lee, J. N., et al., "Compatibility of mammalian cells on surfaces of poly(dimethylsiloxane)," *Langmuir*, Vol. 20, Dec. 21, 2004, pp. 11684–11691.

CHAPTER

# 8

# Pharmacokinetic-Pharmacodynamic Models on a Chip

Jong Hwan Sung[1] and Michael L. Shuler[1, 2]

[1]School of Chemical and Biomolecular Engineering
[2]Department of Biomedical Engineering
Cornell University, Ithaca, NY 14853

## Abstract

Drug development is a risky and expensive process. With recent developments in microfluidic technology, the possibility of utilizing novel microfluidic systems in the drug-development process is being pursued. The microfluidic technology could have a more profound impact if it is combined with a mathematical modeling approach such as physiologically based pharmacokinetic (PBPK) and pharmacodynamic (PD) models for a "body-on-a-chip" approach. The body-on-a-chip was conceived as a physical realization of PBPK models, and its utilization in toxicity testing has been demonstrated at a proof-of-concept level using naphthalene. Currently, we are developing a body-on-chip system to test cancer drugs. Here we review the concept of pharmacokinetic and pharmacodynamic (PK-PD) models and discuss the potential of using microfluidic systems that can physically realize the PK-PD models. Integration of PK-PD models and body-on-a-chip technology to build pharmacokinetic-pharmacodynamic models on a chip could substantially improve current drug-development processes.

Key terms

micro cell culture analog ($\mu$CCA)
in vitro metabolism–dependent toxicity
microfluidics
cancer chemotherapy
body-on-a-chip

## 8.1 Introduction

Drug development is a complex process, requiring knowledge of pharmacology, cell and molecular biology, chemistry, and mathematics. Also it takes many years to complete drug development, and only 1 in 10 drugs entering clinical trials reaches the market, making the drug-discovery process a high-risk enterprise [1]. All these aspects render drug development expensive, both in terms of cost and time. It is estimated that the cost of drug development is approximately $900 million per new drug [2] and can be a high as $2 billion, depending on the type of drug and the developing firm [3]. In addition to the cost, pharmaceutical companies are suffering from decreasing productivity with a decreasing number of new drugs each year. Based on a report by Kola and Landis, the average success rate in all therapeutic areas is approximately 11%, and success rates in some therapeutic areas are even lower, such as a 5% success rate for oncology drugs [2]. In this report, the analysis of the underlying causes of attrition reveals that a significant portion of drug candidates fail due to lack of efficacy or unforeseen toxicity.

Typically, the drug-development process consists of three stages: preclinical in vitro testing, animal testing, and human clinical trials. In vitro testing typically involves a cell-based multiwell plate system (96-well plate or 384-well plate). Cells are exposed to drug candidates, and the response is analyzed. Though a cell-based assay system can provide human cell-specific responses to the drug candidate, it is an incomplete model as it does not provide the responses due to multiorgan interactions. In addition, cell lines cultured in vitro often do not provide the same responses as the cells in vivo, which makes the cultured human-cell model incomplete. While the whole-body response to a drug can be obtained from clinical trials, testing drug candidates in humans is met with serious challenges in resources such as time and cost. Animal models using rodents and nonhuman primates have been used as an alternative approach to human clinical trials. However, animal models are also expensive and carry ethical issues. More importantly, extrapolating from animal to human does not always guarantee successful prediction of human response to a drug candidate. Therefore, it is of paramount importance to develop an improved model system that is predictive of human response. This is especially true with oncology drugs, where animal models are known to be unpredictive of human response [4].

A mathematical modeling approach like pharmacokinetic-pharmacodynamic (PK-PD) modeling has made significant contributions to the drug-development process. Pharmacokinetics, which relates the administered dose of a drug to its concentration profiles in plasma and at a target site, evaluates drug absorption, distribution, metabolism, and elimination (ADME) characteristics. On the other hand, pharmacodynamics examines the pharmacological effect of a drug at a given concentration at the target site. By combining the two, PK-PD models aim to predict the time course of pharmacological effect after a given dosing regimen. The role of PK-PD models in the drug-discovery process has been emphasized in several review papers [1, 5–8], but the impact of a PK-PD modeling approach has been limited due to several obstacles, including the problems of finding optimal parameters and inaccurate predictions due to unknown mechanism of action.

Recently, the advance in micro- and nanofabrication technology has enabled development of a new research area called microfluidics [9], and microfluidic systems with novel designs have been shown to be ideal for attacking problems in cell biology, tissue engineering, and drug discovery [10–12]. The microscale nature of such systems makes them appropriate for addressing questions more relevant to the physiological situation in vivo [13]. Although microfluidics is a useful and powerful technology in its own right, it has great potential for improving the drug-development process when combined with a mathematical modeling approach. A carefully designed microfluidic system can compensate for some of the drawbacks of PK-PD modeling systems, and a quantitative mathematical analysis by PK-PD modeling can guide the design of a physiologically realistic microfluidic system as well as help researchers interpret the experimental results with greater insight. Currently, a mathematical approach, combined with microfluidic systems, has been primarily applied to the simulation of transport phenomenon [14–16], and there are few examples of mathematical models for the pharmacological effect of a drug on cells in a microfluidic system. However, active research efforts in the areas of PK-PD modeling have great potential for integration with microfluidic systems to achieve a synergistic outcome.

In this chapter, we begin by explaining the basic concept of pharmacokinetic-pharmacodynamic modeling. Different types of PK-PD models are described, from conventional pharmacokinetic models to more complicated pharmacokinetic models and an integrated approach for PK-PD modeling. The methodology of setting up the model is described, with our main focus being on the PBPK model. In the following section, a micro cell culture analog ($\mu$CCA) currently being developed in our lab is described to illustrate how PK-PD modeling and microfluidics technology can be combined to examine drug effects on a human body. Limitations of the current $\mu$CCA technology are discussed, together with other current, ongoing research on microfluidic-based technologies that can improve the current $\mu$CCA systems.

## 8.2 Pharmacokinetic-Pharmacodynamic Modeling

### 8.2.1 Basic concept

The term *pharmacokinetics* refers to the prediction of time-dependent concentrations of a substance in a living system [17]. It examines drug absorption, distribution, metabolism, and elimination, collectively known as ADME, which determine the disposition of the drug within an organism. A pharmacokinetic model attempts to predict the concentration profile of a drug in the blood or at the target site from a given dose, as well as the values of pharmacokinetic parameters such as clearance rate and half-life. Predicting pharmacokinetics prior to the clinical phase of drug development has been shown to be useful, especially in the selection of dosages in clinical studies, which has resulted in significant time savings [18]. Several commercial PK-modeling software packages are available on the market, including GastroPlus, SIMCYP, and Cloe PK [19–21].

Pharmacodynamics refers to the time course of a drug's pharmacological effect at the target site for a given target concentration. In short, pharmacokinetics studies what the body does to the drug, whereas pharmacodynamics characterizes what the drug does to the body [22]. The basis of pharmacodynamics is that the pharmacological effect of a drug is a function of its concentration at the target site. The pharmacological effect of a drug is modeled using drug concentration as a variable, whose exact form will differ depending on the type and pharmacological effect of the drug. Pharmacokinetics and pharmacodynamics characterize different realms of pharmacology, but when the two are integrated, PK-PD models can provide a description of the time course of a drug's effects in response to a dosage regimen [5].

### 8.2.2 Pharmacokinetic model

Providing a full description for the process of building a pharmacokinetic model is beyond the scope of this chapter, and for more complete information, we recommend excellent reviews on the methodology and application of PBPK models [17, 23–25]. Here we provide a brief summary of the methodologies for developing a pharmacokinetic model, with a focus on physiologically based pharmacokinetic (PBPK) models.

#### 8.2.2.1 Classical PK model

Based on the complexity of the model, pharmacokinetic models can be classified into different levels of pharmacokinetic models. The empirical model lies at the lowest level of the hierarchy of pharmacokinetic models and is described simply by a sum of exponential terms, as follows [26]:

$$C(t) = \sum C_i \cdot e^{-\lambda_i \cdot t} \tag{8.1}$$

It can be used to describe a typical drug-elimination process and may be good for interpolation, but generally empirical models are not useful for extrapolation because the coefficients do not have any physiological interpretations [26]. A classic form of this empirical model is a two-compartment model, which is shown in Figure 8.1(a). This model assumes two compartments with an intercompartmental interaction, and drug

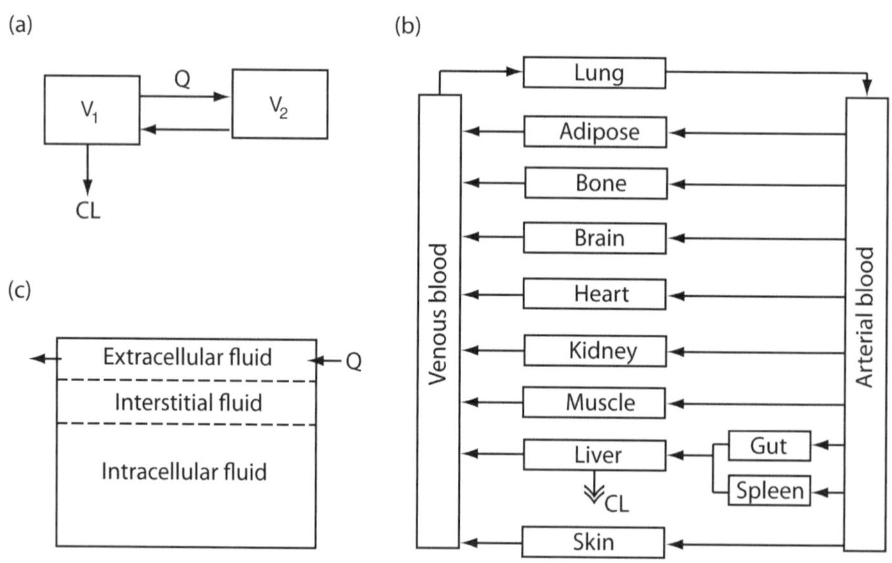


**Figure 8.1** Pharmacokinetic models. (a) Two-compartment PK model. $V_1$ and $V_2$ are volumes of corresponding compartments, $Q$ is a flow rate, and $CL$ is clearance rate. (b) Physiologically based pharmacokinetic (PBPK) model. Separate compartments are assumed for each organ, and blood flows are denoted by arrows. (c) PBPK model for diffusion-limited case. Blood flows through extracellular space and target compounds are transported to intracellular space by diffusion.

elimination occurs in one of the compartments. Mathematically, this model would be expressed as a biexponential model. Although a simple one, it is the conventional form of a pharmacokinetic model and widely used to describe the pharmacokinetic characteristics of a drug [27, 28].

#### 8.2.2.2 PBPK model

A physiologically based pharmacokinetic (PBPK) model utilizes anatomical, physiological, and biochemical information and is therefore more comprehensive and physiologically realistic than the classical PK model. The information needed to develop a PBPK model includes: (1) organ volumes, (2) blood flow rates, (3) enzyme reaction rates, (4) drug-tissue partition coefficients, and (5) membrane permeabilities. As shown in Figure 8.1(b), a separate compartment is assumed for each key organ, and a mass balance equation for each compartment describes the drug-concentration change. The mass balance equation reflects that the time-related change in the drug concentration in a compartment must equal the sum of the rate of drug entering the tissue, the rate of drug leaving the tissue, and the rate of drug elimination within the compartment [see (8.2)]. Elimination includes metabolism of the parent drug.

$$\frac{d(\text{drug concentration in tissue})}{dt} = (\text{rate of drug into tissue}) - (\text{rate of drug out of tissue}) - (\text{rate of drug elimination within tissue}) \quad (8.2)$$

One important assumption of this model is that the distribution can be perfusion (or flow) limited or diffusion limited. In the perfusion-limited case, the limiting step for the distribution of the drug is the rate at which it arrives at the tissue, not the diffusion of the drug into the tissue. This usually applies to hydrophobic drugs that can cross cell membranes easily. In the case of the perfusion-limited model, the above mass balance equation becomes

$$V\frac{dC}{dt} = Q\cdot\left(C_{in} - \frac{C_{in}}{P}\right) - R_e \quad (8.3)$$

where $V$ (in milliliters) is the volume of the tissue, $C$ (in nanomoles per milliliter) is the drug concentration in tissue, $Q$ (in milliliters per hour) is the blood flow rate into the tissue, $C_{in}$ (in nanomoles per milliliter) is the drug concentration entering the tissue, and $P$ is the partition coefficient of the drug between the blood and the tissue. $R_e$ (in nanomoles per hour) is the rate for the elimination reaction, whose exact form depends on the mode of the elimination reaction. In contrast, if it is diffusion limited, which is more likely to occur with hydrophilic drugs that cannot cross the cell membrane easily, the diffusion of the drug into cell becomes the limiting step. In the case of diffusion-limited transport, the situation is a bit more complicated because the diffusion in the extracellular space must be taken into account, as can be seen in (8.4) and Figure 8.1(c).

$$V_e\frac{dC_e}{dt} = Q\cdot(C_{in} - C_{in}) - P_m\cdot(C_e - C_i) \quad (8.4)$$

where $V_e$ is the extracellular space volume, $P_m$ is the membrane permeability, $C_e$ is the free extracellular drug concentration, and $C_i$ is the free, unbound intracellular drug concentration. For a complete model, mass balance equations such as (8.3) and (8.4) are written for each chemical of interest and each organ compartment; then, the model is described by a series of ordinary differential equations that must be solved simultaneously.

A set of ordinary differential equations can easily be solved numerically due to the increasing availability of computing power. However, as mentioned earlier, many parameters are required to build a realistic model. Physiological parameters such as organ sizes and blood flow rates can be found in the literature. Data on enzyme reaction rates can also be found in the literature, but in many cases, they are derived from in vitro experiments, which are not directly applicable to in vivo cases, and the use of scaling factors may be necessary. Methods to extrapolate in vivo clearance rates from in vitro values have been proposed [29, 30], but the applicability of such methods depends on the types of drugs being analyzed. The partition coefficient refers to the distribution of a drug between the tissue and the plasma under steady-state conditions. This value can be found experimentally from tissue incubation [31], or the expected value can be calculated based on the hydrophobicity data of the drug [32]. This method combines the n-octanol:water partition coefficient of a chemical and tissue-composition data to estimate the tissue-plasma partition coefficient. In general, it is difficult to find exact values for enzyme-kinetics data and partition coefficients, which hinders building a physiologically realistic model. Therefore, routines to optimize parameter values from data are often needed to build a realistic PBPK model, which makes an animal PBPK model (e.g., rat) easier to build than the human PBPK model since animal data is more readily available than in the human case.

### 8.2.3 Pharmacodynamic model

A vast array of mechanisms account for the pharmacological effect of drugs, and depending on a drug's mechanism of action, different types of pharmacodynamic models can be developed. According to Mager et al., the types of pharmacodynamic models can be roughly classified as simple direct effect, irreversible, or tolerance models [33]. Each type of pharmacodynamic model can be further classified into different subgroups. Here we summarize briefly only the major model types with a focus on irreversible models and examples of modeling the effect of chemotherapeutic drugs on tumor cells.

#### 8.2.3.1 Simple direct effects

The simplest form of PD model to describe a drug effect is a one that assumes a direct, linear relationship between drug concentration and drug effect, which is expressed as follows:

$$E = S \cdot C + E_0 \tag{8.5}$$

where $S$ is the slope, $E_0$ is the base effect level in the absence of drug, $E$ is the drug effect, and $C$ is the drug concentration. A modified form of this model would be to assume that the drug effect is proportional to the log of the drug concentration, which is expressed as

$$E = S \cdot \log C + E_0 \tag{8.6}$$

These two models are unrealistic in that there is no maximum drug effect, and the drug effect will increase indefinitely as the drug concentration increases. Therefore, the model is valid only in a certain range of drug concentrations. Generally, these models are only valid when the drug effect is less than 20% (linear) or within 20% to 80% (log-linear) of the maximum effect [33]. Also, in case of a log-linear model, the model is not valid when the drug concentration is zero. On the other hand, these models are easy to use, and parameters can be obtained from simple regression analysis. They were successfully implemented in case of diazepam and beta-blocking agents, respectively [34, 35].

Another widely used form of a simple direct effect model is the so-called $E_{max}$ model, also known as the Hill equation, which was originally derived from a drug-receptor interaction [Figure 8.2(a)]. The simplest form of the model is expressed as

$$E = \frac{E_{max} \cdot C}{EC_{50} + C} + E_0 \tag{8.7}$$

$E_{max}$ is the maximum drug effect, and $EC_{50}$ is a concentration at which 50% of the drug effect is seen. This model has some advantages compared to the linear and log-linear models; for example, it can define the maximum drug effect and is still valid for when drug concentration is zero. A more generalized form of the $E_{max}$ model is a sigmoidal $E_{max}$ model:

$$E = \frac{E_{max} \cdot C^{\gamma}}{EC_{50}^{\gamma} + C^{\gamma}} + E_0 \tag{8.8}$$

$E_{max}$ and $EC_{50}$ are the same as in the previous model, and $\gamma$ is a steepness factor for the curve. If $\gamma$ is less than one, the curve becomes smoother, and if $\gamma$ is greater than one, it becomes steeper. The $E_{max}$ and sigmoidal $E_{max}$ models are probably the most classical forms of PD models and have been used in numerous examples [36–39].

All of the simple direct effect models described so far assume that there is no temporal dissolution between drug concentration and drug effect, which means that the maximum concentration and maximum drug effect will occur simultaneously. Therefore, these models are applicable only to the drugs that can achieve rapid equilibrium at the target site. Also, they describe only a reversible drug effect; a decrease in the drug concentration at the target site results in an immediate decrease in the observed drug effect.

#### 8.2.3.2 Irreversible effects

Chemotherapeutic drugs exert irreversible effects on cells, which cannot be modeled with the simple direct effect models described above. The first pharmacodynamic model for the irreversible effects of a chemotherapeutic drug was described in 1971 by Jusko, who modeled the pharmacological effect of phase-nonspecific

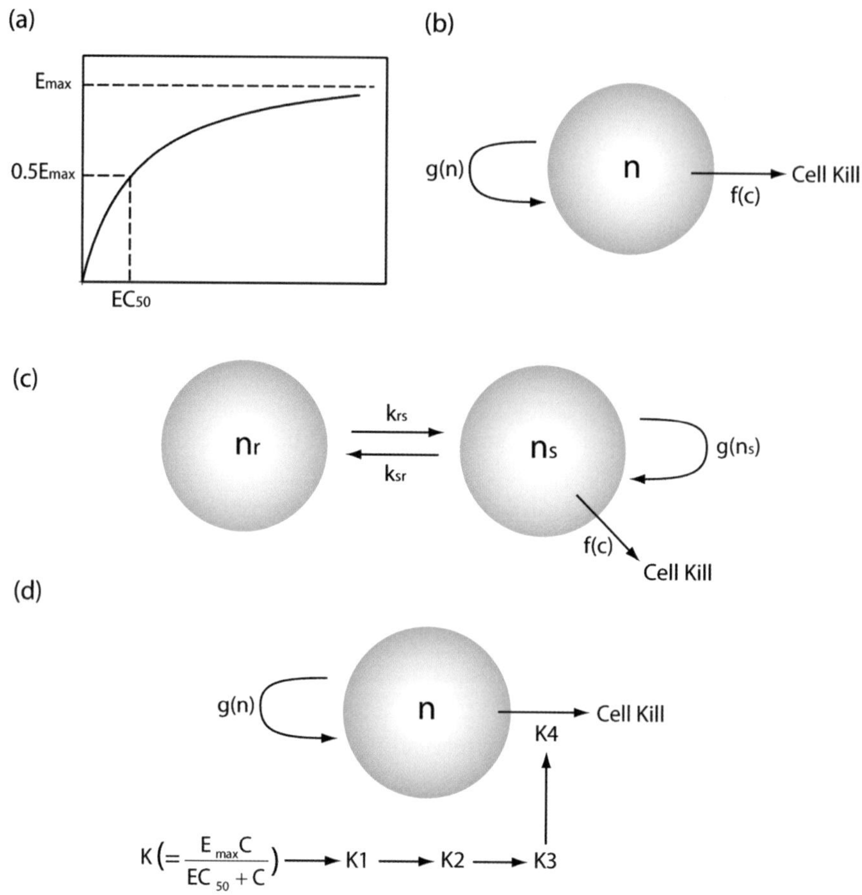


**Figure 8.2** Pharmacodynamic models. (a) Simple direct effect model ($E_{max}$ model). $E_{max}$ is the maximum drug effect, and $EC_{50}$ is the concentration at which half of the maximum drug effect occurs. (b) Cell-proliferation model with phase-non-specific inactivation. Cells proliferate at a rate function [$g(n)$], which is a function of cell number, and are killed at a rate $f(c)$, which is a function of drug concentration. (c) Cell-proliferation model with phase-specific inactivation. Cell population is divided into two groups (sensitive and resistant). Cell growth and killing occurs only in the sensitive population. (d) Transit compartment model. Cell kill is induced via a series of transit compartments (K1~K4).

chemotherapeutic drugs [40]. In this study, Jusko described a cell-proliferation model with irreversible inactivation as follows [Figure 8.2(b)]:

$$\frac{dn}{dt} = g(n) - f(c) \cdot n \tag{8.9}$$

where $n$ is a cell number, $g(n)$ is a proliferation rate of cells in the absence of the drugs, and $f(c)$ describes the effect of the drugs as a function of drug concentration. The cell-proliferation rate is usually described by an exponential model, often with a limitation of maximum cell number.

$$g(n) = k_g \cdot n \cdot (1 - n/n_{max}) \tag{8.10}$$

where $k_g$ (1/hour) is the growth-rate constant, and $n_{max}$ is the maximum cell number. The function for the drug effect $f(c)$ can be in the form of simple direct model discussed above, such as the linear or log relationship, or the $E_{max}$ model. This irreversible model can be an adequate model to describe the action of phase-nonspecific drugs, such as alkylating agents, which act by adding alkyl groups to DNA and thus affect cells regardless of cell-cycle phase. Cisplatin and oxaliplatin are examples of phase-nonspecific drugs.

Many chemotherapeutic agents, however, exert their effect only when the target cell is in a specific phase of the cell cycle, owing to their specific mechanism of action. For example, 5-fluorouracil interferes with DNA synthesis and is therefore known to affect cells in G1/S phase [41]. Paclitaxel (Taxol) is known to inhibit microtubule growth during cell division and therefore to affect cells in G2/M phase only [42]. In such cases, it is necessary to divide the cell population into sensitive and nonsensitive groups and treat them differently. This is known as a cell-cycle-phase-specific model. Mathematically, it is expressed as follows:

$$\frac{dn_s}{dt} = g(n_s) - f(c) \cdot n_s - k_{sr} \cdot n_s + k_{rs} \cdot n_r \tag{8.11}$$

$$\frac{dn_r}{dt} = k_{sr} \cdot n_s - k_{rs} \cdot n_r \tag{8.12}$$

Here, $n_s$ is the number of sensitive cells, and $n_r$ is the number of nonsensitive (or resistant) cells. $k_{sr}$ (1/hour) and $k_{rs}$ (1/hour) refer to the rate of interchange between the sensitive and resistant cell populations. Figure 8.2(c) shows the concept of the cell-cycle-phase-specific model. In many cases, the sensitive population is rapidly growing since most molecular targets of chemotherapeutic agents are related to cell division. Therefore, the proliferation term [$g(n_s)$] is added only to the equation for the sensitive population. The resistant population is generally quiescent and therefore does not have a proliferation term. The two population groups are interchangeable by first-order rate constants. Liu et al. studied the effect of an M-phase-specific drug on the development of cancer by dividing the cell cycle into mitotic phase (M), quiescent phase (G0 phase), and the interphases (G1, G2, S phases), then examining the efficiency of cell synchronization before treatment [43]. In another example, Yano used this model for the analysis of bactericidal kinetics, where bacterial phases were divided into two populations [44].

In many cases, a drug's effect manifests much later than the time of administration, or the time at which the peak concentration occurs. Such a temporal dissolution is evident in case of oncology drugs because it takes days or even weeks for the cytotoxic effect of the drug to be observed. The pharmacodynamic models described thus far assume that the effect of a drug is seen simultaneously with the drug exposure; therefore, they are not adequate to describe delayed drug effects. Sun and Jusko proposed a transit compartment model to characterize the delayed drug effects, where the time delay is modeled via a series of transit compartments, as shown in Figure 8.2(d) [45]. Results show that this model is generally useful for the analysis of pharmacological effects with a cascade of reactions, such as a signal-transduction pathway [33]. One advantage of this modeling approach is that it is possible to simulate the time delay in drug response without mechanistic knowledge of the cascade. An example of equations describing a transit compartment model follows:

$$\frac{dn}{dt} = k_{ng} \cdot n - K4 \cdot n \tag{8.13}$$

$$\frac{dK1}{dt} = \frac{1}{\tau}(K - K1) \tag{8.14}$$

$$\frac{dK2}{dt} = \frac{1}{\tau}(K1 - K2) \tag{8.15}$$

$$\frac{dK3}{dt} = \frac{1}{\tau}(K2 - K3) \tag{8.16}$$

$$\frac{dK4}{dt} = \frac{1}{\tau}(K3 - K4) \tag{8.17}$$

$$K = \frac{E_{max} C}{EC_{50} + C} \tag{8.18}$$

Here $n$ is the cell number, $\tau$ (in hours) refers to the transit time, and $k_{ng}$ (1/hour) is a net growth-rate constant. K4 (1/hour) is the rate function for cell killing, which is related to the drug concentration $C$ via a series of transit compartments (K1 to K4). The transit compartments can also be thought of as the status of the damage on cells. As the cells become more damaged, they progress from K1 to K4, where eventually cell killing is induced. $E_{max}$ and $EC_{50}$ are the maximum cell-killing effect and half-saturation concentration, respectively. In a study by Lobo and Balthasar, three of the models described above (phase-non-specific model, phase-specific model, and transit compartment model) were used to model tumor-cell growth kinetics under the influence of various concentrations of methotrexate [46]. In this study, tumor cells were treated with methotrexate for 24 hours, and the subsequent growth pattern was monitored for more than 20 days. Although methotrexate was removed after the initial 24 hours, cell numbers reached the nadir more than 100 hours after exposure. Comparison of three different mathematical models showed that the transit compartment model described tumor-cell growth kinetics under the influence of methotrexate more accurately than the other two models.

#### 8.2.3.3 Tolerance model

Tolerance models were developed to describe drug effects that diminish after repeated exposure. A well-known example of such a phenomenon is the desensitization of G-protein-coupled receptors (GPCRs) by protein kinases in response to agonists. When a GPCR is activated by ligand binding, G-protein-coupled receptor kinases (GRKs) are activated via a signaling cascade involving G-protein, adenylate cyclase, and cyclic AMP. Activated kinases phosphorylate the receptor, resulting in desensitization [47]. Since activated GPCR causes phosphorylation of itself, the longer GPCR remains active, the more kinases are active, resulting in diminished GPCR activity. Such a phenomenon can be modeled by assuming that the active receptor can be temporarily lost to an inactive pool ($R_i$) by a first-order process ($k_d$):

$$\frac{dR_i}{dt} = k_d \cdot (R - R_i) \tag{8.19}$$

This allows tolerance to evolve in proportion to the original response, which is denoted as $R$ here.

### 8.2.4 Integrated PK-PD modeling

Simulation of a PBPK model gives the concentration profiles of a drug and its metabolites in each organ with a given dose. The pharmacokinetic profile alone can be useful because information such as area-under-curve (AUC) of a drug or metabolites in specific organs can be obtained. For integration of a PBPK model with a PD model, the concentration profile of a drug in a target organ serves as an input for the PD model, yielding a quantitative pharmacological effect of a drug at a given dose. For example, if one builds a PK-PD model for an oncology drug, the concentration profile of a drug in the tumor compartment will serve as an input for a PD model, which simulates the tumor growth in the presence of the drug. Therefore, one can predict the growth kinetics of the tumor for a given dose of the drug. However, as noted earlier, obtaining parameters for the PBPK and PD model is not always straightforward, and parameter optimization may become necessary.

Originally, pharmacokinetic and pharmacodynamic models were considered separate realms, with pharmacokinetic models focusing on the drug-concentration profiles and pharmacodynamic models focusing on the pharmacological response to a drug at a fixed concentration. However, the increasing availability of computing resources since the 1980s has facilitated the complex mathematical calculations required for PK and PD models. Since then, integration of pharmacokinetic and pharmacodynamic models has gained popularity with the expectation of enhancing the efficiency of drug development processes [1, 6–8].

One example of an integrative PK-PD modeling approach is a model for tumor growth in animal models [48]. Pharmacokinetic models of chemotherapeutic agents paclitaxel, 5-fluorouracil, and irinotecan (CPT-11) in mice have been developed and integrated with a pharmacodynamic model to describe tumor growth kinetics and the effects of the drugs on tumor cells using the transit compartment model. The integrated PK-PD model was able to predict the tumor weight progression accurately for 40 days. This example demonstrates a dynamic simulation of the pharmacological effect of a drug because the drug-concentration profile from the pharmacokinetic model serves as a variable in the pharmacodynamic model, thus reflecting the dynamic response of the animal model to changes in drug concentration. We developed a combined pharmacokinetic-pharmacodynamic model for UFT (a combination of uracil and Tegafur, a prodrug of 5-fluorouracil) in rat, which simulated the growth of a tumor in the rat under various dosing strategies [49].

Integrated PK-PD modeling can be versatile and can be applied to dosing and scheduling determination [50], dosing individualization [51], formulation development [52], and toxicity assessment [53]. Currently, use of PK-PD modeling is limited to the early drug-development phase and to research applications only. Before PK-PD models can be utilized more broadly, obstacles in the areas of both PK and PD modeling must be overcome. As noted earlier, physiological and biochemical information is necessary to develop a complete PBPK model. A limiting factor for building a realistic PBPK model is that while some parameters are relatively easy to find, either in the literature or experimentally, some parameters are extremely difficult to estimate, especially for a human model. Although it is possible to optimize the model and find the parameters given sufficient data, data for human parameters is sparse. Indeed, this lack of human data is thought to be the major limitation to the wider use of PBPK models in drug-development processes [26]. It has been reported that a poor prediction by a PBPK model can result from not fully incorporating the processes due to lack of in vitro input parameters [54]. Also, the performance of PBPK models can vary, depending on the class of chemical being simulated. Physiological responses to certain chemical classes are poorly predicted by PBPK models. Therefore, it has been suggested that before the application of PBPK modeling, verification of the model is necessary using a few compounds in the given chemical class [54]. Usually, during the drug-development process, a PBPK model for animals is built first to validate the PK-PD modeling approach. An animal model is easier to build than a human model because animal data is more readily available. If the animal model can predict animal response with reasonable accuracy, then a human model is considered. However, verification in an animal model does not necessarily warrant the development of a human model because extrapolation from an animal model to a human model is not always straightforward.

The obstacles associated with finding realistic parameters are not limited to a PBPK model. When building a pharmacodynamic model, generally parameters are fitted to experimental or clinical data. In the case of an empirical model, fitting the parameters is unavoidable because components of the model do not have a mechanistic basis. However, a mechanism-based PK-PD approach can be built, in principle, using only in vitro data and standard anatomical parameters because physiological and mechanistic description of the model would enable the measurement of parameters from the understanding of the mechanism details [55, 56]. When it is practically difficult to measure parameter values experimentally, a mechanistic basis will enable researchers to predict what types of parameters are necessary for model development. Advances in fundamental biological sciences, such as genomics and proteomics, may expedite the wider application of PK-PD models. However, a mathematical modeling approach has a fundamental limitation in that it cannot capture mechanisms not described in the model; thus, the possibility of an unexpected experimental outcome remains. Therefore, development of an experimental platform that can be used for rapid validation of an integrated PK-PD model would make the drug-development process more cost-effective and less time-consuming. At the same time, once an experimental platform is established, a mathematical modeling approach can guide the design of the system. In addition, mathematical models can be further used for interpretation of experimental results. Microfluidic perfusion cell culture systems offer the possibility of building such a system in a more physiologically relevant way than traditional in vitro cell culture systems.

## 8.3 Micro Cell Culture Analog (μCCA)

As noted earlier, drug development is an inefficient process. Animal trials are expensive, time-consuming, and not particularly effective at predicting human response. While PK-PD models could in principle improve the process, they have had minimal impact thus far, partially due to the difficulty of accurate parameter estimation. In addition, often cell-to-cell and tissue-to-tissue interactions, as well as unknown intracellular reactions, are not captured in PK-PD models. These unknowns lead to a lack of credibility that rule out PK-PD models as primary tools for testing the safety and efficacy of a drug. However, a biological model based on a PK-PD model may offer an in vitro method that can facilitate more efficient drug-development processes. We call this biological model a cell culture analog (CCA). The CCA system was first conceived as a physical realization of a PBPK model. In a PBPK model, separate chambers are assumed for each organ, which are connected by blood flow; in a CCA, separate chambers representing each organ are fabricated on a silicon chip, then connected by channels that mimic the distribution of blood flow. Relevant cell lines are cultured in each chamber to mimic key metabolic or absorption characteristics of that tissue, and cell culture medium is recirculated as a blood surrogate. The CCA attempts to mimic the complex interactions resulting from metabolite exchange between different organs in the human body. Conventional in vitro methods used in drug-screening processes, such as 96-well plate or 384-well plate, have limitations in that they are static systems with a single-cell type and do not replicate the dose dynamics or the complex interactions between organs in the human body, most importantly liver metabolism. With the CCA being a physical realization of a mathematical PBPK model, it is expected that the combined utilization of the CCA and a PK-PD modeling approach can have a synergistic effect, where a PK-PD model could be used in the design of a CCA device and the interpretation of experimental results from it, and a CCA device could experimentally verify simulation results from a PK-PD model. The concept of the integrative approach of PK-PD modeling and a CCA is summarized in Figure 8.3.

Initially, the prototype cell culture analog device used for a proof-of-concept study was described by Sweeney [58]. In this prototype, two milk dilution bottles and a spinner flask were connected using Teflon FEP tubings, and recirculation was achieved using peristaltic pumps [Figure 8.4(a)]. The glass bottles represented the lung and liver compartments, and the "other tissues" compartment was represented by a glass reaction beaker. The system was the first prototype for a multicompartment system with multiple cell types with fluid recirculation. For the lung and liver compartments, rat lung cell line L2 and rat hepatoma cell line H4IIE were used, respectively [Figure 8.4(b)]. As a model drug, naphthalene (oral $LD_{50}$ = 2,200 mg/kg in the rat [59]) was used to test the response

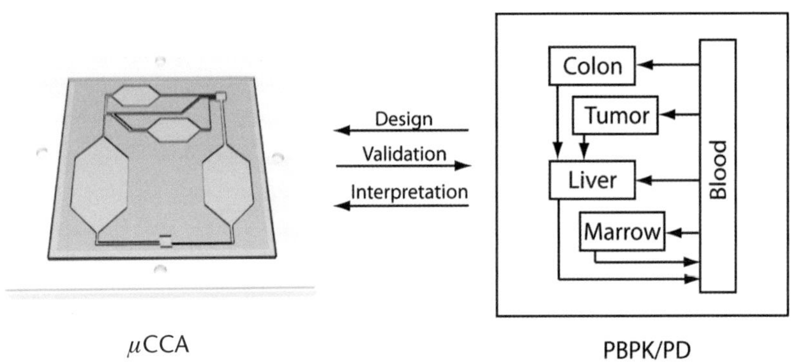


**Figure 8.3** Integrated approach of PK-PD modeling and μCCA. (Figure of a μCCA is modified from [57] with permission.) Each chamber in the mCCA (left) corresponds to colon, tumor, liver, and marrow compartment in the PBPK diagram (right). The chambers are connected with channels that mimic blood flows (arrows in the PBPK diagram). The μCCA provides a method to reproduce a whole-body response to a drug while mimicking pharmacokinetic profiles, whereas PK-PD modeling can be used to analyze and interpret experimental results from a μCCA.

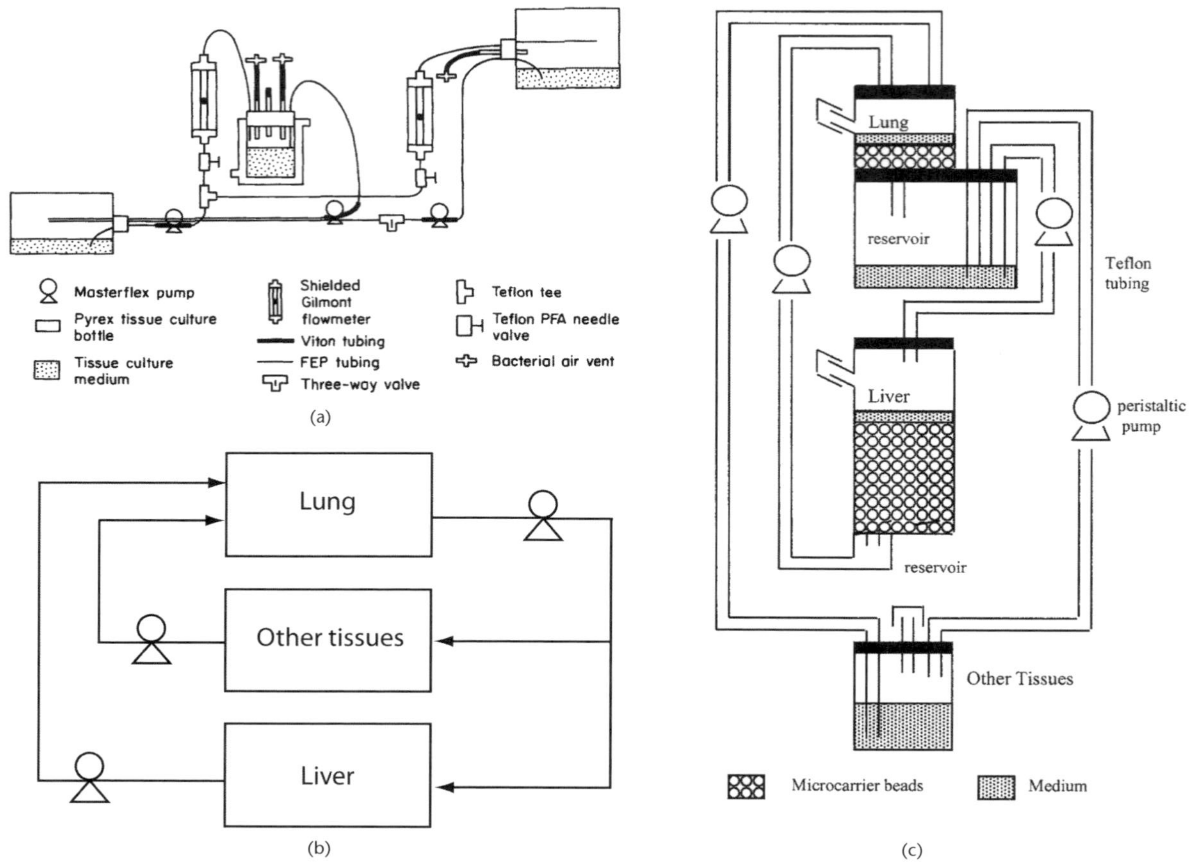


**Figure 8.4** Prototype macroscale cell culture analog devices. (a) First-generation prototype CCA. (Reprinted from [58] with permission.) Culture bottles are connected to one another using tubing and pumps, and the pattern of the blood flow is intended to mimic the blood flow shown in (b). (b) Corresponding PBPK diagram consisting of lung, liver, and other tissues (modified from a figure in [58], with permission). (c) Second-generation CCA with microcarrier beads. (Reprinted from [62] with permission.) The layout of the culture chambers (lung, liver, and other tissues) are the same. Cells are cultured on the microcarrier beads inside the chambers.

of the system. In mice, the nonciliated bronchiolar epithelial cell (Clara cell) of the mouse lung is the primary target of naphthalene toxicity [60]. Experiments with this system successfully showed that the metabolite formed in the liver compartment was responsible for the cytotoxicity observed in the lung cells.

The major limitations of this prototype were the nonphysiological residence times in each chamber and an unrealistic ratio of liver-to-lung cells [61]. Also, the difficulty with stable operation of the device and obtaining time-course data was a limitation of the prototype system. An improved macroscale system was designed to mimic the physiological organ characteristics of a rat more closely than the prototype CCA [62]. The CCA system used cells attached to microcarrier beads, which formed packed beds with cell culture medium recirculating through the beds [Figure 8.4(c)]. The main advantage of packed beds rather than a monolayer cell culture was that a higher cell density was achievable, providing a more realistic liquid-to-cell ratio, and medium residence times were closer to in vivo values. With this second-generation, packed-beds CCA device, several aspects on the device operation were examined, including mixing profiles, naphthalene-distribution profile, and the effect of reactor environment on cell viability. However, when a naphthalene study was performed with the second-generation device, unlike with the previous prototype, there was no detectable change in lung cell viability in response to naphthalene addition to the CCA system [62].

To explain these discrepancies, PBPK models for the two different versions of the CCA system were built and compared [63]. From the simulation of the two PBPK models, it was found that due to a difference in residence times in the two CCAs, levels of naphthoquinones (a naphthalene metabolite) in the system differed by more than two orders of magnitude. Longer residence times in the prototype CCA system allowed time for the formation of naphthoquinones, whereas the shorter residence times in the packed-beds CCA system did not. Based on these results, naphthoquinone was suggested as a possible cause of cell death, rather than naphthalene epoxides as initially thought.

This series of studies involving the development of CCA systems and mathematical PBPK modeling demonstrated that a PBPK model could be used not only to guide the design of a CCA device but also to interpret the experimental results from the CCA. However, there were several limitations in the macroscale systems. First, the liquid-to-cell ratio was not physiologically realistic in the devices. The liquid-to-cell ratio was about 1,000:1 in the first prototype and 5:1 in the second-generation system, compared to the physiological ratio of 1:2. Conventional 2-D monolayer culture methods using standard culture flasks, or wells, also have liquid volumes greater than the volume of the cell layer. As the cells are exposed to an environment that is completely different from their native one, it is not surprising that many cells exhibit a different behavior when separated from their native tissue and cultured by conventional methods [64]. Second, early versions of CCA devices did not have physiologically realistic residence times in organ chambers. For example, the residence time for the liver chamber in the packed-bed CCA device was 1.9 minutes, whereas the actual residence time for the liver in a rat is about 21.3 seconds [63]. A PBPK model is based on physiological parameters such as blood flow rates and organ sizes, and a faithful realization of the PBPK model is achieved only if the residence times in each chamber are close to the actual values because inaccurate residence times in organ chambers could distort a pharmacokinetic-pharmacodynamic response, not correctly reflecting physiological responses in the body.

These limitations of macroscale CCA devices, together with advances in microfluidic technology, spurred the development of a microscale CCA, or $\mu$CCA, which is a miniaturized version of the earlier CCA devices. A cell culture analog device in microscale confers several advantages. (1) The natural length scales in tissue are on the order of 100 $\mu$m from capillary blood flow with cells on the order of 10 to 20 $\mu$m in size; these dimensions can easily be mimicked in the microscale systems. This allows more realistic residence times and relative organ sizes because the flow rate and the geometry of a microfluidic device can be precisely controlled using microfabrication techniques. (2) Being in microscale, fewer cells and less test chemical are required, which enables the use of scarce tissue samples and novel compounds. Authentic tissue samples provide more realistic representations of in vivo environments. A system requiring small amounts of test chemicals would enable high-throughput experiments with newly developed, novel molecules. (3) The small size of the device facilitates operation of multiple devices concurrently, which enables high-throughput assays. Miniaturization also allows integration of the system with other components necessary for drug testing, such as sample analysis and optical detection. (4) Once a master device has been fabricated, subsequent devices can be made very cheaply, reducing the cost of fabrication for mass parallelization.

### 8.3.1 Design of a $\mu$CCA and calculation of flow rates

An experiment with a $\mu$CCA device requires several steps: (1) design and fabrication of the device, (2) cell seeding and assembly of the device, (3) operation of multiple μCCAs, and (4) data acquisition and analysis.

A well-designed microfluidic device can enable novel experiments and reduce experiment time, while a poorly designed device can cause unexpected problems, making the interpretation of experimental results difficult. The principal design criteria for the prototype $\mu$CCA devices are as follows. (1) The ratio of the chamber sizes and the liquid residence times in each compartment should be physiologically realistic. (2) Each chamber should have a minimum of $10^4$ cells to facilitate analysis of chemicals. (3) The hydrodynamic shear stress on the cells should be within physiological values (<2 dyne/cm$^2$ for most cells, except endothelial cells in blood vessels) [65]. (4) The liquid-to-cell ratio should be close to the physiological value (1:2) [66].

Chamber size for each tissue/organ compartment was designed so that the ratio of cells between compartments reflects the ratio in the body. Since this ratio can differ by species, a $\mu$CCA chip may be specific to humans or a particular animal. The absolute sizes of the chambers were determined by the requirement that the smallest cell-containing chamber would hold at least $10^4$ cells. For monolayer cultures, the chambers were typically 30 $\mu$m deep, and deeper chambers (100 $\mu$m) were fabricated when using 3-D tissue constructs. Shear stress was another important criterion; it was maintained at values within physiologically relevant levels for a particular cell type. Residence time was considered the primary constraint because physiologically realistic residence times should capture in vivo reaction kinetics more accurately. The residence time for each organ chamber was determined to be as close as possible to the actual physiological value in humans, and then the flow rate in and out of the chamber was determined simply by dividing the chamber volume by the determined residence time.

After chamber sizes and flow rates were determined, the dimensions of channels connecting the chambers were chosen to meet the desired flow rates. Since the $\mu$CCA mimics the anatomical layout of organs and blood flow, the whole system is a microfluidic network, consisting of organ chambers interconnected by microchannels. A microfluidic network can be analyzed as an analog of an electrical circuit. A pressure drop is equivalent to a voltage drop, a flow rate to an electrical current, and a hydrodynamic resistance to an electrical resistance. In low–Reynolds number hydrodynamics, an equation in exact similarity with Ohm's law for a resistive element can be used to describe the relationship between the pressure drop ($\Delta P$), flow rate ($Q$), and hydrodynamic resistance ($R$) in a channel [67].

$$-\Delta P = R \cdot Q \tag{8.20}$$

To calculate the flow in the microfluidic network, one can use an analog of Kirchoff's current law for electrical circuits, where at each node, the sum of incoming flow rates must be zero. Also, analogous to Kirchoff's voltage law, the sum of potential pressure drops around a loop should equal zero. For example, the $\mu$CCA designed in our lab is illustrated in Figure 8.5(a), with its electrical circuit equivalence shown in Figure 8.5(b). For laminar flow through a rectangular-shaped channel with a high aspect ratio ($h > w$ or $w > h$), the fluidic resistance is determined by the following equation [11]:

$$R = \frac{12 \mu L}{w h^3} \tag{8.21}$$

where $\mu$ is the fluid viscosity, $L$ is the channel length, $w$ is the channel width, and $h$ is the channel height. If the length and width of the channels in a $\mu$CCA are known, then the pressure drop across the channel can be calculated by multiplying the hydrodynamic resistance and the flow rate. Through trial and error, the channel dimensions were determined iteratively so that the pressure drops across all the channels and chambers satisfied the condition of Kirchoff's laws. Once all the dimensions for $\mu$CCA were determined, the design was drawn using AutoCAD software (Autodesk Inc., San Rafael, California).

## 8.3.2 Fabrication of a $\mu$CCA

### 8.3.2.1 Materials

#### *8.3.2.1.1 Equipment*

- Clean room
- AutoCAD (Autodesk)
- GCA/MANN 3600F Optical Pattern Generator (Ultratech, San Jose, California)
- Contact aligner (AB-M HTG 3HR Contact Proximity Aligner, Hybrid Technology Group)
- Inductively coupled plasma ion etcher (Plasmatherm 770, Plasma-therm, St. Petersburg, Florida)
- Tencor P10 profilometer (KLA Tencor, Milpitas, California)
- Gasonics Aura 1000 Asher (Novellus, San Jose, California)

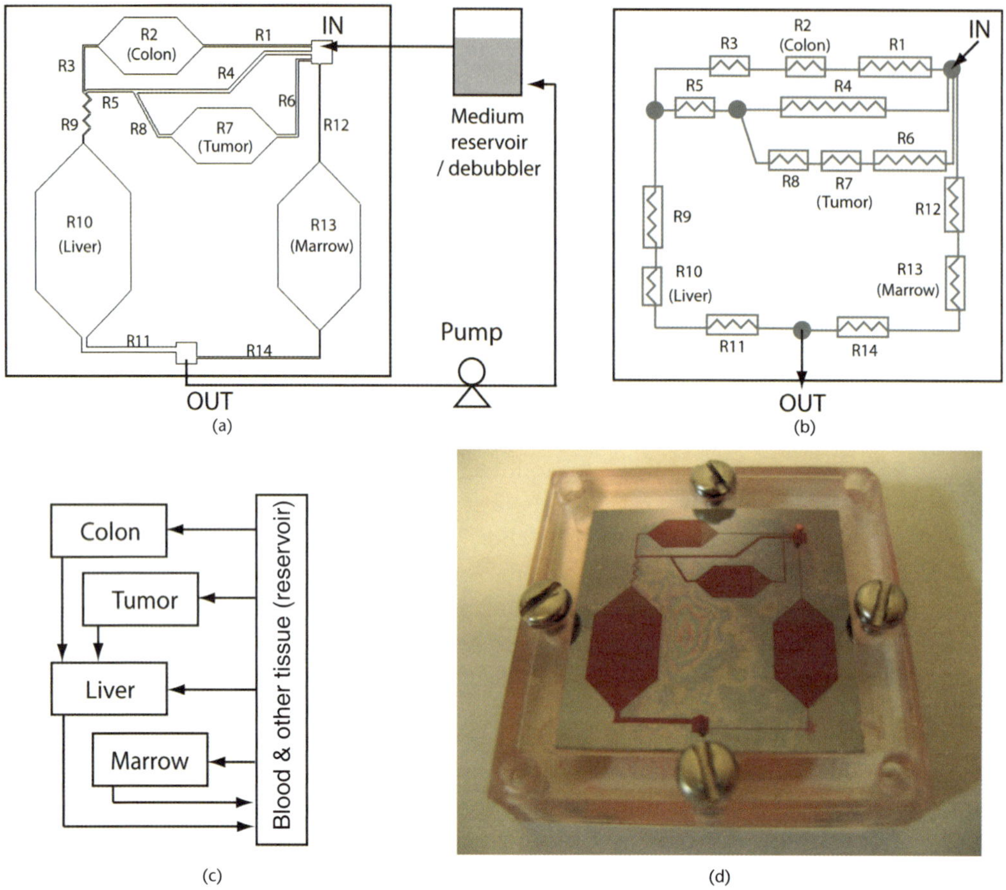


**Figure 8.5** Microfluidic network in analogy with an electrical circuit. (a) Schematic diagram of a $\mu$CCA device. A diagram of a $\mu$CCA device for studying the effect of drugs on colon cancer is shown. Medium flows from the reservoir into the inlet, through channels and chambers, and back to pump and the reservoir. (b) Electrical circuit equivalent of $\mu$CCA, where each element causing hydrodynamic resistance is drawn as a resistor (denoted as R). Microchannels have narrower widths, thus cause greater resistance than chambers. Circles are the nodes where more than two flows are joined. (c) A diagram of corresponding PBPK model. (d) A picture of $\mu$CCA with red dye for visualizing the channels and chambers.

- K&S 7100 dicing saw (Kulicke & Soffa, Fort Washington, Pennsylvania)

##### *8.3.2.1.2 Reagents*

- P20 primer (Shipley, Marlborough, Massachusetts)
- Shipley 1813 photoresist (Shipley)
- MIF 300 developer solution (AZ Electronic Materials, Branchburg, New Jersey)

#### 8.3.2.2 Methods

1. The designed pattern is exposed and developed on a chrome-coated glass mask using GCA/MANN 3600F Optical Pattern Generator.

2. A single-side-polished, 350-$\mu$m-thick, 3″-diameter silicon wafer is primed with P20 primer by spin coating at 3,000 rpm for 30 seconds.
3. Shipley 1813 photoresist is spin-coated on the wafer at 3,000 rpm for 30 seconds.
4. The pattern is transferred from the mask to the wafer by exposure to UV light (405 nm wavelength) for 2.5 seconds, using a contact aligner.
5. The wafer is developed for 1 minute by immersing in MIF 300 developer solution.
6. Etching is performed in an inductively coupled plasma ion etcher (Plasmatherm 770) to the desired depth. The depth of etching can be checked during the etching process using a Tencor P10 profilometer (KLA Tencor).
7. The photoresist is removed using Gasonics Aura 1000 Asher (Novellus). If necessary, the second layer is fabricated using the same steps.
8. After the fabrication is complete, the wafer is cut to a size of 1″ × 1″, using K&S 7100 dicing saw (Kulicke & Soffa) with a S1235 blade. The fabrication process is summarized in Figure 8.6(a).

### 8.3.3 Cell seeding and assembly of the device

#### 8.3.3.1 Materials

##### *8.3.3.1.1 Equipment*

- Plasma cleaner (Harrick Plasma, Ithaca, New York)
- Peristaltic pump (Watson-Marlow, Wilmington, Massachusetts)
- Biosafety cabinet
- Mammalian cell culture incubator
- Pipette man
- Autoclave

##### *8.3.3.1.2 Reagents and supplies*

- PharMed BPT tubes (Cole Parmer, Vernon Hills, Illinois, 0.25 mm inside diameter, 95706-12)
- Poly-D-Lysine (Sigma-Aldrich, St. Louis, Missouri, 50 mg, MW70000-150000 P6407)
- Human plasma fibronectin solution (Chemicon International, Billerica, Massachusetts, FC010-5mg)
- Pipette tips
- Cell culture dishes (Fisher Scientific, Pittsburgh, Pennsylvania, Corning culture dish, 60 × 15 mm, 08-772-21)
- McCoy's 5a Medium, modified (Sigma-Aldrich, St. Louis, Missouri, catalog no. M4892 10X1L)
- ATCC-formulated Eagle's Minimum Essential Medium (ATCC, Manassas, Virginia, 30-2003)
- Trypsin [Invitrogen, Carlsband, California, 0.05% (1×) with EDTA 4Na, liquid, 25300-062]
- Microcentrifuge tubes (VWR, West Chester, Pennsylvania, 1.5 mL)
- Ethanol (Sigma, more than 99.5%)

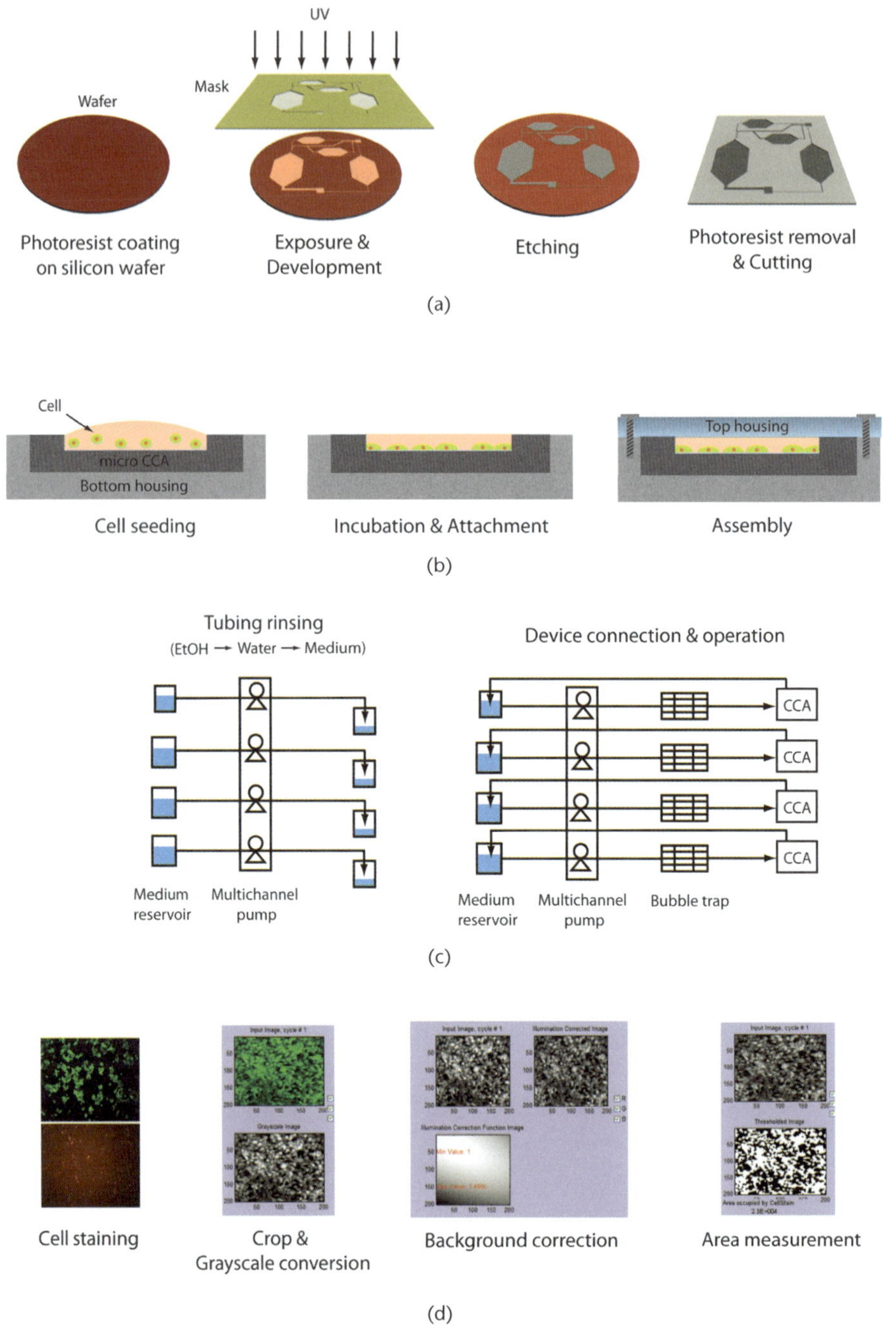


**Figure 8.6** Flow diagram for $\mu$CCA experiment. (a) Fabrication of a $\mu$CCA chip. Photoresist is spin-coated on a silicon wafer, and the $\mu$CCA pattern is exposed on the wafer. After developing and etching, photoresist is removed, and the wafer is cut into separate pieces. For simplicity, only one pattern was drawn on the silicon wafer in the figure. In practice, multiple patterns are fabricated on a single wafer. (b) Cell seeding and assembly. A cell suspension is seeded in a chamber and incubated overnight for attachment. After cells are attached, top housing piece is closed and sealed. Only one chamber is drawn for simplicity. In practice, multiple cell lines are cultured in different chambers, and the cell-seeding process is repeated for each chamber. (c) Operation of a $\mu$CCA. Tubing pieces are rinsed thoroughly to remove residual air bubbles, and a $\mu$CCA is connected to tubing. Six to eight chips are run simultaneously for up to 96 hours. In the figure, only four chips were drawn for simplicity. (d) Cells are stained with fluorescent stain, and images are analyzed using an image-processing software, which involves background correction for nonuniform illumination and measurement of cell area. Pictures were captured from software CellProfiler (Broad Institute, Cambridge, Massachusetts).

- Silicone sheets (Grace Bio Labs, Bend, Oregon, press-to-seal silicone, SWS-S-0.1)
- Eight-well strip plate (VWR, 62409-001)
- Teflon-lined silicone cover (BioTech Solutions, Vineland, New Jersey, 996050MR-96)

#### 8.3.3.2 Methods

1. Cut 0.5 mm thickness silicone gaskets to the size of a μCCA chip. Holes are cut out at the positions of chambers. The gasket is placed on a μCCA chip before autoclaving. The purpose of this gasket is to prevent spilling of cell culture medium later when a cell suspension is placed on the chip.
2. Prior to cell seeding in a μCCA, all of the components are sterilized. PharMed BPT tubings, silicon chips with gaskets, silicone cover for medium reservoir, and polycarbonate housings are autoclaved. Plexiglas housings are sterilized by air plasma cleaning.
3. To facilitate adhesion of cells, 0.1 mg/mL Poly-D-Lysine (Sigma-Aldrich) solution is placed onto the chip through the holes in the silicone gasket, and incubated for 1 hour at room temperature.
4. After rinsing with DPBS, the chip is coated with fibronectin solution (50 μg/mL) for 2 hours at 37°C.
5. Cells are trypsinized from culture flasks and diluted to a desired cell density. Then, 5 to 10 μL of the cell suspension is pipetted through the holes in a silicone gasket onto a chamber in the silicon chip.
6. Repeat step 5 for each cell line that needs to be cultured on the μCCA.
7. The silicon chip is then incubated overnight until cells attach firmly to the chip surfaces.
8. After incubation, the silicone gasket on the μCCA is carefully removed using sterile forceps. The μCCA is placed inside the recess in the custom-made bottom-housing piece. The bottom-housing frame was machined from either 5-mm–thick polycarbonate or Plexiglas (poly-methylmetacrylate, or PMMA) with a 1 mm recess in the center for placing the μCCA. A silicone sheet 0.5-mm-thick is placed between the μCCA and the bottom frame to provide better sealing.
9. A drop of cell culture medium is placed on top of the silicon chip, removing all air bubbles on the surface
10. The top-housing frame is closed in a tilted manner to avoid capturing air bubbles, and the top- and the bottom-housing frames are secured with screws. The top housing was machined from 3-mm-thick polycarbonate or Plexiglas. It consists of inlet and outlet holes for circulation of medium and four holes on the sides for screws. The cell-seeding process is shown in Figure 8.6(b).
11. Before connecting a μCCA to a pump, the tubing needs to be thoroughly rinsed with ethanol for at least 1 hour to remove any residual air bubbles, followed by rinsing with sterile water and cell culture medium for another hour.
12. Medium reservoirs are prepared by filling an individual well of eight-well strip plate with 200 μL medium, then sealing with a Teflon-lined silicone cover.

13. The $\mu$CCA device is then connected to PharMed tubing, of which the other end is connected to a medium reservoir. The whole connected system ($\mu$CCA chips, medium reservoirs, and a peristaltic pump) is placed inside a cell culture incubator.
14. A peristaltic pump (Watson-Marlow) is started to recirculate medium between the $\mu$CCA and the reservoir for 3 days. Figure 8.6(c) shows the schematic diagram of the system setup. Gas exchange (oxygen supply and $CO_2$ removal) is achieved mainly through the walls of the tubing connecting the $\mu$CCA and the medium reservoir.

### 8.3.4 Data acquisition, anticipated results, and interpretation

At the end of an experiment, 5 $\mu$M calcein AM and ethidium homodimer-1 in DPBS solution is circulated into a $\mu$CCA using a peristaltic pump for 1 hour. Cells are examined under a microscope through the top cover frame. Fluorescent images of cells are taken, and image-processing software is used to measure the area occupied by the live and dead cells [Figure 8.6(d)]. The image-processing steps consist of converting an image file into grayscale and correcting for background illumination to create more uniform images. The area occupied by the cells (live or dead) is determined, and the cell viability is calculated as the ratio of the number of live cells to the number of total cells (live and dead). Additional analysis of circulating medium is also possible, using high-performance liquid chromatography (HPLC) or mass spectrometry (MS). For example, the amount of a drug left in the circulating medium can be measured, which would equate to a plasma concentration of a drug in an animal or human model, and can be compared with the simulation result of a PBPK model. Cells cultured in the device can also be recovered after disassembling the device, which can be further analyzed for molecular processes, such as transcription and translation, or cell-cycle analysis.

As a physical counterpart of a PK-PD model, experimental results obtained from a $\mu$CCA can be directly compared with a PK-PD model simulation. For example, the growth of tumor cells in a $\mu$CCA can be compared with the predicted growth of tumor cells in a PK-PD model with a given dose of a chemotherapeutic agent. Obviously, for a direct comparison, a PK-PD model for a $\mu$CCA has to be built. Scaling down a PK-PD model for humans to a model for a $\mu$CCA is relatively straightforward. Anatomical parameters (size of organs and blood flow rates) are replaced with chamber sizes and medium flow rates, and other parameters, such as enzyme kinetic parameters and partition coefficients, can be easily measured in a $\mu$CCA. Being able to compare directly between the simulation results of a PK-PD model and the experimental result from a $\mu$CCA enables improving the design of a $\mu$CCA with PK-PD modeling, verification of simulation results with a $\mu$CCA, and analysis of experimental results.

In a study by Viravaidya [68], a $\mu$CCA device was used to examine naphthalene toxicity as a model system (Figure 8.7). Since the majority of naphthalene toxicity affects the lung, a rat lung cell line (L2) was cultured in the lung compartment, and a rat or human hepatoma cell line (H4IIE or HepG2/C3A) was cultured in the liver compartment. The $\mu$CCA had two additional compartments for fat and other tissue, which did not contain cells in this particular study. In this study, naphthalene was added to the circulating media in a $\mu$CCA device, and the viability of L2 (lung) and HepG2/C3A (liver) cells was tested by MCB staining for GSH. The results showed that after naphthalene was added to the medium, the GSH levels in lung and liver cells continuously decreased for the dura-

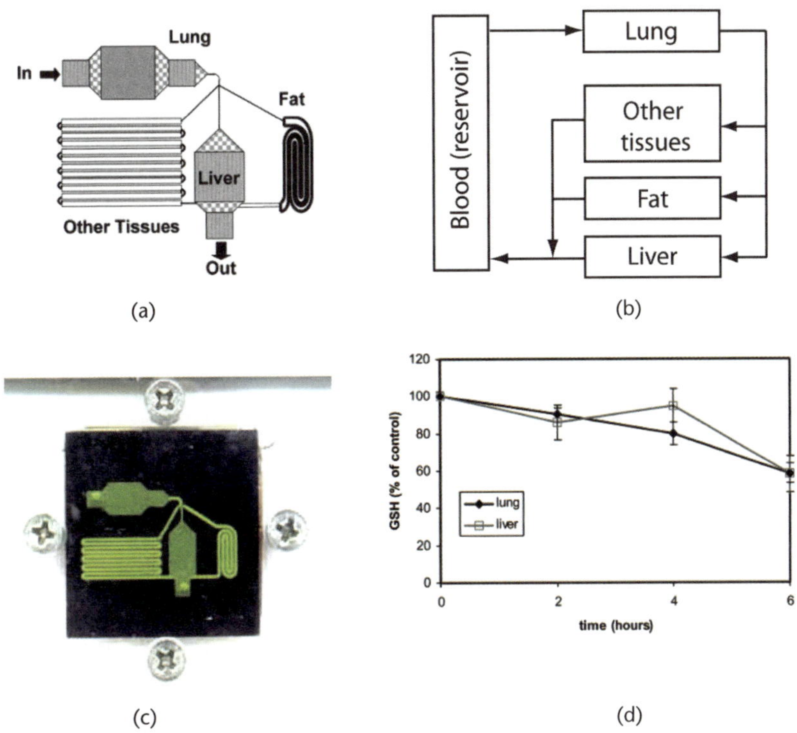


**Figure 8.7** μCCA for probing naphthalene toxicity. (a) Schematic diagram of the μCCA device. (Reprinted from [68] with permission.) The lung, other tissue, fat, and liver compartments were fabricated in the μCCA device. (b) Diagram of a corresponding PBPK model. (c) Picture of the μCCA device with dye for visualizing the flow. (Reprinted from [68] with permission.) (d) Typical GSH responses of L2 (lung) and HepG2/C3A (liver) cells in μCCA devices after treatment with naphthalene at various times. (Reprinted from [68] with permission.)

tion of the experiment, demonstrating the toxic effect of naphthalene on the cells. To verify that the HepG2/C3A (liver) cells were responsible for the observed cytotoxicity of the naphthalene in the mCCA, control experiments were performed where the liver chamber was empty. When HepG2/C3A (liver) cells were not present in the μCCA device, the cytotoxic effect of naphthalene was not observed, as manifested by the high level of GSH in L2 (lung) cells. Further experiments with several metabolites of naphthalene, including naphthoquinone, naphthol, and naphthalene dihydrodiol, revealed that naphthoquinone was responsible for the toxicity of naphthalene in the μCCA with a dose-dependent response [68].

### 8.3.5 Discussion and commentary

#### 8.3.5.1 Operation of a μCCA

Appropriate design of the microfluidics systems is essential to perform an effective μCCA experiment. Some of the issues that researchers face when designing and building a microfluidic perfusion cell culture system for a single compartment have been reviewed elsewhere [11, 69, 70], and here we provide a summary of those issues as well as our own experiences from developing the μCCA device.

Cells in a microfluidic device are exposed to shear stress from the medium flowing past a layer of cells [70]. This effect of shear stress becomes greater in the case of miniaturization because the amount of shear stress changes in proportion to the change in tangential velocity across the height, as (8.22) shows:

$$\tau = -\mu \frac{dv}{dx} \tag{8.22}$$

where $\tau$ is the shear stress, $\mu$ is the fluid viscosity, $v$ is the flow velocity, and $x$ is the height from the bottom of the channel. Given the same average velocity, the shear stress will increase as the height of the channel decreases. The physiologically optimal level of shear stress varies depending on the cell type. In some cases, the presence of a shear stress is desirable; for example, endothelial cells require some level of shear stress to develop properly [71]. On the other hand, the appropriate shear-stress level for endothelial cells can be detrimental or lethal to other cell types [72]. Even when no phenotypic or morphological changes to the cells are observed, the shear stress can affect other processes, such as recombinant protein productivity [73]. In addition, the presence of cells in a microchannel can also have an effect on the flow, which in turn affects the cells. A mathematical simulation by Sugihara-seki showed that the axial spacing between neighboring adherent cells affects the distribution of the stresses on them, which results in drastic variations of the fluid forces with the axial spacing and the relative positions with respect to their neighboring cells [74]. In designing a $\mu$CCA, we ensure that the calculated shear stress is within the physiologically acceptable range (<2 dyne/cm$^2$) [65]. However, in some cell types, we have observed that the attached cell morphology was affected by flow, even at relatively low flow rates, as shown in Figure 8.8.

To address the effect of shear stress on attached cells in $\mu$CCA devices, the same cells (HepG2/C3A) were seeded in two chambers with different flow rates. Cells in the liver chamber show rounded morphology due to the higher flow rate in the liver chamber, whereas the cells in the marrow chamber are well spread and firmly attached to the surface. The calculated shear stress in the liver chamber is less than 1 dyne/cm$^2$, but the flow altered the morphology of the attached cells. In a control experiment where no flow was introduced, cells in two chambers showed no difference in their morphology (not shown). These changes in morphology indicate possible changes in the physiology of the cells.

Also, mass transfer should be considered when designing a microfluidic device for cell culture. Since flow inside a microchannel is a laminar flow characterized by a low Reynolds number, diffusion is the dominating mechanism of transport. This can be advantageous; for example, chemical treatment of only a part of a single cell has been possible by utilizing the laminar flow nature of a microfluidic device [75]. On the other hand, in a $\mu$CCA, a homogeneous environment is more desirable, and the lack of convective mixing can pose a problem. When designing a $\mu$CCA, careful consideration must be given to ensuring sufficient mixing because two flows coming from separate microchannels into a single chamber may not mix completely, and the resulting flow into a chamber may not be homogeneous. Consequently, cells in the chamber may be inadvertently exposed to different concentrations of nutrients and growth factors, depending on their positions within the chamber, and therefore show different responses to seemingly equivalent experimental conditions. In a $\mu$CCA, arrays of baffles

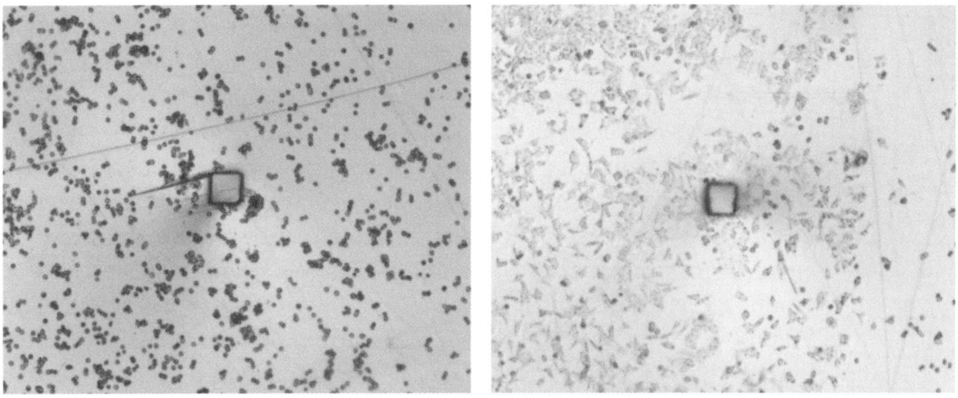

(a) (b)

**Figure 8.8** Effect of shear stress on cell morphology. (a) Cells in the liver chamber, and (b) cells in the marrow chamber.

were fabricated at the entrance of each chamber in order to facilitate mixing, which was verified with mathematical simulation of fluid flow [76].

Microfluidic devices typically have a much larger surface-area-to-volume ratio compared to macroscale bioreactors. While this characteristic confers a more physiologically realistic, tissue-like environment on the microfluidic devices, it can also increase protein adsorption or absorption of small molecules. This problem has been acknowledged especially in the case of polydimethylsiloxane (PDMS) due to its ability to absorb small, hydrophobic molecules [77]. The inner surface of a microfluidic device provides the area for cell attachment; thus, biocompatibility can also be an issue for such a system. Currently, PDMS is by far the most popular material due to its advantages, including high gas permeability, optical transparency, biocompatibility, and ease of fabrication. However, the hydrophobic nature of a PDMS surface may cause unwanted adsorption of hydrophobic drugs or proteins, eventually resulting in the loss of the molecules. The loss of hydrophobic molecules due to surface adsorption is particularly problematic for systems with long-term recirculation.

Packaging and sealing of a microfluidic device is also an important issue as only a small amount of medium is used for the perfusion, and any leak would result in a rapid depletion of circulating medium. For a μCCA, we have used Plexiglas (PMMA) and polycarbonate housings for sealing. Plexiglas provides a transparent surface for visual inspection but is not appropriate for sterilization by autoclaving. This can be a serious drawback for long-term recirculation cell culture, although we have operated such systems for up to 96 hours with sterilization by oxygen plasma cleaning. Polycarbonate is more resistant to high temperature, thus is compatible with autoclaving, but polycarbonate has autofluorescence, which can interfere with analysis using fluorescent biomarkers [78].

In terms of the operation of microfluidic devices for perfusion cell culture, sterilization is an important issue, especially when considering a long-term perfusion cell culture. Autoclaving is generally the most effective and the easiest method for sterilization, but it may not be compatible with some microfluidic components. In such cases, we used an oxygen or air plasma treatment for surface sterilization, which is somewhat less effective than autoclaving. There are some alternative methods for sterilization, such as exposure to UV light [79] and flushing the device with ethanol [80]. We also found that acidified ethanol (pH 2, HCl) is an effective sterilization method. Another obstacle to the consistent operation of a microfluidic device is the formation of bubbles inside a device. Bubbles within in a microchannel can obstruct the fluid flow and kill cells by preventing adequate supply of medium, or they can cause damage to the cells directly at the gas-liquid interface. To minimize bubble formation, several strategies have been suggested, including flushing the system with a high-pressure flow of medium or DPBS prior to an experiment, prepriming the system with low-surface-tension liquids such as ethanol, and operating the system under high pressure [69]. We have utilized a bubble trap installed upstream of a microfluidic device to prevent bubbles from entering the main device [81]. When operating a $\mu$CCA device, we utilize many of the aforementioned strategies. In addition, the medium reservoir serves as a debubbler so that bubbles formed in the tubing stay in the reservoir, where excess gas is vented, and are not transported into the $\mu$CCA. By utilizing these strategies, we can routinely operate the $\mu$CCA system for 96 hours.

#### 8.3.5.2 Authenticity of cells

The most important limitation of the $\mu$CCA is the limited authenticity of cells cultured in the system. By authenticity we refer to the ability of cultured cells to mimic in vivo cellular responses. Limited authenticity is an issue with all in vitro systems. The cell lines cultured in the $\mu$CCA device are established cell lines, which are easy to culture and maintain but have only limited authenticity. For example, a HepG2/C3A cell line is a hepatoma cell line with detectable P450 enzyme activity, but the level of P450 activity is low compared to more authentic cells, such as primary hepatocytes [82]. To mimic the human response more accurately in the $\mu$CCA device, more authentic cells or tissues should be used. This poses some challenges in terms of device operation and maintenance of cell viability because more authentic cells or tissues are generally more difficult to handle. Potential solutions include 3-D tissue-engineered constructs or the use of either biopsy tissues from a patient or tissue slices from an animal. For example, tumor tissue from surgery can be used in a $\mu$CCA. The microscale nature of the device requires a small volume of tissue; it would be possible to perform multiple experiments from a limited amount of tissue. Also, the ability to test a drug on the patient's actual tissue will foster the development of a patient-individualized treatment [83].

Regarding the authenticity of cells for mimicking the human response to foreign compounds, the liver is probably the most important organ because most of the metabolism takes place in the liver. Because of its importance, hepatic tissue engineering for development of a bioartificial liver or drug-screening system has gained attention recently [84]. Several different types of in vitro systems have been developed and used for biotransformation studies, including isolated perfused livers, liver tissue slices, primary hepatocytes, immortalized cell lines, microsomes, and recombinant systems [82]. However, despite the large variety of available in vitro systems, there is no perfect system to study biotransformation. For example, primary hepatocytes are a widely used in vitro system because they maintain many of the in vivo liver-specific functions. However, they are subject to a loss of the liver-specific functions in a long-term culture, especially P450 enzyme activity. Several culture methods have been developed to improve the retention of the liver-specific functions of hepatocytes, including collagen gel sandwich culture, addition of nutrients, and coculture with supporting cell types [82]. Indeed, coculture of hepatocytes with rat liver epithelial cells has been shown to improve the retention of liver-specific functions [85]. By coculturing hepatocytes with supporting cells, researchers aim to mimic the physiological in vivo situation where hepatocytes interact with supporting cells to maintain the differentiated state. The communication of hepatocytes with other cell types is achieved through complex signaling mechanisms, including direct cell-to-cell contact, interaction through the extracellular matrix (ECM), and soluble signaling factors. It has been acknowledged that the role played by ECM can be crucial for maintaining liver functions, and 3-D liver-tissue constructs have gained significant attention as a method for improving the liver-specificity of in vitro culture systems. Research efforts to mimic the physiological and environmental cues include Matrigel cultures, collagen sandwich cultures [86], 3-D membrane bioreactors [87], and 3-D spheroidal cultures of primary hepatocytes [88]. All of these strategies have yielded systems with improved liver functions, but no single in vitro system can yet mimic the liver function to the physiological in vivo level for any sustained period.

Microscale systems to improve the authenticity of hepatocytes cultures were recently developed by several research groups [84]. With advances in microfabrication technology, it has become easier to control the microenvironment for hepatocyte cultures so that the microscale system mimics the physiological microenvironment more closely than macroscale systems. For example, in a study by Khetani and Bhatia, a microscale, multiwell cell culture system was developed by a micropatterning technique using elastomeric stencils to create patterns of hepatocytes surrounded by mouse 3T3 fibroblasts, mimicking the architecture of liver structure [89]. By varying the size of hepatocytes clusters, they were able to find the cluster size that yielded maximal hepatocyte functions. Nahmias et al. developed a protocol for micropatterning the self-assembly of liver sinusoid-like structures with micrometer resolution [90].

In addition to the chemical signaling from neighboring cells and surrounding ECM, it is thought that mechanical cues from blood flow play a role in maintaining the differentiated state of hepatocytes in vivo. Microfluidic systems are ideal for building a microenvironment that mimics the mechanical stresses from blood flow. Perfusion through the system at a physiological level can be incorporated because the flow rate and the shear stress in the system can be controlled. Sivaraman et al. developed a microscale, 3-D, perfused tissue unit for an in vitro physiological model of the liver and compared the liver-specific functions of their 3-D microreactor cultures with conventional 2-D collagen sandwich culture and 3-D Biocoat Matrigel by measuring gene-expression levels of phase I and II enzymes, albumin and urea secretion rates, P450 activities, and inducibility by chemical inducers [91]. Primary rat cells cultured in the 3-D microfluidic system were closer to the native liver than cells cultured by 2-D collagen sandwich cultures or Matrigel cultures in multiwell format. Currently, in our lab, we are developing methods for culturing cells in a 3-D hydrogel matrix inside of a μCCA device in an effort the make the environment more physiologically realistic [92]. To test the system, we measured the enzymatic activity of hepatoma cells embedded in a hydrogel inside the μCCA device. The 3-D matrices of Matrigel with embedded cells are placed into the chambers in the μCCA device. The hydrogel-cell mix is introduced into the device in an unpolymerized state so that when the gel fully polymerizes, it will shrink in volume and provide a space for recirculation of media. The 3-D matrix gel is thin enough (less than 100 μm) so that the diffusion into the matrix is rapid and provides oxygen and nutrients for cell survival. The combination of microfluidics and 3-D hydrogel cell culture is a promising solution to the issue of achieving authentic responses from cells cultured in vitro.

#### 8.3.5.3 Noninvasive detection

The enclosed and miniaturized nature of microfluidic devices poses challenges in obtaining information about the status of cells or the microenvironment during an experiment. Noninvasive, analytical techniques are ideal for obtaining information about the status of a microscale system. A fluorescence-based technique is probably the most widely accepted method for monitoring cell viability, gene expression, and biochemical reactions. Several fluorescent compounds are available commercially to monitor the viability of mammalian cells. Generally, these compounds are nonfluorescent outside the cell and become fluorescent upon a chemical reaction inside the cell. An alternative to exogenous fluorescent compounds is a transfected fluorescent reporter. It is a popular method to genetically tag a protein of interest with a fluorescent reporter protein, which fluoresces when the target protein is expressed. In our lab, we developed a stably transfected cell line that expresses GFP in response to estrogen and used the cell line in a $\mu$CCA device to study the effects of endocrine disruptors [93]. Biochemical reactions can be monitored by fluorogenic substrates, which are only weakly fluorescent but yield a strong fluorescent signal upon biochemical reaction. For example, a variety of fluorogenic substrates have been identified and are routinely used to monitor the enzymatic activity of P450 enzymes [94]. A fluorescent technique can also be used to monitor the status of the microenvironment inside a microfluidic device, for example, dissolved oxygen concentration. As discussed previously, we have demonstrated the use of an integrated oxygen sensor for a $\mu$CCA device [66] that utilizes a fluorescent ruthenium complex quenched by collision with oxygen molecules [95].

A fluorescence-based technique is versatile and will find more use in future microfluidics technology. However, it is necessary to design imaging tools specifically for microfluidic devices. Most current research efforts utilize already existing biological imaging tools, such as fluorescent and confocal microscopy. Therefore, each experiment has to be interrupted as the microfluidic cell culture system must be brought outside an incubator for analysis. Although this is not problematic for end point measurements, it may interfere with experiments that require real-time measurement. Some researchers have used a custom-made Plexiglas incubation box fitted to the stage of a microscope for inspection during an experiment [96, 97]. Although such a setup allows researchers to monitor a microfluidic perfusion culture in real time, it is not directly adaptable to a high-throughput screening format. Recently, we have developed an optical imaging system that is more suitable for high-throughput screening and is compact enough to fit inside a $CO_2$ incubator [98]. The system consists of the optical components necessary to detect a fluorescent signal from a microfluidic device, including a high-power light-emitting diode (LED) and a charge-coupled device (CCD) camera as a light source and a detector, respectively. Fluorescently stained cells were monitored at multiple locations in multiple $\mu$CCA chips simultaneously by using an automated moving stage. We used this optical imaging system to monitor the fluorescence staining of cells and subsequent cell death upon ethanol injection into a $\mu$CCA device (Figure 8.9) [57]. Cell

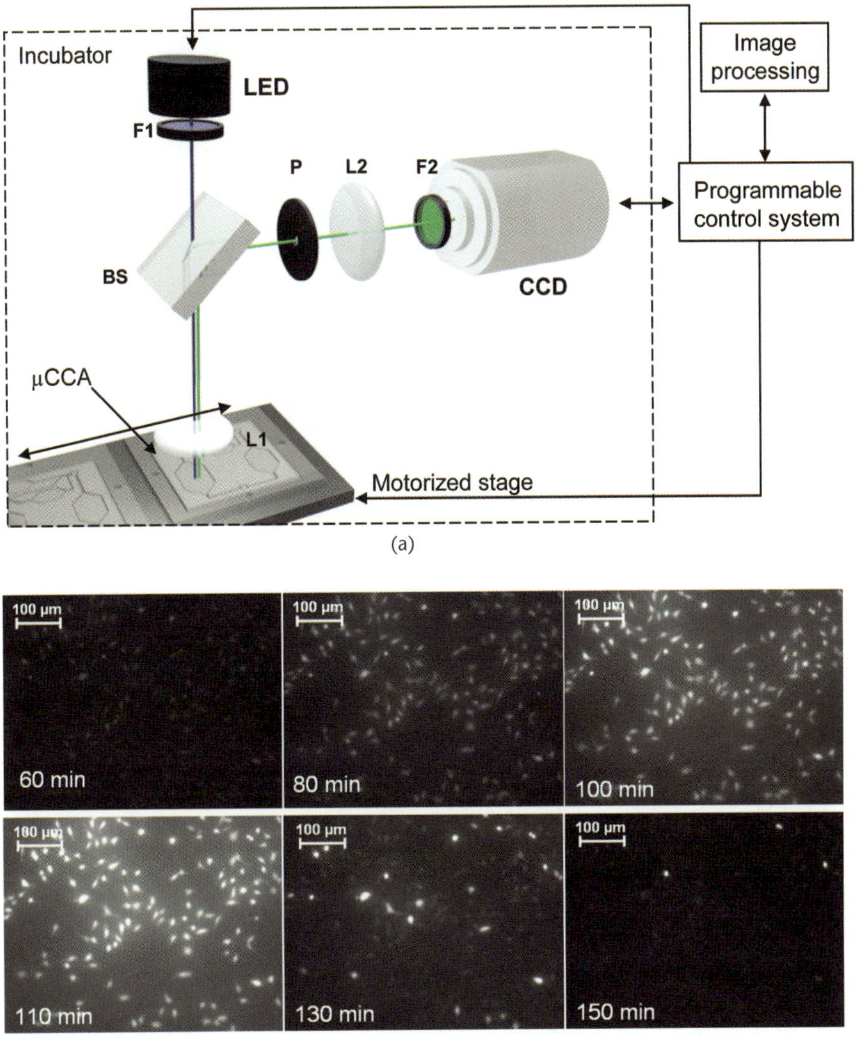


**Figure 8.9** Optical imaging system for $\mu$CCA. (a) Schematic diagram of the system. The system consists of optical components, including the light source (LED), detection module (CCD), filters, and dichroic mirrors. (b) Cell-staining and cell-death experiment data. A 5 $\mu$M Calcein AM solution is introduced into the $\mu$CCA device, and the fluorescent signal becomes stronger as more Calcein AM molecules are hydrolyzed by intracellular esterases. At 90 minutes, the liquid source was switched to 10% ethanol solution, and cells lost viability and the fluorescent signal was rapidly lost. It takes approximately 20 minutes for the switched ethanol solution to go through tubing and enter the $\mu$CCA. (Reprinted from [57] with permission.)

growth inside the $\mu$CCA was also monitored for about 90 hours. The optical detection system is adaptable and can detect a variety of fluorescent signals. In addition to cell viability assays, we have used the imaging system to monitor the P450 enzyme activity of HepG2/C3A cells inside the $\mu$CCA device (unpublished data). Finally, analytical techniques such as liquid chromatography combined with mass spectrometry (LC-MS) [99] have resulted in a greater sensitivity down to the nanomolar or picomolar range of drugs or metabolites. With advances in these analytical techniques, it should be possible to analyze more accurately the small volume of circulating media in a microscale cell culture system. The increasing availability of analytical techniques will provide more information about the status of cells and the environment within a microfluidic system, which will allow the development of more robust mathematical models of the system, fostering the integration of a modeling approach (PK-PD modeling) and an experimental approach ($\mu$CCA).

## 8.4 Application Notes

As demonstrated by the preliminary results using macro- and microscale CCA devices, the pharmacokinetic-pharmacodynamic model on a chip can be a powerful and versatile tool for answering many questions in biology that cannot be addressed with conventional in vitro systems, especially when it is combined with a mathematical modeling approach such as pharmacokinetic-pharmacodynamic models. One potential application of a $\mu$CCA is patient-specific treatment, which has been gaining attention recently [83]. It has been acknowledged that there are often individual variations in patient response, mostly due to inherent genetic variations [100, 101]. This issue, known as pharmacogenetics, mainly focuses on the inherited genetic variation in patients. However, the genetic variation alone does not account for the individual variation in patient response to the same dosing regimen, and phenotypic differences must be considered as well. The individualization of therapeutic strategy based on mathematical modeling has already been proposed [102], but an experimental system for testing the hypothesis is still lacking. The $\mu$CCA device can be an ideal experimental system to identify phenotypic differences within individual patients. For example, the response of patients to a chemotherapeutic agent can vary widely, due to several factors

**Troubleshooting Table**

| Problem | Explanation | Potential Solution |
|---|---|---|
| Circulating medium is depleted after some time | Inadequate sealing | 1. Tighten the screws more.<br>2. Make sure that the top cover frame is as flat as possible to make an even contact with a silicon chip.<br>3. The gasket beneath a silicon chip must be thick enough that the chip and the top cover make a solid contact. |
| There is bacterial contamination of circulating medium | Inappropriate assembly of a $\mu$CCA or preparation of materials | 1. Make sure every component of a μCCA is adequately sterilized.<br>2. Use extra caution for sterility when assembling a $\mu$CCA. |
| Air bubbles appear in microfluidic channels some time after starting the experiment | Presence of bubbles in the tubing or introduction of bubbles into the device through a small gap at connections | 1. Rinse the tubing extensively with ethanol and water before connecting it to a device.<br>2. Make sure that the connections between tubing and device are secure. |
| Cell viability is unexpectedly low | Damage to cells by shear stress; inadequate nutrient transport; insufficient initial seeding cell density | 1. Design the $\mu$CCA so that the flow does not cause excessive shear stress on cells and the mass transport is sufficient to support cell growth.<br>2. Initial cell density can affect the growth of the cells due to cell-to-cell communications. |
| Cell attachment on a silicon chip is not good | Inadequate surface modification | Depending on cell type, different coating materials give the best result. Try collagen, fibronectin, lysine, or a mixture of them. |

such as phenotypic differences in metabolizing enzymes and the genetic constitution of the tumors. Rather than blindly testing a certain therapeutic regimen directly on a patient with cancer, one can take a biopsy sample from the tumor and test the responses of the tumor tissues to various proposed regimens. A $\mu$CCA can be useful in screening combinations of drugs with different ratios and various dosing regimens, which is practically difficult, if not impossible, to do with human clinical trials or animal tests.

Development of a $\mu$CCA that can function as a physical realization of a pharmacokinetic-pharmacodynamic model can work as a validation tool for a drug's proposed mechanism of action, as we have seen in the examples using naphthalene and various chemotherapeutic agents as model drugs [68, 92, 103]. A mathematical model can be developed based on a hypothesis, which can be directly tested with a $\mu$CCA for validation. Additionally, complex drug interactions can be studied using a $\mu$CCA, as drug-drug interaction is an important issue when administering a combination of drugs. A well-known example is the effect of grapefruit juice on a number of drugs through the inhibition of cytochrome P450 enzymes [104]. The combination of a $\mu$CCA and mathematical modeling can be an ideal tool to study the various aspects of drug-drug interaction because it inevitably requires studying multiorgan interaction and the pharmacokinetics/pharmacodynamics of a drug, which are difficult to reproduce with conventional cell-based assay systems.

The rapid development in the fields of biology, chemistry, microfluidics, and analytical/imaging technology will make a more complete form of "body-on-a-chip" possible in near future. With the improvement of the $\mu$CCA and the development of a more realistic, humanlike response, we believe that the potential applications of the $\mu$CCA will not be limited to the aforementioned examples, and the $\mu$CCA will find use in tackling a wide variety of unanswered questions in biology.

## 8.5 Summary Points

- Pharmacokinetics (PK) refers to the concentration profiles of a drug and its metabolites after administration and studies the absorption, distribution, metabolism, and elimination (ADME) characteristics. A pharmacodynamic (PD) model predicts the pharmacological effect of a drug at a given drug concentration. By integrating PK and PD models, the quantitative effect of a drug can be predicted from a given drug dosage.
- A physiologically based pharmacokinetic (PBPK) model, based on the physiological model of the human body, is divided into separate compartments that represent organs.
- As a physical realization of a PBPK model, a micro cell culture analog ($\mu$CCA) can serve as an in vitro experimental platform for testing the effect and toxicity of a drug in a physiologically realistic way, while mimicking multiorgan interactions.
- The integration of PK-PD modeling and a $\mu$CCA can serve as a novel platform for testing drugs and can potentially improve the efficiency of the drug-development process by overcoming the limitations of current in vitro and animal models.

## Acknowledgments

This work was supported in part by the New York State Science and Technology Foundation through the Cornell Nanobiotechnology Center (NBTC), an anonymous gift to Cornell Biomedical Engineering, and a grant from the Army Corp of Engineers (CERL) W9132T-07. J. H. S. gratefully acknowledges support from the Samsung Lee Kun Hee Scholarship Foundation.

### References

[1] Dingemanse, J., and Appel-Dingemanse, S., "Integrated pharmacokinetics and pharmacodynamics in drug development," *Clin. Pharmacokinet,* Vol. 46, No. 9, 2007, pp. 713–737.

[2] Kola, I., and Landis, J., "Can the pharmaceutical industry reduce attrition rates?" *Nat. Rev. Drug Discov.,* Vol. 3, No. 8, 2004, pp. 711–715.

[3] Adams, C. P., and Brantner, V. V., "Estimating the cost of new drug development: Is it really 802 million dollars?" *Health Aff.* (Millwood), Vol. 25, No. 2, 2006, pp. 420–428.

[4] Booth, B., Glassman, R., and Ma, P., "Oncology's trials," *Nat. Rev. Drug Discov.,* Vol. 2, No. 8, 2003, pp. 609–610.

[5] Derendorf, H., and Meibohm, B., "Modeling of pharmacokinetic/pharmacodynamic (PK/PD) relationships: Concepts and perspectives," *Pharm. Res.,* Vol. 16, No. 2, 1999, pp. 176–185.

[6] Karlsson, M. O., et al., "Pharmacokinetic/pharmacodynamic modelling in oncological drug development," *Basic Clin. Pharmacol. Toxicol.,* Vol. 96, No. 3, 2005, pp. 206–211.

[7] Meibohm, B., and Derendorf, H., "Pharmacokinetic/pharmacodynamic studies in drug product development," *J. Pharm. Sci.,* Vol. 91, No. 1, 2002, pp. 18–31.

[8] Perez-Urizar, J., et al., "Pharmacokinetic-pharmacodynamic modeling: Why?" *Arch. Med. Res.,* Vol. 31, No. 6, 2000, pp. 539–545.

[9] Whitesides, G. M., "The origins and the future of microfluidics," *Nature,* Vol. 442, No. 7101, 2006, pp. 368–373.

[10] Li, N., Tourovskaia, A., and Folch, A., "Biology on a chip: Microfabrication for studying the behavior of cultured cells," *Crit. Rev. Biomed. Eng.,* Vol. 31, No. 5–6, 2003, pp. 423–488.

[11] Beebe, D. J., Mensing, G. A., and Walker, G. M., "Physics and applications of microfluidics in biology," *Annu. Rev. Biomed. Eng.,* Vol. 4, No. 2002, pp. 261–286.

[12] El-Ali, J., Sorger, P. K., and Jensen, K. F., "Cells on chips," *Nature,* Vol. 442, No. 7101, 2006, pp. 403–411.

[13] Breslauer, D. N., Lee, P. J., and Lee, L. P., "Microfluidics-based systems biology," *Mol. Biosyst.,* Vol. 2, No. 2, 2006, pp. 97–112.

[14] Zeng, Y., et al., "Mass transport and shear stress in a microchannel bioreactor: Numerical simulation and dynamic similarity," *J. Biomech. Eng.,* Vol. 128, No. 2, 2006, pp. 185–193.

[15] Mehta, K., and Linderman, J. J., "Model-based analysis and design of a microchannel reactor for tissue engineering," *Biotechnol. Bioeng.,* Vol. 94, No. 3, 2006, pp. 596–609.

[16] Roy, P., et al., "Analysis of oxygen transport to hepatocytes in a flat-plate microchannel bioreactor," *Ann. Biomed. Eng.,* Vol. 29, No. 11, 2001, pp. 947–955.

[17] Gerlowski, L. E., and Jain, R. K., "Physiologically based pharmacokinetic modeling: Principles and applications," *J. Pharm. Sci.,* Vol. 72, No. 10, 1983, pp. 1103–1127.

[18] Reigner, B. G., et al., "An evaluation of the integration of pharmacokinetic and pharmacodynamic principles in clinical drug development. Experience within Hoffmann La Roche," *Clin. Pharmacokinet.,* Vol. 33, No. 2, 1997, pp. 142–152.

[19] Agoram, B., Woltosz, W. S., and Bolger, M. B., "Predicting the impact of physiological and biochemical processes on oral drug bioavailability," *Adv. Drug Deliv. Rev.,* Vol. 50, Suppl. 1, 2001, pp. S41–S67.

[20] Willmann, S., et al., "PK-Sim: A physiologically based pharmacokinetic 'whole-body' model," *Biosilico,* Vol. 1, No. 4, 2003, pp. 121–124.

[21] "Population-based pharmacokinetic modeling and simulation," SIMCYP, 2007, www.simcyp.com.

[22] Holford, N. H., and Sheiner, L. B., "Kinetics of pharmacologic response," *Pharmacol. Ther.,* Vol. 16, No. 2, 1982, pp. 143–166.

[23] Suzuki, H., Iwatsubo, T., and Sugiyama, Y., "Applications and prospects for physiologically based pharmacokinetic (PB-PK) models involving pharmaceutical agents," *Toxicol. Lett.,* Vol. 82–83, No. 1995, pp. 349–355.

[24] Rowland, M., "Physiologic pharmacokinetic models: Relevance, experience, and future trends," *Drug Metab. Rev.,* Vol. 15, No. 1–2, 1984, pp. 55–74.

[25] Nestorov, I., "Whole-body physiologically based pharmacokinetic models," *Expert Opin. Drug Metab. Toxicol.*, Vol. 3, No. 2, 2007, pp. 235–249.

[26] Aarons, L., "Physiologically based pharmacokinetic modelling: A sound mechanistic basis is needed," *Br. J. Clin. Pharmacol.*, Vol. 60, No. 6, 2005, pp. 581–583.

[27] Bading, J. R., et al., "Kinetic modeling of 5-fluorouracil anabolism in colorectal adenocarcinoma: A positron emission tomography study in rats," *Cancer Res.*, Vol. 63, No. 13, 2003, pp. 3667–3674.

[28] Baur, M., et al., "Pharmacokinetics of oxaliplatin in patients with severe hepatic dysfunction," *Cancer Chemother. Pharmacol.*, Vol. 61, No. 1, 2008, pp. 97–104.

[29] Iwatsubo, T., et al., "Prediction of in vivo drug metabolism in the human liver from in vitro metabolism data," *Pharmacol. Ther.*, Vol. 73, No. 2, 1997, pp. 147–171.

[30] Shibata, Y., et al., "Prediction of hepatic clearance and availability by cryopreserved human hepatocytes: An application of serum incubation method," *Drug Metab. Dispos.*, Vol. 30, No. 8, 2002, pp. 892–896.

[31] Tsukamoto, Y., et al., "A physiologically based pharmacokinetic analysis of capecitabine, a triple prodrug of 5-FU, in humans: The mechanism for tumor-selective accumulation of 5-FU," *Pharm. Res.*, Vol. 18, No. 8, 2001, pp. 1190–1202.

[32] Poulin, P., and Theil, F. P., "A priori prediction of tissue: Plasma partition coefficients of drugs to facilitate the use of physiologically based pharmacokinetic models drug discovery," *J. Pharm. Sci.*, Vol. 89, No. 1, 2000, pp. 16–35.

[33] Mager, D. E., Wyska, E., and Jusko, W. J., "Diversity of mechanism-based pharmacodynamic models," *Drug Metab. Dispos.*, Vol. 31, No. 5, 2003, pp. 510–518.

[34] Friedman, H., et al., "Pharmacokinetics and pharmacodynamics of oral diazepam: Effect of dose, plasma concentration, and time," *Clin. Pharmacol. Ther.*, Vol. 52, No. 2, 1992, pp. 139–150.

[35] Yamada, Y., et al., "Prediction of therapeutic doses of beta-adrenergic receptor blocking agents based on quantitative structure-pharmacokinetic/pharmacodynamic relationship," *Biol. Pharm. Bull.*, Vol. 16, No. 12, 1993, pp. 1251–1259.

[36] Corey, A., et al., "Pharmacokinetics and pharmacodynamics following intravenous doses of azimilide dihydrochloride," *J. Clin. Pharmacol.*, Vol. 39, No. 12, 1999, pp. 1263–1271.

[37] Mayer, B. X., et al., "Pharmacokinetic-pharmacodynamic profile of systemic nitric oxide-synthase inhibition with L-NMMA in humans," *Br. J. Clin. Pharmacol.*, Vol. 47, No. 5, 1999, pp. 539–544.

[38] Granados-Soto, V., et al., "Relationship between pharmacokinetics and the analgesic effect of ketorolac in the rat," *J. Pharmacol. Exp. Ther.*, Vol. 272, No. 1, 1995, pp. 352–356.

[39] Brown, R. D., Kearns, G. L., and Wilson, J. T., "Integrated pharmacokinetic-pharmacodynamic model for acetaminophen, ibuprofen, and placebo antipyresis in children," *J. Pharmacokinet. Biopharm.*, Vol. 26, No. 5, 1998, pp. 559–579.

[40] Jusko, W. J., "Pharmacodynamics of chemotherapeutic effects: Dose-time-response relationships for phase-nonspecific agents," *J. Pharm. Sci.*, Vol. 60, No. 6, 1971, pp. 892–895.

[41] Wadler, S., et al., "Effects of perturbations of pools of deoxyribonucleoside triphosphates on expression of ribonucleotide reductase, a G1/S transition state enzyme, in p53-mutated cells," *Biochem. Pharmacol.*, Vol. 55, No. 9, 1998, pp. 1353–1360.

[42] Shu, C. H., et al., "Cell cycle G2/M arrest and activation of cyclin-dependent kinases associated with low-dose paclitaxel-induced sub-G1 apoptosis," *Apoptosis*, Vol. 2, No. 5, 1997, pp. 463–470.

[43] Liu, W., Hillen, T., and Freedman, H. I., "A mathematical model for M-phase specific chemotherapy including the G0-phase and immunoresponse," *Math. Biosci. Eng.*, Vol. 4, No. 2, 2007, pp. 239–259.

[44] Yano, Y., et al., "Application of logistic growth model to pharmacodynamic analysis of in vitro bactericidal kinetics," *J. Pharm. Sci.*, Vol. 87, No. 10, 1998, pp. 1177–1183.

[45] Sun, Y. N., and Jusko, W. J., "Transit compartments versus gamma distribution function to model signal transduction processes in pharmacodynamics," *J. Pharm. Sci.*, Vol. 87, No. 6, 1998, pp. 732–737.

[46] Lobo, E. D., and Balthasar, J. P., "Pharmacodynamic modeling of chemotherapeutic effects: Application of a transit compartment model to characterize methotrexate effects in vitro," *AAPS PharmSci*, Vol. 4, No. 4, 2002, p. E42.

[47] Gainetdinov, R. R., et al., "Desensitization of G protein-coupled receptors and neuronal functions," *Annu. Rev. Neurosci.*, Vol. 27, No. 2004, pp. 107–144.

[48] Simeoni, M., et al., "Predictive pharmacokinetic-pharmacodynamic modeling of tumor growth kinetics in xenograft models after administration of anticancer agents," *Cancer Res.*, Vol. 64, No. 3, 2004, pp. 1094–1101.

[49] Sung, J. H., Dhiman, A., and Shuler, M. L., "A combined pharmacokinetic-pharmacodynamic (PK-PD) model for tumor growth in the rat with UFT administration," *J. Pharm. Sci.*, Vol. 8, No. 5, 2009, pp. 1885–1904.

[50] Friberg, L. E., et al., "Models of schedule dependent haematological toxicity of 2′-deoxy-2′-methylidenecytidine (DMDC)," *Eur. J. Clin. Pharmacol.*, Vol. 56, No. 8, 2000, pp. 567–574.

[51] Holford, N. H., "The target concentration approach to clinical drug development," *Clin. Pharmacokinet.*, Vol. 29, No. 5, 1995, pp. 287–291.

[52] Aarons, L., et al., "Role of modelling and simulation in Phase I drug development," *Eur. J. Pharm. Sci.*, Vol. 13, No. 2, 2001, pp. 115–122.

[53] Latz, J. E., et al., "A semimechanistic-physiologic population pharmacokinetic/pharmacodynamic model for neutropenia following pemetrexed therapy," *Cancer Chemother. Pharmacol.*, Vol. 57, No. 4, 2006, pp. 412–426.

[54] Lave, T., and Parrot, N. J. H., "Physiologically based models to predict human pharmacokinetic parameters," in J. B. Taylor and D. J. Triggle, (eds.), *Comprehensive Medicinal Chemistry II*, London: Elsevier, 2007.

[55] Agoram, B. M., Martin, S. W., and van der Graaf, P. H., "The role of mechanism-based pharmacokinetic-pharmacodynamic (PK-PD) modelling in translational research of biologics," *Drug Discov. Today*, Vol. 12, Nos. 23–24, 2007, pp. 1018–1024.

[56] Quick, D. J., and Shuler, M. L., "Use of in vitro data for construction of a physiologically based pharmacokinetic model for naphthalene in rats and mice to probe species differences," *Biotechnol. Prog.*, Vol. 15, No. 3, 1999, pp. 540–555.

[57] Oh, T. I., et al., "Real-time fluorescence detection of multiple microscale cell culture analog devices in situ," *Cytometry A*, Vol. 71, No. 10, 2007, pp. 857–865.

[58] Sweeney, L. M., et al., "A cell culture analogue of rodent physiology: Application to naphthalene toxicology," *Toxicology in vitro*, Vol. 9, No. 3, 1995, pp. 307–316.

[59] Gaines, T. B., "Acute toxicity of pesticides," *Toxicol. Appl. Pharmacol.*, Vol. 14, No. 3, 1969, pp. 515–534.

[60] Plopper, C. G., et al., "Relationship of cytochrome P-450 activity to Clara cell cytotoxicity. I. Histopathologic comparison of the respiratory tract of mice, rats and hamsters after parenteral administration of naphthalene," *J. Pharmacol. Exp. Ther.*, Vol. 261, No. 1, 1992, pp. 353–363.

[61] Shuler, M. L., et al., "A self-regulating cell culture analog device to mimic animal and human toxicological responses," *Biotechnology and Bioengineering*, Vol. 52, No. 1, 1996, pp. 45–60.

[62] Ghanem, A., and Shuler, M. L., "Characterization of a perfusion reactor utilizing mammalian cells on microcarrier beads," *Biotechnol. Prog.*, Vol. 16, No. 3, 2000, pp. 471–479.

[63] Ghanem, A., and Shuler, M. L., "Combining cell culture analogue reactor designs and PBPK models to probe mechanisms of naphthalene toxicity," *Biotechnol. Prog.*, Vol. 16, No. 3, 2000, pp. 334–345.

[64] Cushing, M. C., and Anseth, K. S., "Materials science. Hydrogel cell cultures," *Science*, Vol. 316, No. 5828, 2007, pp. 1133–1134.

[65] Powers, M. J., et al., "A microfabricated array bioreactor for perfused 3D liver culture," *Biotechnol. Bioeng.*, Vol. 78, No. 3, 2002, pp. 257–269.

[66] Sin, A., et al., "The design and fabrication of three-chamber microscale cell culture analog devices with integrated dissolved oxygen sensors," *Biotechnol. Prog.*, Vol. 20, No. 1, 2004, pp. 338–345.

[67] Ajdari, A., "Steady flows in networks of microfluidic channels: Building on the analogy with electrical circuits " *Comptes Rendus Physique,* Vol. 5, No. 5, 2003, pp. 539–546.

[68] Viravaidya, K., Sin, A., and Shuler, M. L., "Development of a microscale cell culture analog to probe naphthalene toxicity," *Biotechnol. Prog.*, Vol. 20, No. 1, 2004, pp. 316–323.

[69] Kim, L., et al., "A practical guide to microfluidic perfusion culture of adherent mammalian cells," *Lab Chip*, Vol. 7, No. 6, 2007, pp. 681–694.

[70] Walker, G. M., Zeringue, H. C., and Beebe, D. J., "Microenvironment design considerations for cellular scale studies," *Lab Chip*, Vol. 4, No. 2, 2004, pp. 91–97.

[71] Nerem, R. M., et al., "The study of the influence of flow on vascular endothelial biology," *Am. J. Med. Sci.*, Vol. 316, No. 3, 1998, pp. 169–175.

[72] Korin, N., et al., "A parametric study of human fibroblasts culture in a microchannel bioreactor," *Lab Chip*, Vol. 7, No. 5, 2007, pp. 611–617.

[73] Keane, J. T., Ryan, D., and Gray, P. P., "Effect of shear stress on expression of a recombinant protein by Chinese hamster ovary cells," *Biotechnol. Bioeng.*, Vol. 81, No. 2, 2003, pp. 211–220.

[74] Sugihara-Seki, M., "Flow around cells adhered to a microvessel wall. I. Fluid stresses and forces acting on the cells," *Biorheology*, Vol. 37, Nos. 5–6, 2000, pp. 341–359.

[75] Takayama, S., et al., "Selective chemical treatment of cellular microdomains using multiple laminar streams," *Chem. Biol.*, Vol. 10, No. 2, 2003, pp. 123–130.

[76] Viravaidya, K., *Development of a Four-Chamber Microscale Cell Culture Analog for Toxicological and Pharmacological Studies*, Ithaca, NY: Cornell University, May 2004, p. xiii, 177 leaves.

[77] Toepke, M. W., and Beebe, D. J., "PDMS absorption of small molecules and consequences in microfluidic applications," *Lab Chip*, Vol. 6, No. 12, 2006, pp. 1484–1486.

[78] Wabuyele, M. B., et al., "Single molecule detection of double-stranded DNA in poly(methylmethacrylate) and polycarbonate microfluidic devices," *Electrophoresis*, Vol. 22, No. 18, 2001, pp. 3939–3948.

[79] Hung, P. J., et al., "Continuous perfusion microfluidic cell culture array for high-throughput cell-based assays," *Biotechnol. Bioeng.*, Vol. 89, No. 1, 2005, pp. 1–8.

[80] Leclerc, E., Sakai, Y., and Fujii, T., "Microfluidic PDMS (polydimethylsiloxane) bioreactor for large-scale culture of hepatocytes," *Biotechnol. Prog.*, Vol. 20, No. 3, 2004, pp. 750–755.

[81] Sung, J. H., and Shuler, M. L., "Prevention of air bubble formation in a microfluidic perfusion cell culture system using a microscale bubble trap," *Biomed. Microdevices*, in press.

[82] Brandon, E. F., et al., "An update on in vitro test methods in human hepatic drug biotransformation research: Pros and cons," *Toxicol. Appl. Pharmacol.*, Vol. 189, No. 3, 2003, pp. 233–246.

[83] Deeken, J. F., et al., "Toward individualized treatment: Prediction of anticancer drug disposition and toxicity with pharmacogenetics," *Anticancer Drugs*, Vol. 18, No. 2, 2007, pp. 111–126.

[84] Nahmias, Y., Berthiaume, F., and Yarmush, M. L., "Integration of technologies for hepatic tissue engineering," *Adv. Biochem. Eng. Biotechnol.*, Vol. 103, No. 2007, pp. 309–329.

[85] Clement, B., et al., "Long-term co-cultures of adult human hepatocytes with rat liver epithelial cells: Modulation of albumin secretion and accumulation of extracellular material," *Hepatology*, Vol. 4, No. 3, 1984, pp. 373–380.

[86] LeCluyse, E. L., et al., "Cultured rat hepatocytes," *Pharm. Biotechnol.*, Vol. 8, 1996, pp. 121–159.

[87] Nussler, A. K., et al., "The suitability of hepatocyte culture models to study various aspects of drug metabolism," *ALTEX*, Vol. 18, No. 2, 2001, pp. 91–101.

[88] Walker, T. M., Rhodes, P. C., and Westmoreland, C., "The differential cytotoxicity of methotrexate in rat hepatocyte monolayer and spheroid cultures," *Toxicol in vitro*, Vol. 14, No. 5, 2000, pp. 475–485.

[89] Khetani, S. R., and Bhatia, S. N., "Microscale culture of human liver cells for drug development," *Nat. Biotechnol.*, Vol. 26, No. 1, 2008, pp. 120–126.

[90] Nahmias, Y., and Odde, D. J., "Micropatterning of living cells by laser-guided direct writing: Application to fabrication of hepatic-endothelial sinusoid-like structures," *Nat. Protoc.*, Vol. 1, No. 5, 2006, pp. 2288–2296.

[91] Sivaraman, A., et al., "A microscale in vitro physiological model of the liver: Predictive screens for drug metabolism and enzyme induction," *Curr. Drug. Metab.*, Vol. 6, No. 6, 2005, pp. 569–591.

[92] Sung, J. H., and Shuler, M. L., "A micro cell culture analog ($\mu$CCA) with 3-D hydrogel culture of multiple cell lines to assess metabolism-dependent cytotoxicity of anti-cancer drugs," *Lab Chip*, Vol. 9, No. 10, 2009, pp. 1385–1394.

[93] Xu, H., Kraus, W. L., and Shuler, M. L., "Development of a stable dual cell-line GFP expression system to study estrogenic endocrine disruptors," *Biotech. Bioeng.*, Vol. 101, No. 16, 2008, pp 1276–1287.

[94] Donato, M. T., et al., "Fluorescence-based assays for screening nine cytochrome P450 (P450) activities in intact cells expressing individual human P450 enzymes," *Drug Metab. Dispos.*, Vol. 32, No. 7, 2004, pp. 699–706.

[95] Bambot, S., et al., "Optical oxygen sensor using fluorescence lifetime measurement," *Adv. Exp. Med. Biol.*, Vol. 361, No. 1994, pp. 197–205.

[96] Prokop, A., et al., "NanoLiterBioReactor: Long-term mammalian cell culture at nanofabricated scale," *Biomed. Microdevices*, Vol. 6, No. 4, 2004, pp. 325–339.

[97] Solent Scientific, 2008, www.solentsci.com.

[98] Tatosian, D. A., Shuler, M. L., and Kim, D., "Portable in situ fluorescence cytometry of microscale cell-based assays," *Opt. Lett.*, Vol. 30, No. 13, 2005, pp. 1689–1691.

[99] Blair, I. A., and Tilve, A., "Analysis of anticancer drugs and their metabolites by mass spectrometry," *Curr. Drug. Metab.*, Vol. 3, No. 5, 2002, pp. 463–480.

[100] Park, D. J., Stoehlmacher, J., and Lenz, H. J., "Tailoring chemotherapy in advanced colorectal cancer," *Curr. Opin. Pharmacol.*, Vol. 3, No. 4, 2003, pp. 378–385.

[101] Relling, M. V., and Dervieux, T., "Pharmacogenetics and cancer therapy," *Nat. Rev. Cancer*, Vol. 1, No. 2, 2001, pp. 99–108.

[102] Gardner, S. N., "Modeling multi-drug chemotherapy: Tailoring treatment to individuals," *J. Theor. Biol.*, Vol. 214, No. 2, 2002, pp. 181–207.

[103] Tatosian, D. A., and Shuler, M. L., "A novel system for evaluation of drug mixtures for potential efficacy in treating multidrug resistant cancers," *Biotech. Bioeng.*, Vol. 103, No. 1, 2009, pp. 187–198.

[104] Uno, T., and Yasui-Furukori, N., "Effect of grapefruit juice in relation to human pharmacokinetic study," *Curr. Clin. Pharmacol.*, Vol. 1, No. 2, 2006, pp. 157–161.

CHAPTER

# 9

# Lab-on-a-Chip Impedance Detection of Microbial and Cellular Activity

Liju Yang,[1] Xuanhong Cheng,[2] Yi-Shao Liu,[3] and Rashid Bashir[3]

[1]Biomanufacturing Research Institute and Technology Enterprise, Department of Pharmaceutical Sciences, North Carolina Central University, Durham, NC 27707
[2]Department of Materials Science and Engineering, Program of Bioengineering, Lehigh University, Bethlehem, PA 18015
[3]Department of Electrical and Computer Engineering, Department of Bioengineering, Micro and Nanotechnology Laboratory, University of Illinois, Urbana-Champaign, Urbana, IL 61801

## Abstract

Lab-on-a-chip type of devices capable of impedance sensing has recently attracted a lot of interest for label-free, real-time, and noninvasive electrical detection of biological activities. In this chapter, we describe four lab-on-a-chip systems for the detection of microbial and cellular activities based on the unique electrical and electrophysiological properties of micro-organisms and mammalian cells. Two of the systems were designed based on impedance monitoring of live micro-organism activities for: (1) monitoring the concentration of bacterial cells during growth, and (2) the detection of *Bacillus anthracis* spore germination. The other two systems were designed for detection of cell concentration by measuring the impedance changes due to their ion release for applications in counting: (1) CD4+ T lymphocytes, and (2) food-borne pathogenic bacterial cells. These microfabricated impedance sensors show great promise in the detection of cells and their metabolic activities with improved simplicity, higher sensitivity, and faster detection time than conventional methods.

Key terms
microfluidic chips
lab-on-a-chip
electrical detection
impedance spectroscopy
dielectrophoresis
bacteria cells
mammalian cells
cell counting
spore germination

## 9.1 Introduction

The advances in microelectromechanical systems (MEMS) technology have allowed scientists to construct novel devices or systems with sizes comparable to biological entities and sensitivity high enough for a wide variety of important biomedical and biological applications. Development of "lab-on-a-chip" types of devices uses MEMS technology to integrate various microfabricated sensors or detection platforms with many of the unit operations associated with sample preparation and presentation, such as separation, mixing, incubation, and concentration. The use of lab-on-a-chip devices for microbial and cellular detection has shown the following advantages over traditional methods: (1) reduction of the sensor elements to the size of a single cell or even smaller, providing a higher sensitivity; (2) reduction of reagent volume and associated cost; (3) reduction of the time to results due to the small volume and high effective concentration; (4) amenability to system miniaturization and portability; and (5) compatibility with large numbers of assays and multiplexed measurements.

Impedance sensing, as one of the principal electrical/electrochemical transductions, is becoming a fertile area for developing methods for a wide range of biological and biomedical applications. Several factors attribute to the popularity of impedance sensing: (1) the distinct electrical properties associated with specific biological entities and/or biological reactions motivate the use of impedance-sensing techniques; (2) impedance measurement is one of the most promising techniques for label-free, real-time, and noninvasive biological detection; and (3) impedance detectors can be easily miniaturized to meet the growing needs of portable systems with an analytical footprint considerably smaller than laboratory-based instruments [1].

The distinct electrical properties of biological cells and their electrophysiology are fundamental for developing impedance-based methods to detect biological activities. Biological cells consist of adjacent structures of materials that have very different electrical properties. The cell membrane consists of a lipid bilayer, where the lipid molecules are oriented with their polar groups facing outwards into the aqueous environment and their hydrophobic hydrocarbon chains pointing inwards to form the membrane interior. The inside of a cell is complex and contains membrane-covered particulates, such as mitochondria, vacuoles, a nucleus, and many charged molecules. While the cell membrane is highly insulating, the interior of the cell is highly conductive. The conductivity of the cell membrane is around $10^{-7}$ S/m, whereas the conductivity of the interior of a cell can be as high as 1 S/m [2].

Based on the electrophysiological and electrical properties of biological cells, three major mechanisms have been explored for the detection and quantification of biological cells using microscale impedance-based measurements:

1. *Making use of the metabolic activity of biological cells:* This is represented by impedance microbiology, which is a technique based on the measurements of the electric impedance change in a medium or a reactant solution resulting from cell metabolism [3, 4]. Based on this principle, a new technique called "impedance microbiology-on-a-chip" has been demonstrated by our group [5]. The idea is to confine a few live bacterial cells into a small volume on the order of nano- to picoliters such that metabolites of these cells are concentrated and detectable by impedance measurement with interdigitated microelectrodes. We have also

successfully developed a microchip for impedance monitoring of spore germination [6].

2. *Making use of the highly ionic cytoplasmic content of the cells:* As the inside of a cell contains many charged molecules and is highly conductive (1 S/m), the impedance change due to the lysis of cells or release of intracellular ions can provide a means to detect biological cells. In this chapter, we will review two microchip-based impedance-detection systems for: (1) enumeration of CD4+ T lymphocytes through cell lysates [7], and (2) detection of bacterial cells based on impedance change from their ion release into deionized (DI) water [1].
3. *Making use of the insulating properties of the cell membrane:* Because of their highly insulating cell membrane, cells attached on an electrode surface effectively reduce the conducting area and hence increase the interfacial impedance. The sensor probes the attachment of cells by measuring the change of the interfacial electrical properties arising from the insulating property of the cell membrane. Many cell-based impedance sensors are based on this mechanism. By culturing cells on microelectrodes and monitoring impedance changes caused by adherent cells, one can quantify changes in the impedance associated with the cell membrane, cell-substrate interaction, and cell-cell separation with exquisite sensitivity and in a noninvasive manner [8, 9]. We constructed a bacterial immunosensor based on this mechanism: antibodies specific to the target bacterial cells are immobilized on an electrode surface, and selective attachment of cells is detected electrically [10].

In this chapter, we describe four impedance-detection systems based on the first two mechanisms described above for monitoring biological metabolic activity and detecting cells.

## 9.2 Lab-on-a-Chip for Monitoring Microbial Metabolic Activity

### 9.2.1 "Impedance microbiology-on-a-chip" for bacterial concentration and detection

One common impedance method for detection of bacterial growth is *impedance microbiology*, which is based on the measurement of changes in electrical impedance of a culture medium or a reaction solution resulting from the bacterial growth. This growth-based impedance technique allows one to distinguish between viable and dead cells and to detect viable bacteria within 24 hours. In 1992, the impedance method was approved by the Association of Official Analytical Chemists International (AOAC) as the first action method for screening *Salmonella* in food samples [11, 12].

In impedance microbiology, the impedance change is typically measured using a pair of electrodes submerged in the growth medium or the reactant solution. The impedance change in the medium is mainly produced by the release of ionic metabolites from live cells. There are two main origins of ion release by bacteria into their growth environment [13]: one is energy metabolism (catabolism) in which bacteria consumes oxygen and sugars and produces carbon dioxide and organic acids; the other is ion exchange through the cell membrane. Ions (such as $K^+$ and $Na^+$) are actively transported across ion channels embedded in the cell membrane, which serves to regulate the membrane potential and the osmotic difference between the interior and exterior of the cell. Between the two origins, energy metabolism is the major path of ion release from cells to the environment, and the ion-exchange process is a small contributor. These released ions cause changes in the ionic composition of the medium and consequently increase the conductivity of the medium. To detect bacteria, the impedance sensor measures the relative or absolute changes in conductance, capacitance, or impedance at regular time intervals during the growth of bacteria at a given temperature. The measured electrical signals are then graphically plotted on the ordinate against the incubation times on the abscissa, producing impedance growth curves. The time at which the decrease in impedance value exceeds a threshold is defined as the detection time, $t_d$. Generally, the impedance threshold is not reached until the bacteria number reaches approximately $10^6$ to $10^7$ cfu/mL (as determined by the plating method). For conventional impedance-microbiological methods, the detection time ranges from about 1 to 8 hours for initial bacterial concentration of $10^7$ to $10^1$ cfu/mL.

Miniaturization of an impedance-detection system into a chip-based device has shown great promise in rapid detection of bacterial growth. Our group was among the first to fabricate integrated silicon-based biochips for impedance detection of microbial metabolism [5, 14, 15]. The basic idea was to confine a few live bacterial cells into a small volume on the order of nano- to picoliters, such that the metabolites of a few live cells in a low-conductivity buffer can be rapidly detected by impedance measurements using interdigitated microelectrodes. To concentrate bacterial cells from a diluted sample into a small volume, we used a technique called dielectrophoresis (DEP), which is the electrokinetic motion of dielectrically polarized particles in nonuniform electric fields [16]. As most biological cells behave as dielectric particles in an external electric field, DEP allows trapping, concentration, and separation of biological cells in a liquid suspension.

#### 9.2.1.1 Methods and devices

##### *9.2.1.1.1 Chip design and fabrication*

The impedance microbiology-on-a-chip device contained three sets of interdigitated microelectrodes and flow channels. Figure 9.1(a) shows the principle of the operation of the DEP-based deviation and capture of bacterial cells in the microchip. One set was for dielectrophoretical deviation of bacterial cells from the main channel into the small channel that leads to the detection chamber. In the detection chamber, one set of electrodes was for DEP capture of bacterial cell into the detection chamber, and the other set of electrodes was for monitoring the impedance change of bacterial growth in the chamber. The detection chamber had a volume of 400 pL. Figure 9.1(b) shows the cross section of the detection chamber with DEP electrodes and impedance-measurement electrodes. Figure 9.1(c) shows the completely packaged microchip.

The microchips were fabricated on 4″ wafers with a (100) surface and a thickness of 500 $\mu$m. Electrodes and channel patterns were made using standard photolithographic technology. The channels were 12 $\mu$m deep and were made by etching the wafer with a hard mask. The DEP electrodes were deposited by sputtering 1,000Å of aluminum onto 2,000Å of silicon dioxide on the bottom of the channel. On the DEP electrodes, another layer of silicon dioxide of 3,500Å was deposited by plasma-enhanced chemical vapor deposition in order to completely isolate the DEP electrodes and prevent electrolysis of the liquid in the channels. The impedance-measurement electrodes and temperature

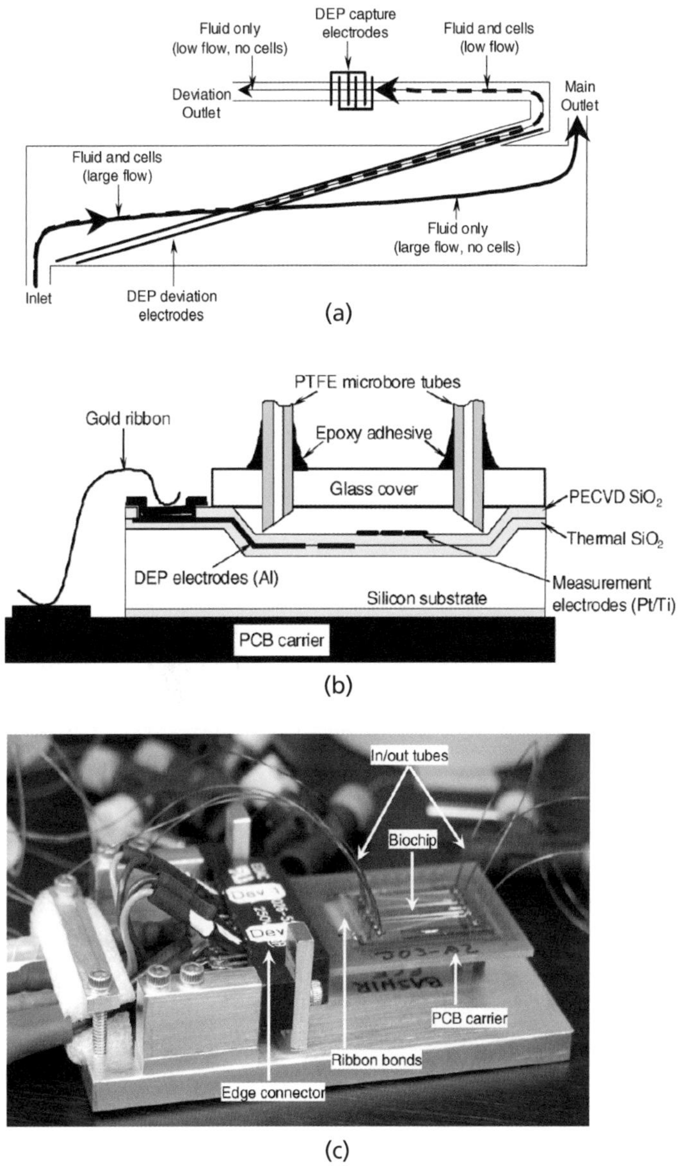


**Figure 9.1** (a) The schematic design of the microchip with DEP deviation electrodes in the main channel and a small channel leading the flow to the DEP capture electrodes in the detection chamber. (b) Simplified cross section of the packaged microchip showing the DEP capture electrodes and the impedance-measurement electrodes in the detection chamber. (c) An image of the packaged microchip connected to the measurement and control system. (Reprinted with permission from the *J. Microelectromechanical Systems* and kind permission from [5].)

sensor were deposited by sputtering 800Å of platinum over a titanium adhesion layer. The chip was assembled with a glass cover with inlet and outlet holes using anodic bonding. Detailed procedure can be found in [5].

#### *9.2.1.1.2 Bacterial cell preparation*

*Listeria monocytogenes v7* was grown in Luria-Bertani (LB) medium at 37°C for at least 16 hours. The cells were harvested and washed by repeated centrifugation and resuspension in sterile LB medium. The cells were then diluted in sterile LB to desired concentrations. When fluorescent cell were needed, 1 mL of the as-grown live cells was stained with green fluorescent dye $DiOC_6(3)$ (3,3'-dihexyloxacarbonanine iodide). Fluorescence-stained cells were washed and diluted in the same way as described above for further use.

#### *9.2.1.1.3 On-chip DEP concentration and impedance detection of metabolism*

All the on-chip experiments were carried out with the chip heated to 37°C. A total volume of 40 $\mu$L of the cell suspension in DI water was injected into the main channel at a flow rate of approximately 1.7 $\mu$L/min. The DEP deviation and capture electrodes were excited with a 16 $V_{pp}$ square signal at 100 kHz. During the injection, the flow rate in the incubation chamber was manually controlled to be between 4 and 10 nL/min. After the sample injection, the DEP deviation electrodes were turned off, and Half-LB medium (HLB, mixing equal parts of LB and DI water) was injected at a flow rate of less than 0.5 $\mu$L/min, while the excitation voltage on the capture electrodes was increased to 20 $V_{pp}$, and the frequency was increased to 3 MHz in order to maximize the DEP forces acting on cells. Once the incubation chamber was filled with HLB, the flow was stopped, and the fluidic channels and microtubes were pinched to seal them completely. The DEP capture electrodes were then turned off. Impedance measurement electrodes were turned on, and the cells were incubated for approximately 12 hours. The impedance was measured with an Agilent 4284A LCR meter (Agilent Technologies Inc., Palo Alto, California) connected to the chip through an Agilent 34970A switching unit fitted with two Agilent 34905A RF multiplexer cards. All the instruments were connected to a computer through a GPIB interface. The impedance measurements and chip temperature were controlled by custom LabVIEW program (National Instruments Corp., Austin, Texas). Impedance was measured at 51 frequencies logarithmically spaced between 100 Hz and 100 kHz, with a 150 mV amplitude. Sinusoidal and square wave DEP signals were generated by Agilent 333120A synthetized signal generators.

### 9.2.1.2 Results and discussion

To demonstrate the complete process of cell concentration and impedance measurement of bacterial metabolism on the chip, samples containing fluorescently labeled *L. monocytogenes v7* cells at concentrations of $2.3 \times 10^5$, $6.8 \times 10^5$, and $8.7 \times 10^4$ cfu/mL were tested. The sample with $2.3 \times 10^5$ cfu/mL cells was injected with the DEP electrodes off, while the other two samples were injected with the DEP electrode activated. When the DEP electrode was off, the cells were not concentrated in the incubation chamber. There was only a probability of approximately 0.09 to find one cell in the chamber. When DEP electrodes were activated, almost all the cells were captured by the DEP electrodes into the detection chamber. Figure 9.2(a) shows a representative image of fluorescently labeled *Listeria* cells concentrated by DEP into the picoliter measurement chamber. Although the actual number of cells collected was not determined, it was visually confirmed that only a very small fraction of cells escaped the DEP deviation and capture processes (no more than approximately 10%). However, during the switch from water to HLB, a more significant fraction of the cells were lost because the DEP

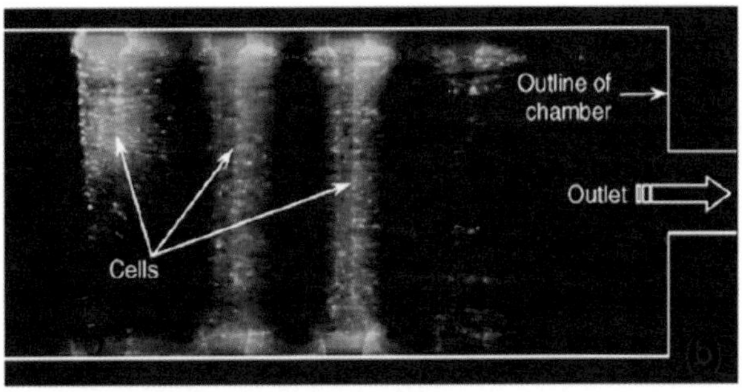


(a)

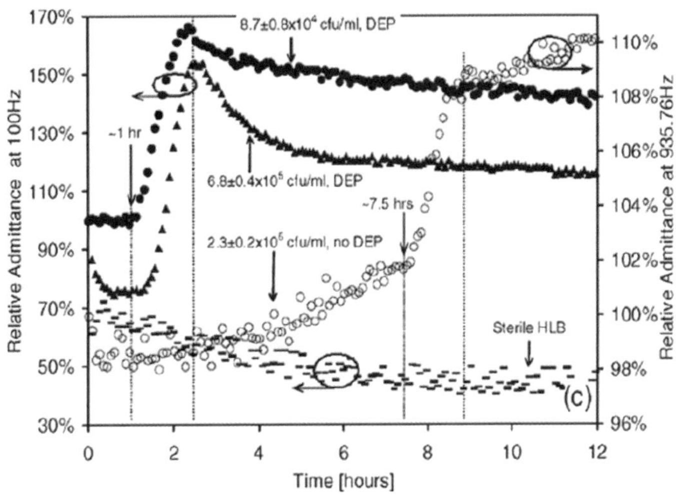


(b)

**Figure 9.2** (a) Fluorescence-labeled *Listeria* cells concentrated by DEP from a suspension of $6.8 \times 10^5$ cfu/mL into the incubation chamber immediately before the start of incubation. (b) Relative admittance change during the incubation of *Listeria* cells injected into the microchip at various concentrations, with and without DEP concentration, plus sterile HLB. Values at $t = 0$ are defined as 100%. (Reprinted with permission from the *J. Microelectromechanical Systems* and kind permission from [5].)

force was weakened by the increased medium conductivity and the fluctuation of the flow rate. Nonetheless, thousands of cells were still collected in the chamber, as shown in Figure 9.2(a). The concentration factor of this chip was between $10^4$ to $10^5$ when the cells in an original sample volume of 40 $\mu$L were concentrated into the 400 pL chamber, provided that 10% to 100% of the cells were captured by DEP. Such DEP concentration technique in microfabricated chips eliminates the need to enrich the bacterial population by long culture steps in conventional cell culture methods. With the dramatic increase of bacterial cell concentration at the locality of the detection chamber, it is expected that the detection time on the impedance growth curve can be effectively reduced, resulting in rapid detection.

The significant reduction in detection time was demonstrated by the comparison of the impedance growth curves of *Listeria* cells in HLB medium on chip with and without DEP concentration [Figure 9.2(b)]. As shown in the figure, the sterile media did not exhibit any clear metabolic signal at any frequency. The bacterial sample containing approximately $6.8 \times 10^5$ cfu/mL with the DEP concentration presented an impedance metabolic signal corresponding to the exponential growth at approximately 1 hour, while the sample containing similar concentration of cells without DEP concentration required approximately 7.5 hours to produce a detectable impedance signal. The results demonstrated that concentration of bacterial cells by the DEP can effectively shorten the detection time needed for the impedance detection of cell metabolic activity.

Miniaturized sensors for impedance-based detection of bacteria integrated with a DEP-based cell-concentration system hold great potential to dramatically reduce the time needed to detect bacteria based on their metabolic activity to hours, which is a great improvement compared with conventional methods that require several days. Such a microscale system also has great potential applications for screening industrial and clinical samples for total bacterial contents.

### 9.2.2 Microfluidic biochips for impedance detection of *Bacillus anthracis* spore germination

*Bacillus anthracis* has long been identified as the causative agent of the disease anthrax. The intentional contamination of seven letters with *B. anthracis* spores in 2001 resulted in 22 cases of anthrax, 5 of which were fatal [17], and focused attention on the detection of spores of *Bacillus anthracis*. *Bacillus anthracis* has a long-term environmental persistence due to the formation of endospores, which develop over a time course of several hours inside a cell exposed to nutrient starvation or other environmental stresses [18]. A dense protein coat, low cytoplasmal water activity, and small, acid-soluble, DNA-binding proteins render the spore highly resistant to dessication, irradiation, chemical oxidation, and other environmental assaults [19]. This resistance renders decontamination of an environment in which endospores are present very difficult and makes detection of low-level spore contamination an important goal. Many detection methods, such as colony morphology, staining of the unique poly-D-glutamate capsule, PCR amplification of specific DNA sequences, and c-phage susceptibility testing, require the outgrowth of vegetative cells before testing; they also often require time-consuming manual steps. Several innovative methods reported in recent years can detect endospores at a threshold of about $10^3$ spores in a sample [20–25]. Most of these methods involve micro- or macroscale PCR analysis or various forms of optical detection. Automated spore-detection systems have used real-time PCR with thermal cycling chambers made from etched and fusion-bonded silicon to carry out the PCR-based detection assays for *Bacillus* spp. and *Yersinia* spp. [26]. It should be noted that most of the reported methods either perform identification of the spores before spores germinate and fail to distinguish between viable and nonviable spores [21], or they detect the pathogens at relatively late vegetative growth phase [20]. In this section, we review a detection method for viable spores by impedance measurement. The detection was performed as early as the spore-germination stage, and a signal was detected only when the spores were viable.

#### 9.2.2.1 Methods and devices

We first examined the concept using macroscale experiments, where we measured the germination of a 5 mL spore suspension, with concentration ranging from $10^7$ to $10^9$ spores/mL, in a rinsed (with sterile DI water) and sterile 15 mL plastic centrifuge tube (430052, Corning Inc., Corning, New York) using a commercial conductivity meter

(6307 microcomputer pH/conductivity meter, Jenco Instruments, San Diego, California). The spores were preheated in a 65°C water bath for 30 minutes before the experiment. Afterwards, germinant solution was added, and the conductivity probe was inserted into the 15 mL centrifuge tube containing the sample. Conductivity values were recorded every minute. Three different concentrations of spores were compared to control experiments (i.e., spores only, DI water only, and germinant only) to find the detection limit. The conductivity probes were calibrated before each experiment. The DI water had a measured conductivity of 2 to 3 $\mu$S/cm, which was within the accepted range and thus demonstrated the sensitivity of the instrument. The experiment was carried out at room temperature [27].

The microfluidic device for on-chip detection of spore germination was constructed as a three layer BioMEMS device. The first layer was a Pyrex (7740, Corning Inc.) substrate with interdigitated electrodes for exerting dielectrophoresis force to capture and concentrate spores and for recording the change of admittance (inverse of impedance) within the solution. The metal electrodes were deposited as 250Å titanium and 350Å gold by evaporation (E-Beam Evaporator, CHA Industries, Fremont, California), followed by a lift-off process. On top of the Pyrex substrate was a 40 $\mu$m PDMS layer with patterned microfluidic channels and chambers for sample delivery and germination detection. The third layer was a thick, 2 mm PDMS slab with microfluidic pathways serving as valves to close off the channels in the thin second layer to enhance the signal strength by forming a closed environment for detection, as well as to constrain spores inside the detection chamber after the dielectrophoresis (DEP) capture force was released. The layout of the chip and the completed chip are illustrated in Figure 9.3(a, b). The Pyrex layer containing the electrode substrate and the hybrid PDMS layer was bonded by surface treatment using oxygen plasma (200W, 15 seconds) and aligned immediately (within 5 minutes). The etch gas was 80% argon and 20% oxygen. Microbore tubings (OD: 0.016", ID: 0.006", Cole Parmer, Vernon Hills, Illinois) were inserted into the punched inlet holes and sealed with 10:1 PDMS for injection of liquids. The cross section of the fabricated biochip and its functions is illustrated in Figure 9.3(c) [28–30].

The electrical measurements on-chip were carried out with an automated recording system. The system included an injector, measuring probes (Micromanipulator Co., Carson, Nevada), an LCR meter (Agilent Technologies), a computer, and a microscope (Eclipse E600FN, Nikon Inc., Melville, New York). Heat-treated spores were injected by the injection system into the microfluidic device mounted on the microscope platform, with a flow rate of 30 $\mu$L/min for 5 minutes, followed by 0.2 $\mu$L/min. The injection system had multiple injection valves and switches to change solutions for delivery of samples and germinant. Electrical recording started right after spores and germinant were delivered into the chip. Data was recorded at 2-minute intervals for 1 hour. Verification of germination after each experiment was done by observing the refractility of *Bacillus anthracis* spores using phase-contrast microscopy. Ungerminated spores are refractile (phase bright) and germinated spores are not (phase gray or phase dark). The experiment was carried out at room temperature.

#### 9.2.2.2 Results and discussion

The results of spore-germination detection with a commercial conductivity meter are illustrated in Figure 9.4(a). This figure shows the results with spore concentrations of

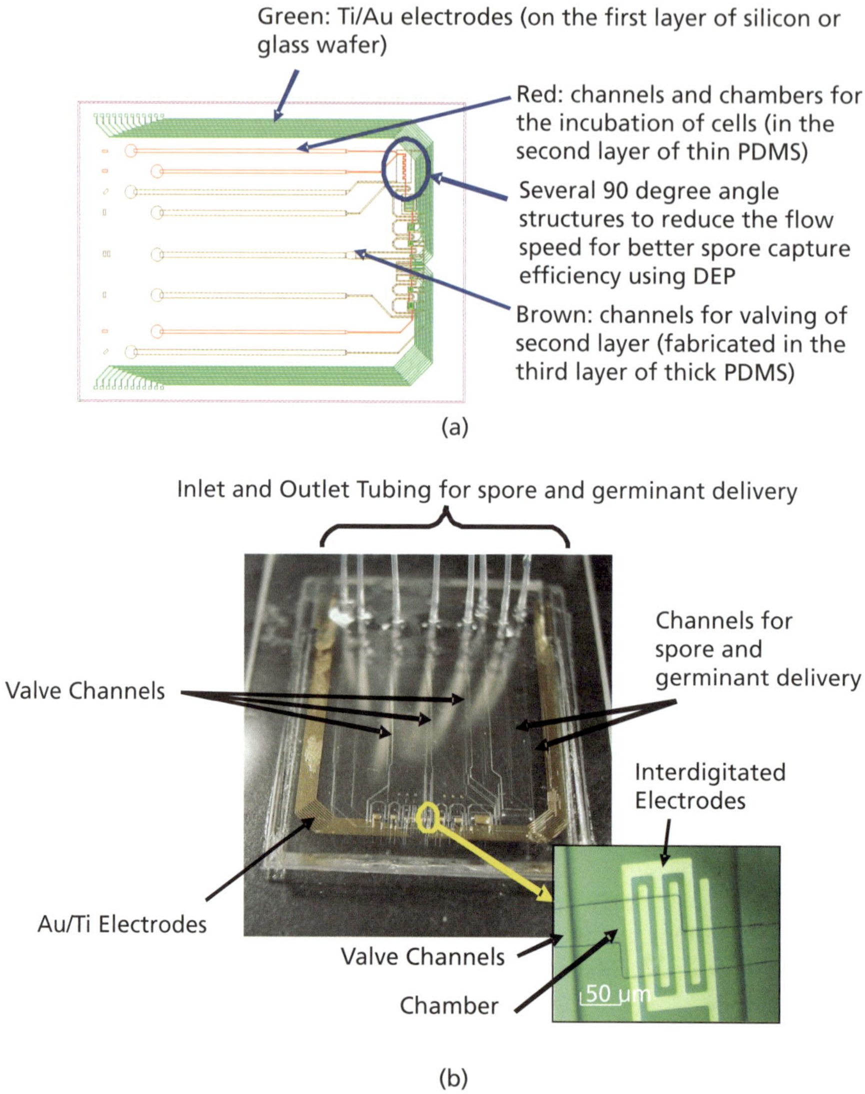


**Figure 9.3** (a) Top-view layout of the microfluidic device, (b) optical image of the completed device, and (c) cross section of the microfluidic device. Spores were heat-activated off-chip, then passed through and captured on interdigitated electrodes by DEP in the desired chamber. (d) Valving is effected by pressurizing the third layer and thus pressing against the second layer channel to form a closed environment for spore germination to take place. The effectiveness of the valves was demonstrated with a solution of safranin dye. (e) Fluorescence microscope image of spores captured within the chamber using DEP forces. 20V peak-to-peak, 100 kHz. Total elapsed time is 1 minute. Activated spores stained with the FITC dye, $DiOC_6(3)$. (Reprinted with permission from *Lab on a Chip* and kind permission from [6].)

$10^7$, $10^8$, and $10^9$ spores/mL. The most pronounced result is observed at a concentration of $10^9$ spores/mL, where a net increase of conductivity occurs in the range of 5 to 7 $\mu$S/cm. An immediate increase in conductivity suggests that germination began in the first 2 minutes after germinant was added and finished within 20 minutes, while control experiments showed no significant change in conductivity over time. We consid-

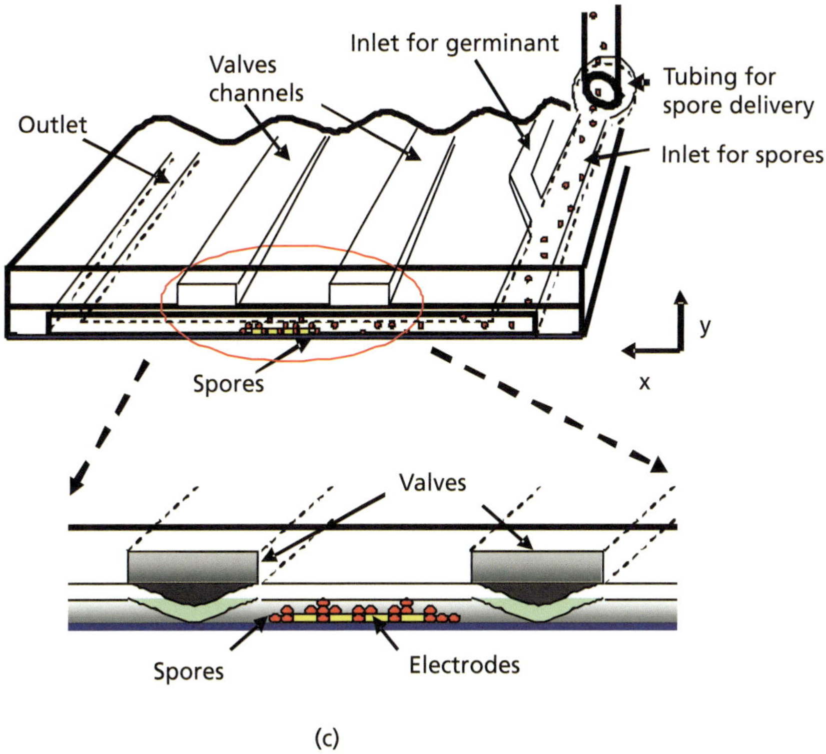


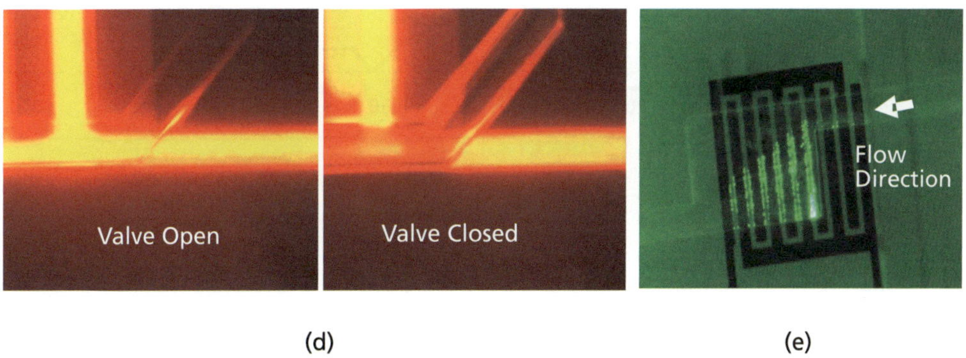


**Figure 9.3** (continued)

ered the $10^9$ spores/mL concentration as the one that can be safely detected and used this value for the design of the microscale assay.

In microfluidic device experiments, spores were delivered to the target chamber (0.1 nL, $100 \times 100 \times 10$ $\mu$m) in a carrier stream of DI water. The spores were captured at the edge of the embedded electrodes by dielectrophoretic forces induced at 20V and 100 kHz. With a flow rate of 0.2 $\mu$L/min (peak flow velocity: 40 cm/min), 90% of the spores in the carrier stream were captured by DEP [31]. Figure 9.3(e) shows the capture of spores by DEP forces, 20V peak to peak, 100 kHz.

We first germinated the spores without actuating the built-in valves. Measurement of admittance started right after the germinant solution had fully replaced the DI water.

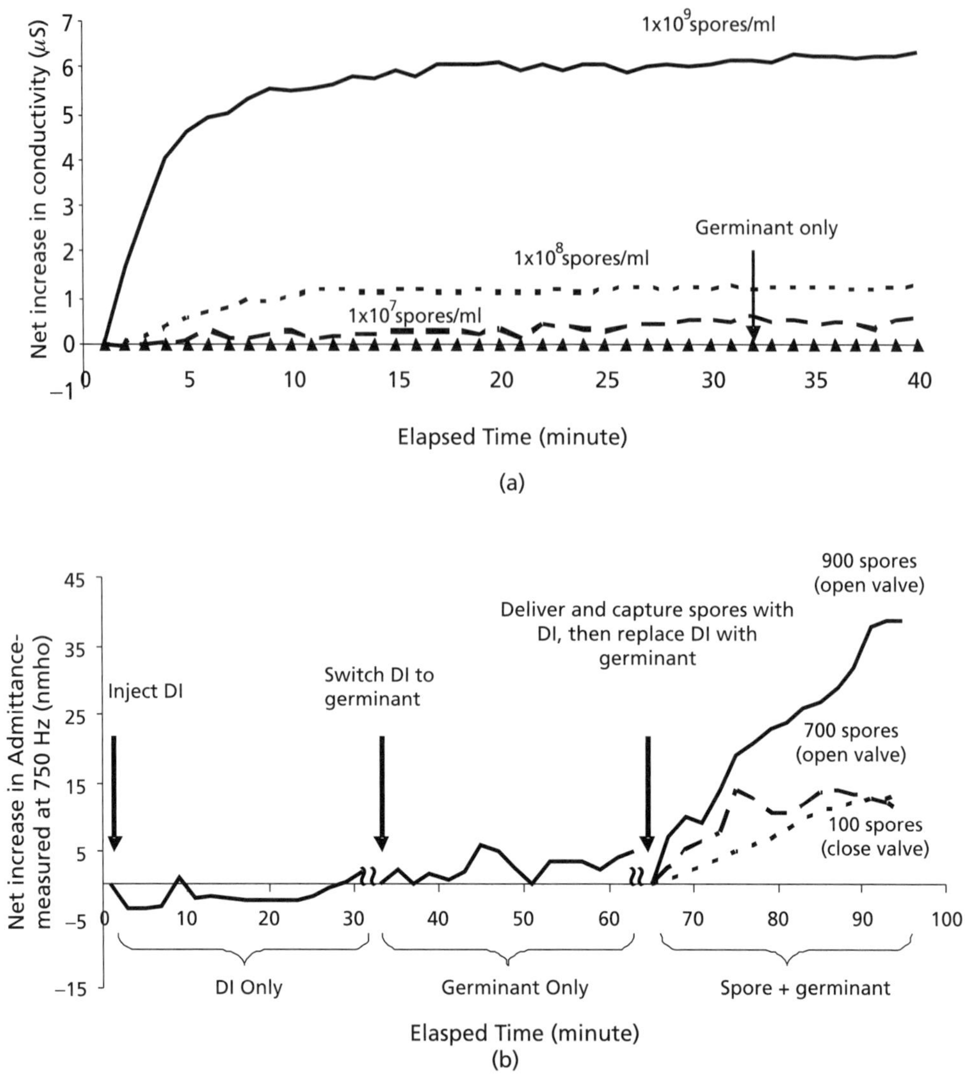


**Figure 9.4** (a) Impedance curves of the spore germination in the macroscale germination experiments with samples containing $10^9$, $10^8$, and $10^7$ spores/mL. The control sample contains germinant solution only. (b) Representative on-chip experimental results with about 100, 700, and 900 spores in a 0.1 nL chamber (graphs of data with 700 and 900 spores were open-valve experiments, while the graph with 100 spores was from a closed-valve experiment). The results from serially performed control experiments are also shown. (Reprinted with permission from *Lab on a Chip* and kind permission from [6].)

The results showed an increase in admittance upon spore germination [Figure 9.4(b)]. In comparison to the control experiment with germinant only, the samples with approximately 700 spores and approximately 900 spores presented significant increases in admittance, starting 2 minutes after the experiment began. There was no significant increase in admittance in the control experiments. Fluctuations in admittance occurred because the chamber was open to the flow stream without the actuated valve. However, when spores started to germinate, enough ions were released to overcome the baseline fluctuation and showed a significant increase in admittance. The detection limit with this experiment proved to be a few hundred spores in a 0.1 nL chamber.

To overcome the problem of fluctuations in admittance, the designed microfluidic valves were actuated to isolate the germination chamber from the rest of the system. The channels in the third layer of the microfluidic device (the PDMS slab) were pressurized with germinant solution to close the valves in the second layer [the PDMS membrane; see Figure 9.3(c, d)]. We used germinant solution to close the valves in order to avoid admittance perturbation due to ion permeation from the valve channel into the germination chamber through the PDMS layer. Two sets of sensing probes were used to measure two identical chambers (0.1 nL) simultaneously, with one serving as a control chamber and the other as the experimental chamber. A comparison of the results from these two identical chambers determined that the chamber with spores and germinant had a significant increase in admittance when germination was taking place, whereas the control chamber showed no significant difference from the control experiments without germinant. The sample with 100 spores showed an immediate increase in admittance and reached an admittance increase of approximately 10 nmho in 20 minutes. It is clear that only when spore germination was taking place did a significant change in the measured signal occur. (There was a slight admittance increase in the control chamber in the germination experiment due to the influence of ions permeating the PDMS from the experimental chamber.) Based on a comparison of the admittance changes with 700 spores in the open-valve experiment and with 100 spores in the closed-valve experiment, results were very similar, indicating that the isolation valves also enhanced the sensitivity of the admittance measurements. This implies that a lower detection limit could be achieved by including isolation valves in the chip design. The detection limit was less than 100 spores in a 0.1 nL chamber ($10^9$ spores/mL). Theoretically, this limit could be reduced to 10 spores or even 1 spore with smaller-sized chambers.

In this study, we demonstrated a method for automatic and rapid electrical detection of germination of viable spores within a microfluidic biochip. The microfluidic device includes special design features that facilitate spore capture and isolation as well as electrodes for spore concentration and impedance measurements. The limit of detection was shown to be a few hundred spores in a 0.1 nL chamber without use of the isolation valves. The detection limit was reduced to fewer than 100 spores in a 0.1 nL chamber when the chamber was isolated by closing the isolation valves. The detection limit can be further lowered by using a smaller capture-measurement chamber. The detection time is as short as 2 hours from heat activation of a suspected organism, which makes the impedance-based detection method a promising candidate for an on-site environmental diagnostic platform.

## 9.3 Lab-on-a-Chip for Impedance Detection of Cell Concentration Based on Ion Release from Cells

### 9.3.1 Microchips for impedance detection of CD4+ T lymphocytes

Although multiple miniaturized platforms exist for cell counting in suspension, such as flow cytometry and Coulter counters [32–35], methods to enumerate attached cells within microfluidic devices are limited. Optical-microscopy-based cell detection, although straightforward, remains dependent on a stable light path and lensing, filtering, and focusing mechanisms that add cost and complexity to detection. In addition, optical detection tends to be low throughput due to the small detection area available at a single time. A valuable complement to optical microscopy is surface impedance sensing to enumerate cells attached on a substrate [36–38]. However, in the impedance-sensing method, a near unity of cell coverage on the electrode surface is required to generate a detectable signal.

To address the need for sensitive detection of a small number of cells attached on a relatively large surface area or in a large volume, we introduce in this chapter an approach called "cell lysate impedance spectroscopy." In this approach, surface-bound cells are lysed in a microfluidic channel, and the bulk conductance changes are measured through surface-patterned electrodes and impedance spectroscopy. As the intracellular ion content is relatively constant in each cell type, the number of released ions measured electrically is indicative of the cell number. Using immunoaffinity-isolated CD4+ T cells as an example, we demonstrate here that bulk solution conductance increases proportionally to the number of cells in the microdevice. In addition, this method has a detection threshold of 20 cells/$\mu$L, which is sufficiently useful for many clinical and research applications that require cell counting.

#### 9.3.1.1 Methods and devices

Microfluidic devices were fabricated by bonding two pieces of glass slide with a PDMS gasket that is 50 $\mu$m thick and contains an opening window of 5 cm × 4 mm. The PDMS gaskets were prepared by curing spin-coated PDMS on a transparency slide, followed with hand-cutting windows of desired sizes. Interdigitated (IDT) gold electrodes were patterned on the bottom slide by standard photolithography [Figure 9.5(a)], facing to the microfluidic channel side during assembly. Two holes were drilled on the cover slide and bonded with PDMS ports to form fluid inlets and outlets. Assembled devices were then functionalized with a monoclonal CD4 antibody and primed with PBS containing 1% BSA and 1 mM EDTA [7]. CD4+ T lymphocytes from healthy donors were captured in the microfluidic chip by flowing cultured peripheral blood mononuclear cells (PBMC) into the device at 5 $\mu$L/min. This flow rate is optimal for efficient and specific capture of CD4+ T lymphocytes [7], and the duration of sample injection determines the total number of captured cells.

To minimize background conductance, we use the following operational sequence to lyse the captured cells: First, extracellular ions present in the microchannel were washed out using a low-conductive washing solution containing 8.5% sucrose and 0.3% dextrose at a flow rate of 20 $\mu$L/min until impedance signals were stable. This ion-free solution has been found to maintain viability of mammalian cells. Next, a low-conductive solution containing 2% sucrose and 0.07% dextrose was flowed in at a flow rate of 10 $\mu$L/min for 1 minute for controlled cell lysis. The lysing solution was formulated such that cell lysis occurred after a complete replacement of the washing solution by the lysing solution. Cells were then kept in the lysis solution in a static state for another 10 minutes to allow cell lysis to reach a steady state.

Impedance measurements were taken using an Agilent 4284 LCR meter (Agilent Technologies). The microelectrode devices were connected to the LCR meter through platinum probes. The impedance-measurement process was automated by custom LabVIEW (National Instruments Corp.) virtual instruments and GPI B interface. Imped-

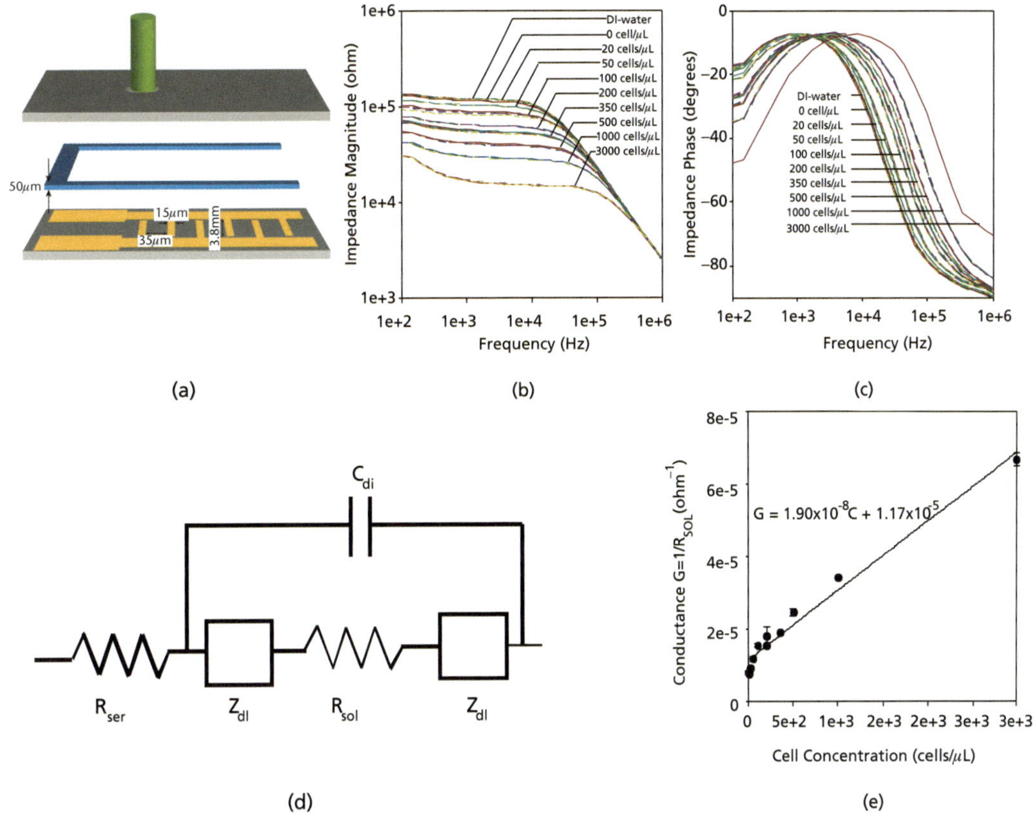


**Figure 9.5** Microfluidic devices used in this study and impedance spectroscopy measurement using off-chip cell lysate. (a) The devices are composed of two glass slides with micropatterned IDT electrodes and a PDMS gasket. (b) Impedance magnitude and (c) phase spectra of DI water and cell lysate with different starting cell concentrations measured on the IDT device. Three to five scans were performed at each cell concentration in the frequency range between 100 and $10^6$ Hz. (d) An equivalent circuit used in our study to model the electrode/electrolyte system for extracting bulk solution conductance, $1/R_{sol}$, which directly correlates with cell ion release [1]. (e) A linear relationship between measured bulk solution conductance (solid dots) and cell concentration is observed, and the best fits are shown as solid lines. Error bars indicate the standard deviation from three to five continuous measurements within a single device. (Reprinted with permission from *Lab on a Chip* and kind permission from [7].)

ance spectra were measured in the frequency range of 100 Hz to 1 MHz with a frequency increase factor of 1.5 and amplitude of 250 mV.

#### 9.3.1.2 Results and discussion

To test the detection sensitivity of ions released from primary cells using impedance spectroscopy, we first lysed PBMCs of known concentrations in Eppendorf tubes with DI water and measured the impedance of the lysate using the microfluidic device with coplanar interdigitated microelectrodes (IMEs) as shown in Figure 9.5(a). Figure 9.5(b, c) shows the spectra of impedance magnitude [Figure 9.5(b)] and phase [Figure 9.5(c)] as a function of frequency for cell concentrations ranging from 0 to 3,000 cells/μL measured using the IMEs. We observed that each spectrum has two regions, a constant-impedance region in the frequency range from 100 Hz to 10 kHz and an impedance-decreasing region when frequency is greater than 100 kHz. With increasing cell concentrations, there is a consistent decrease in impedance magnitude in the low-frequency range and a shift of phase peak to higher frequency. This suggests strongly that semiquantitative measurement of cell number can be achieved through ion release.

To understand solution conductance as a function of cell number, we carried out modeling studies to extract bulk conductance from the impedance spectra. Electrodes in an electrolyte solution can be modeled using an equivalent circuit as shown in Figure 9.5(d) [5, 7], where $C_{di}$ is the dielectric capacitance (it contains dielectric contributions from all the materials surrounding the electrodes, including the solution), $R_{sol}$ is the bulk solution resistance (charge transport across the bulk solution), $Z_{dl}$ is the interfacial impedance (the so-called Warburg impedance that accounts for the change in the ionic gradient at the interface), and $R_{ser}$ is the resistance of the on-chip wiring. The interfacial impedance can be expressed as

$$Z_{dl} = 1/\left[(j\omega)n\,B\right] \tag{9.1}$$

where $j = \sqrt{(-1)}$, and $n$ and $B$ are parameters dependent on the properties of the electrolytes and the electrodes. This is the simplest model that would properly fit the measured data over the whole frequency range at all times. Solution bulk conductance $G_{sol}$ is simply the reciprocal of $R_{sol}$.

By applying the circuit model to the group of curves in Figure 9.5(b, c), bulk conduction was extracted and plotted as a function of cell concentration [solid dot in Figure 9.5(e)]. It is observed that solution conductance increases linearly with the number of cells, confirming our hypothesis that ion release and solution conductance change are indicative of cell number. Moreover, using ion release to detect cells appears to be extremely sensitive and can detect as few as 20 cells/μL in an ion-free solution. The slope of the conductance curve represented measurement sensitivity of the microchip, and the sensitivity of the IMEs was about $1.90 \times 10^{-8}$/cell $\mu$L.

After impedance measurement using off-chip lysate, we then studied how impedance changes in the process of cell capture and on-chip cell lysis. CD4+ T cells were captured from cultured peripheral blood mononuclear cells, followed with washing with PBS buffer and then the low-conductive washing solution. Afterwards, the low-conductive cell-lysing solution was introduced into the microfluidic device, and cells were allowed to lyse for 10 minutes. At the end, the reference spectra were obtained with DI water. In the entire process, impedance spectra were acquired continuously and reflected the bulk solution resistance. Take the impedance magnitude at 760 kHz, for example [Figure 9.6(a)], where maximum separation occurs between the impedance magnitude curves in Figure 9.5(b), it remains in the low kilo-ohm range when cells are in PBS due to the high ionic concentration of biological buffers. It increases dramatically to above 10 kΩ upon introduction of the low-conductive washing solution. When cells are kept in the low-conductive washing solution in a static state, impedance magnitude decreases slightly, likely due to cell ion release in a low-conductive environment. After injection of the ion-free lysing solution, an initial impedance jump is noticed because the lysing solution has a lower conductivity than the washing solution. This is followed

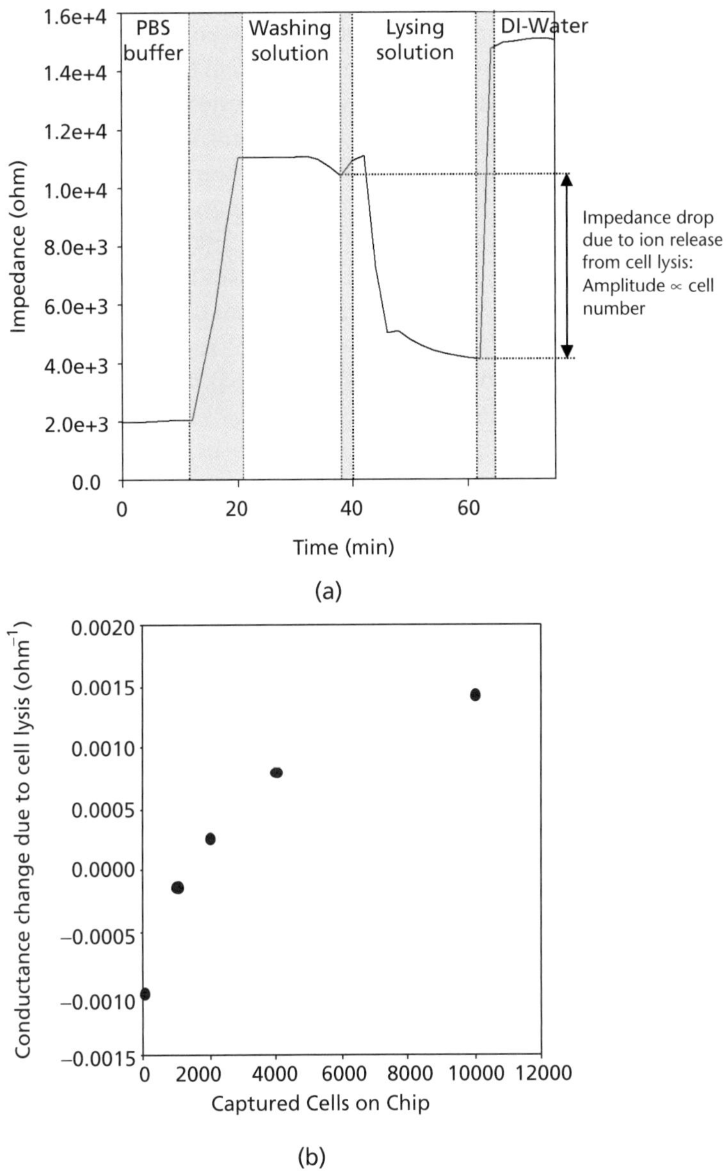


**Figure 9.6** (a) Impedance measurement at 760 Hz in the process of cell capture and on-chip lysis. The respective incubation steps are labeled on top of the graph, and the shaded areas between these labeled steps are transient states during solution exchanges. The impedance drop before and 10 minutes after injecting the lysing solution is associated with cell lysis and used as a cell-numbers indicator. (b) Conductance change in the process of on-chip cell lysis versus the number of cells captured within microfluidic devices. Bulk solution conductance was extracted from the impedance spectra, and conductance drop before and 10 minutes after flowing in the lysing solution was taken as the indicator to count cell. This conductance change increases continuously with the number of cells captured within the microfluidic chip, suggesting immobilized cells can be counted by electrical measurement of their ion release. Nonlinearity of the relationship may arise from incomplete diffusion of ions within the measurement time. Each data point in the plot represents measurement from one device. (Reprinted with permission from *Lab on a Chip* and kind permission from [7].)

with an abrupt drop of impedance and a subsequent slower impedance decrease. This two-stage impedance drop during cell incubation in the lysing solution matches optical observation of lysed cell numbers in the same solution (data not shown), suggesting that the decrease in impedance magnitude arises from lysis of the captured cells.

Following the same data-fitting procedure as described above, bulk conductance was extracted from the impedance spectra and the conductance change before and 10 minutes after flowing in the lysing solution was taken as a result of complete ion release from captured cells. When we compare this conductance change to manual cell counts within the microfluidic devices [Figure 9.6(b)], it is evident that the bulk conductance change is proportional to the number of captured cells. The results successfully demonstrate that cells can be detected and counted within a microfluidic device through the impedance/conductance measurement of cell lysate.

In conclusion, impedance spectroscopy can be used to detect mammalian cells immobilized in a microfluidic device through their ion release. The microdevice helps to confine the ions in a small volume for sensitive measurement. Not only is the approach useful for terminal cell counting, but it also holds the promise to study live-cell activities through their ion exchange with the environment.

### 9.3.2 Interdigitated microelectrode chip for impedance detection of bacterial cells

The conductivity of bacterial suspensions has been used to study the electrical properties of bacterial cell-surface and related cell-surface interfacial physiology [39, 40]. It can also be used to quantify the concentration of bacterial cells in suspensions. We present here a simple and rapid impedance method to detect bacterial cells in suspensions using interdigitated microelectrodes.

When bacterial cells are suspended in DI water, there are two possible ways for the bacterial cells to alter the impedance of DI water. One is via the charged nature of bacterial cell surfaces. The bacterial cell walls contain various acidic groups such as carboxyl, phosphate, and amino groups [41]. Generally, there is a higher concentration of anionic groups than of cationic groups, which results in a negative cell-wall charge at neutral pH. This charge is compensated for by counterions that penetrate into the porous cell wall and, to a minor extent, by coions that are expelled from it, thereby conferring electrostatic charge to the cell periphery [39, 40, 42]. The charge density of the bacterial cell wall can be as high as 0.5 to 1.0 $C/m^2$ [39]. The conductivity of the DI water is in a range from as low as about 1 to 2 $\mu S/cm$ to up to about 10 to 15 $\mu S/cm$. When bacterial cells are suspended in low-conductive DI water and reach a sufficient concentration, they can alter the conductivity of the suspension because of their cell-wall charges. The other way for bacterial cells to alter the conductivity of DI water is via the ion release from bacterial cells. When bacterial cells are suspended in a solution, such phenomena may occur as leakage of ions through the cytoplasmic membrane and negative adsorption of electrolyte or ion uptake into the cytoplasm and specific adsorption of ions. These processes can influence the conductivity of the bulk solution [39]. When bacterial cells are suspended in DI water, they experience an osmotic shock. In response to the fluctuations in environmental osmolarity, cells adjust their intracellular solute concentrations in order to maintain a constant turgor pressure and ensure continuation of cellular activity. Other properties of cells, such as cell size and buoyant density, can also be altered in response to the osmotic shock [43]. The charges on the cell wall, the release of ions, and other responses to the osmotic shock in combination account for the impedance change in DI water with suspended bacteria.

Here, we used *Salmonella typhimurium*, a Gram-negative, food-borne, bacterial pathogen, as an example to demonstrate impedance detection of bacterial cells in suspensions. *Salmonella* cell suspensions in DI water and phosphate buffered saline (PBS) solution were studied over a wide range of frequencies. Bacterial cells suspended in DI water with different cell concentrations were shown to have different electrical impedance spectral responses. In a certain frequency range, impedance of the cell suspension is directly proportional to the cell concentration, which can be used to quantify bacterial cells in a label-free, inexpensive, and simple fashion.

### 9.3.2.1 Methods and device

#### *9.3.2.1.1 Device*

The device for electrical impedance measurements consists of a silica or glass chip patterned with an array of interdigitated microelectrodes (IMEs) and a microchamber (~25 $\mu$L capacity) right above the electrode area formed by silicone rubber, as shown in Figure 9.7(a). The interdigitated microelectrodes were fabricated on a flat silica or glass sub-

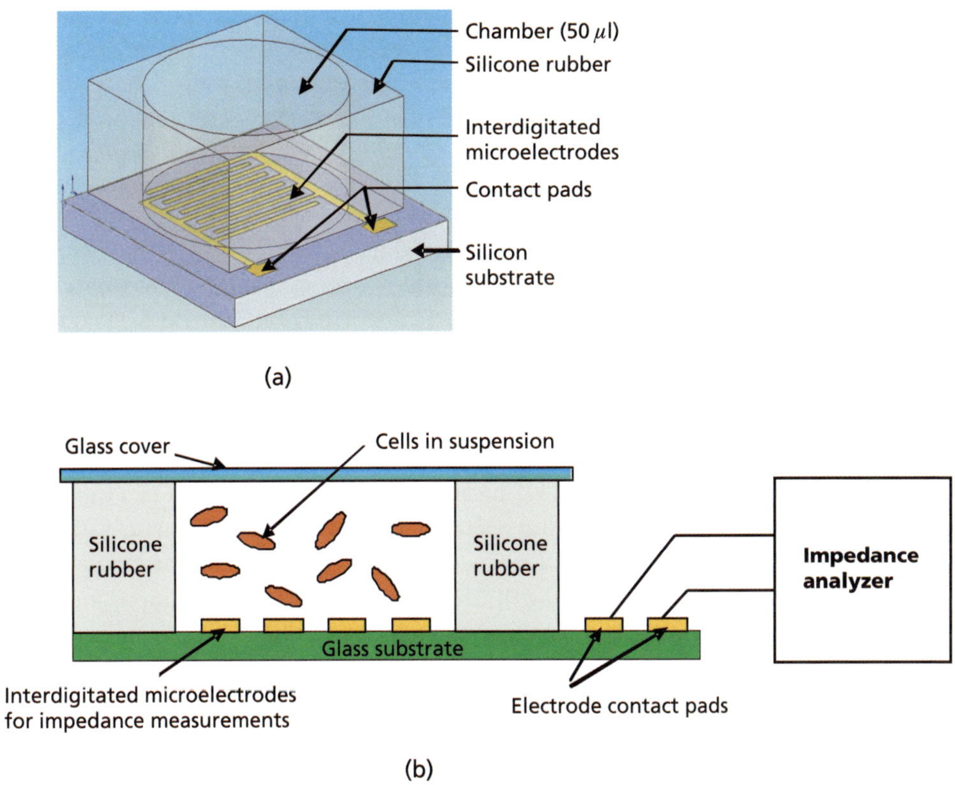


**Figure 9.7** (a) The device for impedance detection of bacterial cells based on ion release from cells into buffers. It consists of a chamber formed by silicone rubber and a set of interdigitated microelectrodes at the floor of the chamber. The chamber capacity is 25 $\mu$L. The gold interdigitated microelectrode consists of 50 pairs of finger electrodes with 15 $\mu$m of digit electrode width and 15 $\mu$m interdigit space. (b) Schematic setup for electrical impedance spectroscopic measurements of impedance of bacterial cell suspensions in DI water or PBS. (Reprinted with permission from *Talanta* and kind permission from [1].)

strate using standard lithographic microfabrication technology. The IME consists of a pair of microband arrays of digit electrodes that mesh with each other. The width of digit electrodes and the interdigit space can be in the range of microns to nanometers, with a total of tens to hundreds of pairs of finger electrodes. The two sets of microelectrodes are used as the two poles in a bipolar impedance-measurement setup. The IMEs used in this experiment have a total of 50 pairs of finger electrodes with each having a width of 15 $\mu$m and a space of 15 $\mu$m. IMEs are also commercially available.

The chamber was made by punching a hole in a piece of silicon rubber using a standard puncher of a desired size. The silicon rubber was then glued to the chip using epoxy with the chamber appropriately aligned with the electrode area.

#### *9.3.2.1.2 Preparation of bacteria cells*

Stock culture of *Salmonella typhimurium* was purchased from Carolina Biological Supply Company (Burlington, North Carolina). The culture was grown in brain-heart-infusion (BHI) broth (TEKnova, Hollister, California) at 37°C for 16 to 18 hours. The cells were centrifuged (Eppendorf, Westbury, New York) at 6,000 × g for 2 minutes. After removal of the supernatant, the cell pellet was resuspended in sterilized DI water or PBS. The cells were washed three times with DI water or PBS in order to get rid of residues from the growth medium. Then, they were serially (1:10) diluted with DI water or PBS to desirable concentrations for further experiments.

The traditional plating method was used to determine the viable cell number in the stock cell suspension prepared in the above step. The cells suspension was serially (1:10) diluted with DI water. Then, 100 $\mu$L of appropriate dilutions were plated onto XLT4 agar plates (Difco, Sparks, Maryland). Colonies were counted after incubation of the plates at 37°C for 24 hours. Generally, the cell numbers in the stock cell suspension averaged about $10^9$ cfu/mL.

#### *9.3.2.1.3 Electrical impedance spectroscopy (EIS)*

Impedance measurements were performed using an IM-6 impedance analyzer (Zahner-Elektrik Gmbh & CoKG, Kronach, Germany) with the IM-6/THALES software. Figure 9.7(b) shows the schematic impedance-measurement setup. For impedance measurements, 20 $\mu$L of each sample was placed into the microchamber and covered with a glass cover. One of the two microband array electrodes was connected to the test and sensing probes, and the other was connected to the reference and counter electrodes on the IM-6 impedance analyzer. EIS measurements were carried out in a frequency range from 1 Hz to 100 kHz. Bode (impedance and phase versus frequency) diagrams were recorded. Impedance at a fixed frequency was measured using the capacitance-potential (C/E) program at 1 kHz with an amplitude of ±50 mV. Impedance data were recorded at every minute. All tests were performed at room temperature.

Simulation was performed using the SIM program. From each measured spectrum, 50 data points were automatically selected by the software as the input, and the fitting curves were generated using an equivalent circuit model.

### 9.3.2.2 Results and discussion

#### *9.3.2.2.1 Impedance spectra of bacterial cell suspensions in DI water and PBS*

Figure 9.8 presents the Bode impedance spectra of *Salmonella typhimurium* cell suspensions in (a) DI water and (b) PBS, along with their equivalent circuits and best-fitting

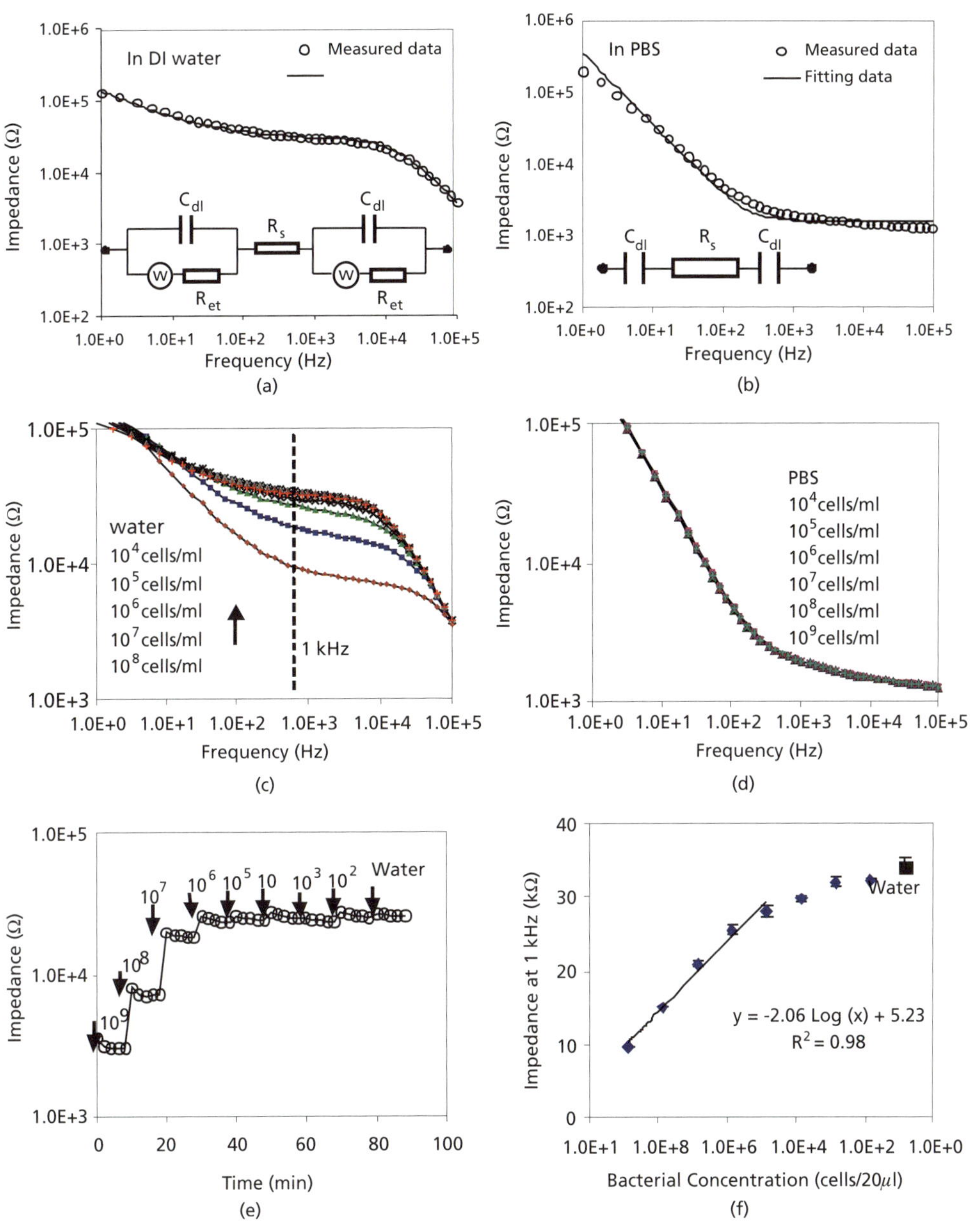


**Figure 9.8** (a, b) Impedance spectra of *Salmonella* cell suspensions in (a) DI water and (b) PBS, together with their fitting curves and the equivalent circuits. *Salmonella* concentration: $1.93 \times 10^6$ cfu/mL. (c, d) Impedance spectra of *Salmonella* suspensions in (c) DI water and (d) PBS, with the cell concentrations in the range of $10^4$ to $10^9$ cfu/mL, along with samples of water and PBS as controls. (e) Typical impedance responses to the samples with different concentrations of cells when they were measured at a fixed frequency of 1 kHz. (f) The linear relationship between the logarithmic value of the concentration of *Salmonella* cells and the impedance measured at 1 kHz. Error bars are standard deviations of three to five measurements. Impedance spectra were measured in the frequency range of 1 Hz to 100 kHz with an amplitude of ±50 mV. (Reprinted with permission from *Talanta* and kind permission from [1].)

spectra. For *Salmonella* cell suspension in DI water, the measured spectrum [Figure 9.8(a), blank dots] is a typical Bode plot for a system in which the polarization is due to a combination of kinetic and diffusion processes. Based on the general electrical-equivalent model of an electrochemical cell [44] and the behavior of the IME microelectrode [45], the measured spectrum can be modeled by an equivalent circuit that consists of the ohmic resistance ($R_s$) of the solution between two electrodes, double-layer capacitance ($C_{dl}$), electron-transfer resistance ($R_{et}$), and Warburg impedance ($Z_w$) around each electrode. The agreement between the measured data and the fitting spectra (solid line) indicated that the equivalent circuits provided a feasible, if not unique, model to describe the impedance characteristics of *Salmonella* suspensions in DI water. Using this circuit model, the simulated values of $C_{dl}$, $Z_w$, $R_{et}$, and $R_s$ were 884.9 pF, 147.2 kΩ/s$^{0.5}$, 13.54 kΩ, and 491.2Ω, respectively, with the mean error of modulus impedance of 0.6%. The spectrum and the circuit model suggest that electrochemical reactions occur on the IME electrodes and that the cells may have released some electrochemical active composites to the DI water.

For *Salmonella* suspensions in PBS, the impedance spectrum [Figure 9.8(b), blank dots] shows two domains: a double-layer region in the low-frequency range from 1 Hz to approximately 500 Hz, and a resistive region in the frequency range from approximately 500 Hz to 100 kHz. The electrical impedance behavior of the cell suspension in PBS can be represented by the equivalent circuit of the IME system in aqueous solutions reported previously [45–47]. In this circuit model, two identical double-layer capacitances ($C_{dl}$) of each set of the IME are connected to the medium resistance ($R_s$) in series. $C_{dl}$ dominates the impedance in the low-frequency range (double-layer region), whereas $R_s$ dominates the impedance in the high-frequency range (resistive region). By simulation, the values of $C_{dl}$ and $R_s$ were 892.8 nF and 1.62 kΩ, respectively, with the mean error of modulus impedance of 3.0%. In the cell suspension in PBS, the impedance spectrum does not show any characteristics related to electrochemical active parameters, which implies cells may not release active electrochemical species into PBS.

Figure 9.8(c, d) shows the Bode impedance spectra of *Salmonella* suspensions in (c) DI water and (d) PBS solution with different cell concentrations from $10^4$ to $10^9$ cfu/mL. It is observed that the suspensions with different cell concentrations in DI water each have a distinct impedance response in the frequency range from 100 Hz to 10 kHz, whereas impedance spectra of *Salmonella* suspensions in PBS were identical in the full frequency range. The results verified that when bacterial cells were suspended in low-conductive DI water, they could alter the conductivity of the suspension.

#### *9.3.2.2.2 Quantifying bacterial concentration in DI water by impedance*

As the solution impedance decreases with the increasing cell concentration in DI water within a certain frequency range, we can estimate the cell concentration in DI water using the impedance value at a fixed frequency. As the best representative frequency, 1 kHz was used to investigate the relationship between impedance values and cell concentrations in DI water suspensions. Figure 9.8(e) shows typical impedance responses at 1 kHz to samples of different bacterial concentrations. When the bacterial concentration decreased from $10^9$ cfu/mL to $10^8$, $10^7$, and $10^6$ cfu/mL, impedance of the suspensions significantly increased from 3.13 ± 0.26 kΩ to 7.29 ± 0.17, 18.7 ± 0.28, and 24.4 ± 0.58 kΩ. Figure 9.8(f) shows the plot of the impedance values as a function of the bacterial concentrations. There is a linear relationship between the impedance and the loga-

**Troubleshooting Table**

| Problem | Explanation | Potential Solution |
|---|---|---|
| There is fluid leakage | The chip is not well assembled | Use a new and well assembled chip. |
| DEP does not capture cells or spores | The connection between the DEP electrodes and the DEP signal generator does not work | Check the connection or change to a new chip. |
| There is no signal in impedance measurements | There is a bad connection between the chip and the impedance instrument | Check the connections. |
| Impedance baseline with deionized water is not consistent when a different vial of water is injected into the device | Deionized water changes its conductance after it exposes to air | Prepare multiple vials of water in prerinsed clean tubes and sterilize the batch together. Discard any water vials that have been left open for more than 30 minutes. |
| Impedance baseline with deionized water from the same vial is not consistent | Impedance is affected by ion release from the polymer material for chip fabrication | Soak the whole chip in deionized water over night, or flush the channels with deionized water for an extended period (hours) until the baseline stabilizes with continuous injection of deionized water. |
| Cells don't lyse or lyse too quickly using the present lysis solution | Different cells have different tolerances for the hypotonic environment | For each cell type, the lysis solution needs to be optimized such that during the solution exchange from isotonic to hypotonic solution, the cells remain intact but will lyse rapidly after the lysis solution flow stops. |
| Impedance does not change after the germinant is injected | Spores germinate before reaching the detection chamber | One-time use of the device will prevent germination due to chemical residues. A separate inlet for germinant delivery is also recommended to avoid pregermination. |

rithmic value of the cell concentration in the range from $10^4$ to $10^8$ cells/20 $\mu$L ($10^6$ to $10^{10}$ cfu/mL). The linear regression equation is $Z$ (k$\Omega$) = $-2.06$ Log $C$ (cells/20 $\mu$L) + 5.23 with $R^2 = 0.98$. The detection limit was calculated to be $6.9 \times 10^4$ cells/20 $\mu$L ($3.45 \times 10^6$ cfu/mL).

In this section, we have demonstrated a new, simple, and rapid method to detect bacterial cells by measuring the impedance properties of their suspensions in DI water using interdigitated microelectrodes. This method does not require any label or amplification steps. It can be used as an alternative approach to quantify bacterial cells in suspensions to impedance microbiology. The detection limit of this method is comparable with many other label-free immunosensors for detection of pathogenic bacteria using different transducer techniques, including QCM immunosensors for detection of *Salmonella* with detection limits of $3.2 \times 10^6$ cfu/mL and $9.9 \times 10^5$ cfu/mL [48, 49], SPR immunosensors for the detection of *Salmonella enteritidis* and *Listeria monocytogens* with detection limits of $10^6$ cfu/mL [50], a SPR sensor for detection of *E. coli* O157:H7 with a detection limit of $10^7$ cfu/mL [51], and an electrochemical impedance immunosensor for the detection of *E. coli* O157:H7 with a detection limit of $10^6$ cfu/mL [10]. To afford this method with selectivity, we have recently implemented magnetic separation prior to the impedance detection [1]. To further improve the detection limit of this approach, a concentration step that enriches the small number of bacterial cells into a microdetection chamber would be very useful.

## 9.4 Conclusion

Advances in microfabrication have paved the way for miniaturization of many traditional detection platforms into microdevices or chips. Impedance sensing as a principal electrical transducer technique is one of the best-suited measurement approaches that can be incorporated into microchips. We have demonstrated four microchip-based systems that have been successfully used for monitoring microbial and cellular activities and for detecting bacterial and mammalian cells. The microscale impedance-based methods have shown advantages in improved sensitivity, reduced quantities of costly reagents, reduced detection time, and flexibility in integration with other electrical methods such as DEP.

## 9.5 Summary Points

- The microfabricated systems described here demonstrate the application of microscale lab-on-a-chip devices for the detection of biological activities with high sensitivity and shortened assay time.
- Microscale "impedance microbiology" can be realized in a microchip format, which can be integrated with an electrical concentration step, through dielectrophoresis (DEP), for rapid detection of bacterial cells based on their metabolic activity.
- The DEP concentration in microchip "impedance microbiology" eliminates the traditional cell-growth-enrichment step and thus can significantly save assay time, which is a great improvement compared with conventional methods that require several days.
- Automatic and rapid electrical detection of the germination of viable spores can be achieved. DEP concentration of a low number of spores in the ultrasmall detection chamber (0.1 nL) within the microfluidic biochip can improve the detection limit down to fewer than 100 spores and significantly reduce the detection time compared with traditional methods.
- Microfluidic chip-based impedance spectroscopy can be designed to detect both mammalian and bacterial cells by monitoring the impedance change in the culture buffer as a result of ion release from the cells. The microdevice helps to confine the ions in a small volume for sensitive impedance measurement.

## Acknowledgments

L. Yang acknowledges the funding from the Golden LEAF Foundation and the state of North Carolina through the Biomanufacturing Research Institute & Technology Enterprise (BRITE) Center for Excellence. X. Cheng and R. Bashir acknowledge support from the NIH/National Institute of Biomedical Imaging and Bioengineering under Grant No. P41 EB002503 (BioMEMS Resource Center, PI: Professor M. Toner). R. Bashir acknowledges research support through the Center for Food Safety Engineering at Purdue University, funded through a cooperative agreement with the Agricultural Research Service of the U.S. Department of Agriculture, project number 1935-42000-035. The authors also thank the staff and facilities of the Birck Nanotechnology Center at Purdue University.

## References


[1] Yang, L., "Electrical impedance spectroscopy for detection of bacterial cells in suspensions using interdigitated microelectrodes," *Talanta*, Vol. 74, 2008, pp. 1621–1629.

[2] Pethig, R., and Markx, G. H., "Application of dielectrophresis in biotechnology," *TIBTECH*, Vol. 15, 1997, pp. 426–432.

[3] Silley, P., and Forsythe, S., "Impedance microbiology—a rapid change for microbiologists," *J. Applied Bacteriology*, Vol. 80, 1996, pp. 233–243.

[4] Wawerla, M., et al. "Impedance microbiology: Applications in food hygiene," *J. Food Protection*, Vol. 62, 1999, pp. 1488–1496.

[5] Gomez, R., Morisette, D. T., and Bashir, R., "Impedance microbiology-on-a-chip: Microfluidic bioprocessor for rapid detection of bacterial metabolism," *J. Microelectromechanical Systems*, Vol. 14, 2005, pp. 829–838.

[6] Liu, Y.-S., et al., "Electrical detection of germination of model *Bacillus anthracis* spores in microfluidic biochips," *Lab on a Chip*, Vol. 7, 2007, pp. 603–610.

[7] Cheng, X., et al., "Cell detection and counting through cell lysate impedance spectroscopy in microfluidic devices," *Lab on a Chip*, Vol. 7, No. 6, 2007, pp. 746–755.

[8] Giaever, I., and Keese, C. R., "Micromotion of mammalian cells measured electrically," *Proc. Natl. Acad. Sci. USA*, Vol. 88, 1991, pp. 7896–7900.

[9] Giaever, I., and Keese, C. R., "A morphological biosensor for mammalian cells," *Nature*, Vol. 366, 1993, pp. 591–592.

[10] Yang, L., Li, Y., and Erf, G. F., "Interdigitated array microelectrode-based electrochemical impedance immunosensor for detection of *Escherichia coli* O157:H7," *Analytical Chemistry*, Vol. 76, 2004, pp. 1107–1113.

[11] Gibson, D. M., Coombs, P., and Pimbley, D. W., "Automated conductance method for the detection of salmonella in foods—collaborative study," *J. AOAC International*, Vol. 75, 1992, pp. 293–302.

[12] AOAC, "Salmonella in food, automated conductance methods: AOAC official method 991.38," *Official Methods of Analysis of AOAC International*, 16th ed., Gaithersburg, MD: Association of Official Analytical Chemists International, 1996.

[13] Owicki, J., and Parce, J., "Biosensors based on the energy metabolism of living cells: The physical chemistry and cell biology of extracellular acidification," *Biosensors and Bioelectronics*, Vol. 7, 1992, pp. 257–272.

[14] Gomez, R., Bashir, R., and Bhunia, A., "Microscale electronic detection of bacterial metabolism," *Sens. Actuators B: Chemical*, Vol. 86, 2002, pp. 198–208.

[15] Gomez, R., et al., "Microfluidic biochip for impedance spectroscopy of biological species," *Biomedical Microdevices*, Vol. 3, 2001, pp. 201–209.

[16] Pohl, H. A., *Dielectrophoresis*, Cambridge, U.K.: Cambridge University Press, 1978.

[17] Jernigan, J. A., et al., "Bioterrorism-related inhalation anthrax: The first 10 cases reported in the United States," *Emerging Infectious Diseases*, Vol. 7, No. 6, 2001, pp. 933–944.

[18] Straiger, P., and Losick, R., "Molecular genetics of sporulation in bacillus subtilis," *Annual Review of Genetics*, Vol. 30, 1996, pp. 297–341.

[19] Nicholson, W. L., et al., "Resistance of bacillus endospores to extreme terrestrial and extraterrestrial environments," *Microbiology and Molecular Biology Reviews*, Vol. 64, 2000, pp. 548–572.

[20] Hartley, H. A., and Baeumner, A. J., "Biosensor for the specific detection of a single viable *B. anthracis* spore," *Analytical and Bioanalytical Chemistry*, Vol. 376, No. 3, 2003, pp. 319–327.

[21] Ryu, C., et al., "Sensitive and rapid quantitative detection of anthrax spores isolated from soil samples by real-time PCR," *Microbiology and Immunology*, Vol. 47, 2003, pp. 693–699.

[22] Lee, J., and Deininger, R. A., "A rapid screening method for the detection of viable spores in powder using bioluminescence," *Luminescence*, Vol. 19, 2004, pp. 209–211.

[23] Welkos, S. L., et al., "A microtiter fluorometric assay to detect the germination of *Bacillus anthracis* spores and the germination inhibitory effects of antibodies," *J. Microbiological Methods*, Vol. 56, 2004, pp. 253–265.

[24] Hamouda, T., Shih, A. Y., and Baker, J. R., Jr., "A rapid staining technique for the detection of the initiation of germination of bacterial spores," *Letters in Applied Microbiology*, Vol. 34, No. 2, 2002, pp. 86–90.

[25] Kiel, J. L., et al., "Growth medium for the rapid isolation and identification of anthrax," *Proc. SPIE*, Vol. 4036, 2000, pp. 92–102.

[26] Hindson, B., et al., "APDS: The autonomous pathogen detection system," *Biosensors and Bioelectronics*, Vol. 20, 2005, pp. 1925–1931.

[27] Titball, R. W., and Manchee, R. J., "Factors affecting the germination of spores of *Bacillus anthracis*," *J. Applied Bacteriology*, Vol. 62, 1987, pp. 269–273.

[28] Chang, W. J., et al., "Hybrid poly(dimethylsiloxane) (PDMS)/silicon biochips for bacterial culture applications," *Biomedical Microdevices*, Vol. 5, 2003, pp. 281–290.

[29] Agirregabiria, M., et al., "Fabrication of SU-8 multilayer microstructures based on successive CMOS compatible adhesive bonding and releasing steps," *Lab on a Chip*, Vol. 5, 2005, pp. 545–552.

[30] Baek, J. Y., et al., "A pneumatically controllable flexible and polymeric microfluidic valve fabricated via in situ development," *J. Micromech. Microeng.*, Vol. 15, 2005, pp. 1015–1020.

[31] Yang, L., et al., "A multifunctional microfluidic system for dielectrophoretic concentration coupled with immuno-capture of low number of *Listeria* monocytogenes," *Lab on a Chip*, Vol. 6, 2006, pp. 896–905.

[32] Gawad, S., Schild, L., and Renaud, P., "Micromachined impedance spectroscopy flow cytometer for cell analysis and particle sizing," *Lab on a Chip*, Vol. 1, 2001, pp. 76–82.

[33] Kruger, J., et al., "Development of a microfluidic device for fluorescence activated cell sorting," *J. Micromech. Microeng.*, Vol. 12, No. 4, 2002, pp. 486–494.

[34] Huh, D., et al., "Use of air-liquid two-phase flow in hydrophobic microfluidic channels for disposable flow cytometers," *Biomedical Microdevices*, Vol. 4, No. 2, 2002, pp. 141–149.

[35] Koch, M., Evans, A. G. R., and Brunnschweiler, A., "Design and fabrication of a micromachined coulter counter," *J. Micromech. Microeng.*, Vol. 9, No. 2, 1999, pp. 159–161.

[36] Lundien, M. C., et al., "Induction of MCP-1 expression in airway epithelial cells: Role of CCR2 receptor in airway epithelial injury," *J. Clin. Immunol.*, Vol. 22, No. 3, 2002, pp. 144–152.

[37] Giaever, I., and Keese, C. R., "Monitoring fibroblast behavior in tissue-culture with an applied electric-field," *Proc. Natl. Acad. Sci. USA*, Vol. 81, No. 12, 1984, pp. 3761–3764.

[38] Ehret, R., et al., "Monitoring of cellular behaviour by impedance measurements on interdigitated electrode structures," *Biosensors and Bioelectronics*, Vol. 12, No. 1, 1997, pp. 29–41.

[39] Van Der Wal, A., et al., "Conductivity and dielectrical dispersion of gram-positive bacterial cells," *J. Colloid Interface Sciences*, Vol. 186, 1997, pp. 71–79.

[40] Wilson, W. W., et al., "Status of methods for assessing bacterial cell surface charge properties based on zeta potential measurements," *J. Microbiological Methods*, Vol. 43, 2001, pp. 153–164.

[41] Carstensen, E. L., and Marquis, R. E., "Passive electrical properties of microorganisms. III. Conductivity of isolated bacterial cell walls," *Biophys. J.*, Vol. 8, 1968, pp. 536–548.

[42] Mozes, N., and Rouxhet, P. G., "Microbial hydrophobicity and fermentation technology," in R. J. Doyle and M. Rosenberg, (eds.), *Microbial Cell Surface Hydrophobicity*, Washington, D.C.: American Society for Microbiology, 1990, pp. 75–105.

[43] Baldwin, W. W., et al., "Changes in buoyant density and cell size of *Escherichia coli* in response to osmotic shocks," *J. Bacteriology*, Vol. 170, No 1, 1988, pp. 452–455.

[44] Bard, A. J., and Faulkner, L. R., *Electrochemical Methods: Fundamentals and Applications*, New York: John Wiley & Sons, 2001.

[45] Van Gerwen, P., et al., "Nanoscaled interdigitated electrode arrays for biochemical sensors," *Sens. Actuators B: Chemical,* Vol. 49, 1998, pp. 73–80.

[46] Laureyn, W., et al., "Nanoscaled interdigitated titanium electrodes for impedimetric biosensing," *Sens. Actuators B: Chemical*, Vol. 68, 2000, pp. 360–370.

[47] Yang, L., et al., "Interdigitated microelectrode (IME) impedance sensor for the detection of viable salmonella typhimurium," *Biosensors and Bioelectronics*, Vol. 19, 2004, pp. 1139–1147.

[48] Park, I., and Kim, N., "Thiolated salmonella antibody immobilization onto the gold surface of piezoelectric quartz crystal," *Biosensors and Bioelectronics*, Vol. 13, 1998, pp. 1091–1097.

[49] Park, I., Kim, W., and Kim, N., "Operational characteristics of an antibody-immobilized QCM system detecting salmonella spp," *Biosensors and Bioelectronics*, Vol. 15, 2000, pp. 167–172.

[50] Koubova, V., et al., "Detection of foodborne pathogens using surface plasmon resonance biosensors," *Sens. Actuators B: Chemical*, Vol. 74, 2001, pp. 100–105.

[51] Fratamico, P. M., et al., "Detection of *E. coli* O157:H7 using a surface-plasmon resonance biosensor," *Biotechnology Techniques*, Vol. 12, 1998, pp. 571–576.

CHAPTER

# 10

# Controlling the Cellular Microenvironment

George Eng,[1] Milica Radisic,[2] and Gordana Vunjak-Novakovic[1]

[1]Department of Biomedical Engineering, Columbia University, New York, NY
[2]Institute of Biomaterials and Biomedical Engineering, Department of Chemical Engineering and Applied Chemistry, University of Toronto, Toronto, Ontario

## Abstract

The methods described here were developed to enable studies of cell behavior, interaction, and functional assembly on a small scale under conditions that are both biologically relevant and highly controllable. Our focus is on systems designed to study cells surrounded by other cells and tissue matrix, subjected to both the molecular and physical signals. Three different technologies are included that enable microenvironmental control of cells: (1) surface patterning for spatially defined cell coculture, (2) microfluidic patterning for the formation of tissue organoids, and (3) microarray systems for 3-D patterning of embryonic stem cells. An additional technology is presented for inducing cell alignment and elongation by combined effects of substrate topology and electrical-field stimulation. The application of each method is illustrated by experimental data obtained in our previous studies and discussed with respect to the scope of its application and possible extensions to other cell types and biological questions.

Key terms
bioreactor
electrical stimulation
hydrogel, microarrays
microfluidics
microprinting
stem cells
substrate
topography

## 10.1 Introduction

During development, tissues emerge from coordinated sequences of cell proliferation, differentiation, and functional assembly. These sequences are orchestrated by spatial and temporal gradients of regulatory factors that originate from the surrounding cells, extracellular matrix, and external environment. Likewise, stem-progenitor cells participating in the repair and regeneration of adult tissues respond to the entire milieu of the specific state of injury or disease. In native environments, cells are surrounded by other cells and matrix and are subjected to cascades of molecular and physical regulatory factors. In contrast, cells cultured in conventional culture dishes are deprived of their native surroundings, maintained by periodic medium change, and are not subjected to physical factors that are critically important to most cells. Clearly, cell behavior under these conditions will differ from that seen in biologically relevant settings and will not be predictive of the situations in vivo. The utility of data obtained under such culture conditions for any application of interest—from basic research of stem cell biology to translation into animal models—will be rather limited.

With rapid advances in stem cell biology and the need to translate scientific advances into medical applications, there is great interest in the development and utilization of "biomimetic" culture environments. The premise is that to unlock the full potential of stem cells, one would need to study the cells in dynamic culture environments resembling those that the cells encounter in vivo. At the same time, the multitude of factors of interest (including the types, levels, and combinations of regulatory signals; multiple sources and phenotypes of the cells; ranges of pathological conditions to be treated) drives the development of small-scale, modular, microarray-type culture systems that allow high-throughput investigation. Recent advances in tissue engineering may in fact change the way we conduct cell and tissue experiments by providing *culture systems that combine the controllability of in vitro technologies with the biological fidelity normally seen in whole animals.*

In the general context of designing biologically relevant, yet controllable, culture environments, we describe here two cell culture platforms:

1. Microenvironmental control of cell-cell interactions with three related sets of methods:
    i. Surface patterning for cell coculture
    ii. Microfluidic systems for cardiac organoid formation
    iii. 3-D patterning of embryonic stem cells
2. Interactive use of substrate topography and electrical stimulation for the control of cell alignment

These highly engineered, small-scale platforms enable the interaction of stem cell biology and tissue engineering at multiple levels for a variety of cell types and applications—from regenerative medicine to the screening of environmental factors and studies of disease. In general, these systems provide contexts for cell cultivation somewhere between a standard well plate and an animal model. We present the methods used to assemble and operate each of these culture platforms, summarize representative experimental data, and discuss further extensions to research studies and applications that may be of interest.

# 10.2 Microenvironmental Control of Cell-Cell Interactions

## 10.2.1 Surface patterning for cell coculture

One of the simplest methods to study the interactions between two different types of cells involves binary surface patterning [1]. First, a cell-repulsive material is patterned on a substrate. Type 1 cells are seeded, resulting in cell attachment to the bioadhesive, but not to the biorepulsive, surface. The bioresistant surface is transformed to become bioadhesive through the application of a soluble agent that is not toxic to the already attached cells. Then, Type 2 cells are seeded to attach to free bioadhesive surfaces but not the surfaces already occupied by Type 1 cells.

This method allows for two different cell types to be patterned in any 2-D configuration relative to each other, with any size or shape of the individual field occupied by one or the other cell type and with resolutions of cell patterning close to that of the size of an individual cell. An additional advantage is that such patterns can be generated using biocompatible materials with minimal exposure to chemicals. This can be an important tool for studying cell-cell interactions and "community" cell behaviors, with or without the application of additional molecular regulatory factors.

### 10.2.1.1 Materials

| Equipment | Reagents | Disposables |
|---|---|---|
| Clean room | Polydimethylsiloxane (Slygard 184, Essex Chemical) | Petri dish (10 cm) |
| Spin coater (Laurell, WS-400B) | Hyaluronic acid (Sigma, H5388) | Pipette tips |
| Plasma cleaner (Harrick Plasma PDC-002) | Fibronectin (Sigma, F1141) | Tongue depressor (VWR, 57100-252) |
| Silicon master template (University Wafer) | Type I collagen (BD 356236) | Glass slides |
| Dessicator chamber (Nalgene, 5310-0250) | Curing agent (Slygard 184, Essex Chemical) | Mouse embryonic stem cells (R1 line ATCC, SCRC-1036) |
| Curved forceps (VWR, 82027-392) | PBS (Invitrogen) | AML12 murine hepatocytes (ATCC, CRL-2254) |
| Biological safety cabinet | Alconox detergent (VWR, 21835-032) | NIH-3T3 fibroblasts (ATCC, CRL-1658) |
| Automated pipette man (Drummond DP-200) | | |
| Fluorescent microscope | | |

### 10.2.1.2 Methods

The schematic in Figure 10.1 outlines the complete patterning process with three key steps: fabrication of the PDMS mold, patterning of the glass surface, and seeding of two different cell populations.

#### *10.2.1.2.1 Preparation of the PDMS mold*

1. Weigh 30g of silicone elastomer into a disposable container. Then, add 3g of curing agent.
2. Use a tongue depressor or other disposable mixing tool to thoroughly mix the PDMS, taking care to scrape the sides and bottom of the container. Mix for 3 minutes.
   *Note:* Incomplete mixing will result in semiliquid PDMS, which will ruin any silicon master.
3. Place the mixed PDMS into a dessicator chamber and expose to vacuum. Wait 2 minutes until you see the bubbles rise to the surface of the PDMS. Then, slowly open to room atmosphere. Repeat these steps if there are still residual bubbles.

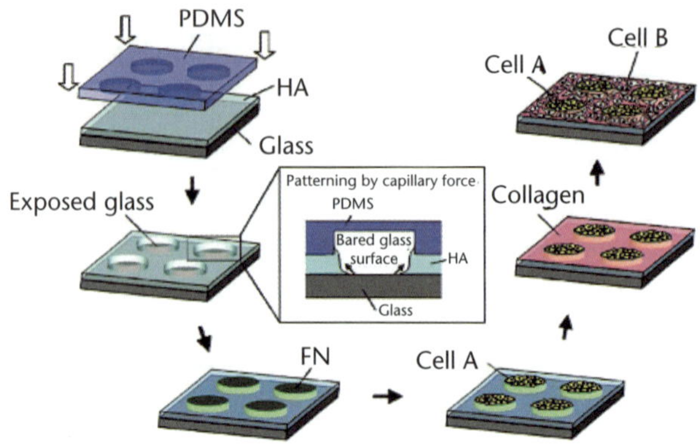


**Figure 10.1** Surface patterning for cell coculture by using capillary force lithography and layer-by-layer deposition. HA solution was applied to a glass slide by spin coating, and a PDMS mold was immediately placed on the layer of HA. In regions of exposed glass, HA receded, and these regions were coated with FN, where primary cells (Type 1 cells) could be selectively adhered. Subsequently, the HA surface was complexed with collagen, allowing for the adhesion of secondary cells (Type 2 cells). (Reproduced with permission from [1], Figure 1.)

*Note:* Degassing is important for the formation of PDMS from the silicon master with proper features.

4. Place silicon master into a petri dish, and position it at about 1 to 2 cm from the outside edge. If there are multiple pieces, try to arrange them with a 1 to 2 cm distance between pieces.

   *Note:* This is to ensure easier removal of the cured PDMS mold without breaking the fragile silicon masters.
5. Bake at 60°C for 2 hours.
6. Cut 1 cm around the silicon master using a scalpel. Gently peel the PDMS from the petri dish, pulling slowly, using a curved forceps.
7. Carefully separate the PDMS from the silicon master, applying even pressure. Any sudden increase in pressure may cause breakage of the silicon.

#### *10.2.1.2.2 Hyaluronic acid patterning*

1. Place a glass slide inside the plasma cleaner.
2. Close the valve and turn on the vacuum until it decreases to 10 Torr. Then, turn on the cleaner power, and open the valve slowly while keeping the pressure around 10 Torr.

   *Caution:* Without additional attachments, the plasma cleaner valve can be quite sensitive.
3. Once a pink glow is established, let the glass slide be cleaned for 10 seconds.
4. Remove the glass slide from the plasma cleaner using a forceps.

   *Caution:* Be careful to touch only the edges of the slide.

   *Caution:* Once plasma-cleaned, the glass slide must be used as soon as possible (ideally within 1 hour).
5. Add 1 mL of HA solution (5 mg/mL in water) to the center of the glass slide, place in the spin coater, and run at 1,500 rpm for 10 seconds.
6. Clean the PDMS mold with soap and water (Alconox powder detergent 1% in DI $H_2O$), followed by a rinse with ethanol to remove any dust or PDMS debris. Let dry.

7. Hold PDMS mold features down, touching only the sides, and gently place on the thin HA film.

   *Caution:* Be careful to not slide the mold over the surface.
8. Let the mold sit undisturbed for 12 hours.
9. Peel the PDMS away from the slide by lifting it at one edge. Try to pull evenly until released and visualize the pattern under a microscope.

   *Caution:* It is important to pull slowly and to fully remove the PDMS stamp because any PDMS that gets permanently attached will ruin the patterns.

##### *10.2.1.2.3 Cell seeding*

1. Sterilize HA patterned slides with overnight exposure to UV light within a biosafety cabinet.
2. Pipette 1 mL of fibronectin (100 $\mu$g/mL in PBS) onto each slide to cover the entire surface evenly for 20 minutes. Repipette up and down to make sure the surface maintains even fibronectin coating.
3. Wash with PBS, gently pipetting 10 mL on top of the glass slide three times.
4. Add Type 1 cells (10 mL at a concentration of $1 \times 10^6$ cells/mL in culture medium); allow cells to adhere for 8 hours.
5. Check under a microscope for morphological changes in the cells.

   *Note:* The cells should appear attached and partially elongated, depending on the cell type.
6. Aspirate culture media and wash with fresh media (10 mL, two times), taking care to not wash away attached cells.
7. Pipette collagen on the slide (1 mL at a concentration of 500 $\mu$g/mL), and incubate for 20 minutes.
8. Aspirate and replace with Type 2 cells (10 mL at a concentration of at $1 \times 10^6$ cells/mL).
9. Allow the cells to adhere for 8 hours, check morphology, and wash excess cells as in step 3 of this section.

#### 10.2.1.3 Data acquisition, anticipated results, and interpretation

This technique provides the ability to attach two different cell populations to any 2-D pattern. If the cell types are morphologically different, the two populations can be distinguished. If this is not the case, cell-membrane dyes can be used to track the two cell types and monitor the changes in cell patterning. In the example shown in Figure 10.2, mouse embryonic stem cells and AML12 murine hepatocytes were labeled with a green fluorescent dye (CFSE), and 3T3 fibroblasts were labeled red (PKH26). The right panel displays patterned cell culture using the HA/collagen patterning system. Both ES cells [Figure 10.2(a)] and AML12 murine hepatocytes [Figure 10.2(c)] adhered to the HA bare regions where fibronectin was absorbed. The surrounding cells were 3T3 fibroblasts. Morphologically, the cells in Figure 10.2(a) are distinct, and their identity can easily be determined. However the AML12 fibroblast coculture demonstrates fewer morphological differences, and the two cell-membrane dyes were necessary to facilitate pattern recognition. In both cases, the patterns were reproducible over a large area and remained stable for 1 week [1].

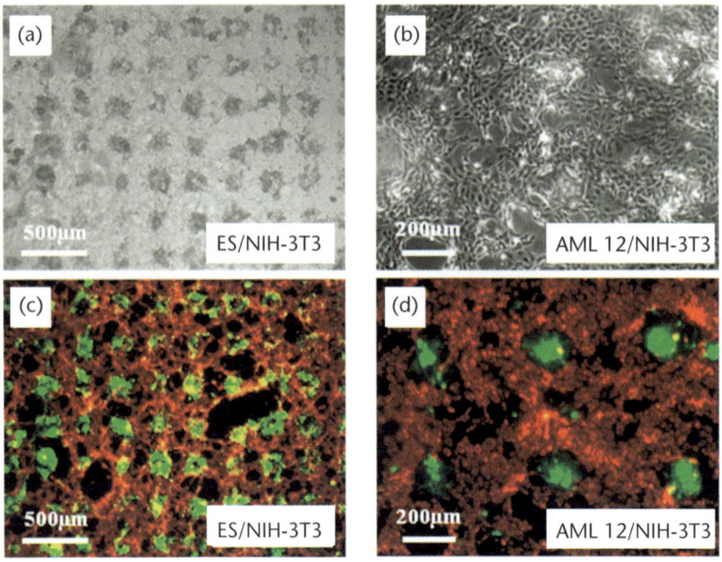


**Figure 10.2** Patterned cell culture and patterned coculture on HA/collagen surface. (a) After 3 days of culture, ES cells formed dense spherical aggregates and were clearly distinct from the surrounding fibroblasts monolayer. (b) In light microscopy, the cocultured AML12 hepatocytes and NIH-3T3 fibroblasts were difficult to distinguish. (c, d) Fluorescently stained primary cells (green) and secondary cells (red) were easily visualized for ES/NIH-3T3 and AML 12/NIH-3T3 cocultures at 3 days of culture. (Reproduced with permission from [1], Figure 7.)

#### 10.2.1.4 Discussion and commentary

The simple method for surface patterning of the cells described above allows studies of cell-cell interactions under well-defined conditions. The patterns can be designed to replicate some aspects of the tissue architecture in a way suitable for high-throughput cell culture experiments. Many previous studies of cellular cocultures utilized a random seeding method, where the proportions of cell types may be varied, but the specific geometry and orientation of the cells was not tightly controlled. Precise positioning of the cells is indeed necessary to study exactly how cells interact with each other. Common variables that can be changed include the numbers of cells of each type "printed" within the topographical features on the substrate, the distances between the cell types, and specific geometries unique for specialized tissues.

### 10.2.2 Microfluidic systems for cardiac organoid formation

The tissue architecture of the heart is highly organized at many hierarchical levels. Cardiomyocytes are arranged in a linear manner with extensive cell-cell communications through gap junctions. To promote the elongation, alignment, and coupling of cardiac myocytes, a microfluidic-based patterning technique was used. Native extracellular matrix material, hyaluronic acid, was used as a template polysaccharide to generate cardiac organoids composed of multiple cardiomyocytes functioning synchronously. After 3 days of culture, the cells started to detach from the substrate and form contractile strands. This method represents a general technique to generate small, but functional,

**Troubleshooting Table**

| Problem | Explanation | Possible Solution |
|---|---|---|
| Indistinct HA pattern formation | Possibly too little capillary force between the solution and PDMS | Increase plasma cleaner time |
| Indistinct cell pattern formation | Cell density too low | Increase cell concentration in solution |
| PDMS pieces attached to glass slide | Cross-linking of PDMS to glass | Allow less time for drying |

cardiac tissue organoids for potential application in biologic studies, in drug and toxicology screening, and as diagnostic models of normal and pathological cardiac function.

#### 10.2.2.1 Materials

| Equipment | Reagents | Disposables |
|---|---|---|
| Spin coater (Laurell, WS-400B) | Polydimethylsiloxane (Slygard 184, Essex Chemical) | Pipette tips |
| Plasma cleaner (Harrick Plasma PDC-002) | Hyaluronic acid (Sigma, H5388) | Scalpel blades |
| Video microscope (Nikon) | Fibronectin (Sigma, F1141) | Petri dishes (10 cm) |
| Clean room facility | High glucose DMEM (Invitrogen, catalog no. 11965) | Glass slides |
| Pipetteman | Heat-inactivated fetal bovine serum (FBS) (Gibco, catalog no. 10082) | Silicon master template (University Wafer) |
| Forceps (curved ends) | Penicillin (Gibco) | |
| Incubator | 95% ethanol (Pharmco, catalog no. 111000190) | |
| | Hank's Balanced Salt Solution (Invitrogen) | |
| | HEPES (Invitrogen) | |
| | FBS (Invitrogen) | |
| | L-glutamine (Invitrogen) | |
| | Trypsin (Invitrogen) | |
| | Alconox detergent (VWR, 21835-032) | |

#### 10.2.2.2 Methods

The preparation of cardiac organoids proceeds in four steps: (1) microfluidic patterning to fabricate the device, (2) isolation of cardiomyocytes, (3) seeding of cardiomyocytes into microfluidic channels, and (4) cultivation of seeded cells until they detach from the substrate and form organoids.

##### *10.2.2.2.1 Microfluidic patterning*

1. Prepare a PDMS mold using the methods described in Section 10.2.1.2.1.
2. Clean glass slides with ethanol and DI water.

   *Note:* Using a new package of slides that have not been manipulated will provide a better result.
3. Clean the prepared PDMS molds by washing with soap and water (Alconox powder detergent 1% in DI $H_2O$).
4. Rinse with ethanol (95% 5 mL, three times).
5. Place the mold with features up into the plasma cleaner.
6. Close the valve and turn on the vacuum until it reaches 10 Torr.
7. Turn on the cleaner power, and open the valve slightly, keeping the pressure around 10 Torr.

   *Note:* Without additional attachments, the valve can be quite sensitive.

8. Once a pink glow is established, let clean for 60 seconds.
9. Remove the glass slides from the plasma cleaner using a curved forceps. Be careful to touch only the edges.

   *Caution:* Once plasma-cleaned, the samples must be used fresh.
10. Allow the slides to sit for 5 minutes in a biological hood.
11. Place the PDMS mold features down onto the glass slide. Watch to ensure proper adhesion of the PDMS to the glass slide.

    *Note:* Proper sealing is visualized by a slight darkening of the interface between the two substrates.
12. Pipette 200 $\mu$L of HA solution, or as much as is needed depending on the pattern size, onto one side of the PDMS mold, allowing the capillary action to pull the solution through.

    *Note:* It is very important that the solution is applied to only one side; if applied to both, the air will become entrapped, and there will be no movement of the solution.
13. Let the slide dry undisturbed for 12 hours at room temperature in the biological safety cabinet.
14. Remove the PDMS pattern by carefully peeling from one side, lifting slowly.

    *Caution:* Do not slide the pattern.

#### *10.2.2.2.2 Cardiomyocyte isolation*

1. Isolate cardiomyocytes according to an institutionally approved IACUC protocol and under supervision of the committee for animal care.
2. Remove hearts from neonatal rats.

   *Note:* Take care to avoid the peritoneum and lung tissue.
3. Quarter the ventricles and incubate overnight at 4°C in a 0.06% w/v solution of trypsin in Hank's Balanced Salt Solution (HBSS).
4. Digest with collagenase II in HBSS at 0.1% w/v for 3 minutes at 37°C. Do this multiple times (four to six), and centrifuge the digestions at 750 rpm for 4 minutes.
5. Pool in DMEM with 4.5 g/L glucose with 10% FBS, 10 mM HEPES and 2 mM L-glutamine, and 100 U/mL penicillin.
6. Preplate the cells for 1 hour to enrich for cardiomyocytes. Use the suspension of nonattached cells.

   *Note:* Please see [6] for more details about the cell-isolation procedure.

#### *10.2.2.2.3 Cardiomyocyte seeding*

1. Make 5 mL of 5 mg/mL solution of hyaluronic acid in DI water. Weigh out the appropriate amount of HA, and add to a 15 mL conical tube, avoiding the edges of the tube if possible.

   *Note:* HA is slow to dissolve, so leave on an orbital shaker overnight.
2. Sterilize HA solution with UV light treatment in a biological safety hood overnight.
3. Place the glass slides with HA patterns into a petri dish and rinse for 2 minutes with PBS. Sterilize with UV light for 10 minutes.
4. Add isolated cardiomyocytes at $1.5 \times 10^6$ cells in 2 mL of culture media.
5 Replace the media 5 hours later and every 2 days subsequently.

   *Note:* Be careful to not disturb the cells during the attachment process.

#### 10.2.2.3 Data acquisition, anticipated results, and interpretation

The lanes formed by micropatterning provided the surface guidance for the cells to align and subsequently couple for contractile function. Cardiomyocytes preferentially attached to fibronectin-coated regions immediately adjacent to the HA lanes [Figure 10.3(a)]. By 24 hours of culture, cells that attached to the adhesive regions had elongated. Individual contractile cells were apparent at day 2. Due to the force exerted by the cardiomyocytes, the patterns began to condense and formed linear aggregates. At day 4, the organoids pulled themselves from the substrate. By day 6 of culture, there was a clear progression from single adherent cells to organoids with multiple cardiomyocytes [Figure 10.3(b)].

Video image analysis using ImageJ software (available at http://rsb.info.nih.gov/ij) provided quantification of the contractile function of the cardiomyocytes. The video was recorded at 25 frames per second and exported into a set of image files. For each beating unit, a landmark was made in the initial image, and the change in distance from the initial landmark to subsequent images provided the measure of contractile shortening and lengthening. The change in contractility followed previously reported results and had a unique profile. Contractile function developed in culture, and the amplitude of contractility at day 6 was sevenfold greater than at day 3 [2].

The organoids were composed of multiple cells expressing Troponin I, a marker commonly used to validate the identity of the contracting cells as cardiomyocytes. As seen in Figure 10.3(b), inset, these cells have well-developed contractile apparatus. A key property of this microfluidic platform is the ability to prompt the cells to couple with each other and beat synchronously and to form functional units by day 4 in culture [2].

(a)

(b)

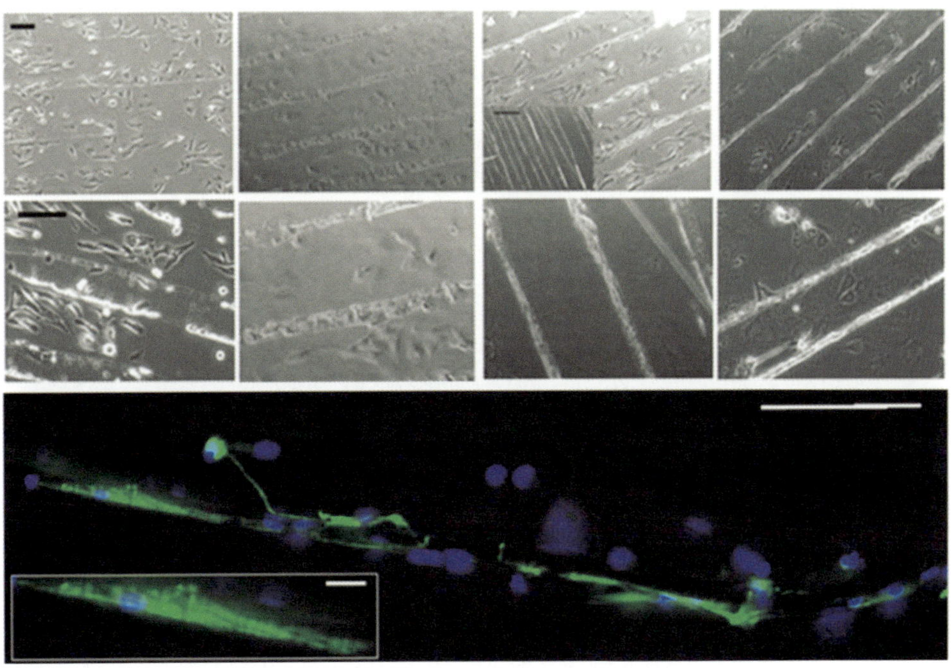

**Figure 10.3** Formation of cardiac organoids. (a) Progression of cardiac organoid formation on HA patterned surfaces. Day 4 inset image taken shows several millimeter-long cardiac organoids. Scale bars: 100 $\mu$m. (b) Expression of cardiac Troponin I. Scale bar =100 $\mu$m; inset scale bar = 15 $\mu$m. (Reproduced with permission from [2].) (See Figures 3 and 6 in [2]).

#### 10.2.2.4 Discussion and commentary

The ability to control cell alignment, orientation, and coupling on a microscale enables better models to grow tissue constructs and use them as in vitro models of cardiac function. This technique utilizes biocompatible materials to organize cardiomyocytes into functional units that can be used for drug screening, diagnostic purposes, cell studies, and disease models. Harvesting the functional units does not require additional steps or the use of enzymes or chemicals. The patterning process is simple and robust, utilizing the capillary action through microfluidic channels. Importantly, tissue organoids are large enough to allow functional studies and small enough to allow high-throughput studies.

### 10.2.3 3-D patterning of embryonic stem cells

While surface patterning of different cell types enables studies of cell-cell interactions, it is limited to the cultivation on a flat substrate, whereas cells normally reside in a 3-D microenvironment. The method described here utilizes the techniques of soft lithography to generate microarray systems suitable to study cell-cell interactions in three dimensions. This is a relatively new and exciting area for microtechnology that allows the researcher to better mimic the spatial configurations cells encounter in their native in vivo environments by creating cell-hydrogel constructs.

The method still utilizes PDMS and photolithography, but the paradigm is different, and the PDMS mold has relatively deep features (≈50 $\mu$m). Many groups have used poly(etheylene glycol) methacrylate, a bioresistant, biocompatible material that readily forms stable hydrogels by photopolymerization. The process involves the deposition of a small volume of hydrogel solution onto a substrate, followed by careful placement of the PDMS mold, which shapes the liquid into the desired form. The entire construct is then exposed to UV light, which passes through the PDMS and crosslinks the solution into a hydrogel. The PDMS is removed, and the microstructures consisting of cells in hydrogel are left on the substrate. Cells can be either confined within the hydrogel or patterned using the hydrogel as a negative mold. We describe here a method for cultivating "embryoid bodies" of controllable size.

#### 10.2.3.1 Materials

| Equipment | Reagents | Disposables |
|---|---|---|
| Clean room facility | Polydimethylsiloxane (PDMS) (Slygard 184, Essex Chemical) | Glass slides (Fisher Scientific) |
| Chemical hood | | |
| Spin coater (Laurell, WS-400B) | Fibronectin (Sigma, F1141) | Petri dish |
| Plasma cleaner (Harrick Plasma PDC-002) | Trimethylsilyl methacrylate (Sigma, 347493) | 10 mL pipette |
| UV light (EFOS Ultracure, UV spot lamp) | Poly(ethylene glycol) diacrylate (Sigma, 437441) | |
| Silicon master template (University Wafer) | 2-hydroxy-2methylpropiophenone (Sigma, 405655) | |
| Flow cytometry (BD FACSCalibur) | $H_2O_2$ (Sigma, H1009) | |
| Confocal microscopy | $H_2SO_4$ (Sigma, 320501) | |
| 200 $\mu$L pipetteman (Eppendorf 022472054) | 3-(trichlorosilyl)propyl methacrylate (TPM) | |
| | Di$H_2O$ | |

#### 10.2.3.2 Methods

##### *10.2.3.2.1 Methacrylation of glass slides*

1. Prepare the PDMS mold as described in Section 10.2.1.2.1.
2. Place a glass slide inside the plasma cleaner.
3. Close the valve and turn on the vacuum until it decreases to 10 Torr. Then, turn on the cleaner power, and open the valve slowly while keeping the pressure around 10 Torr.
   *Caution:* Without additional attachments, the plasma cleaner valve can be quite sensitive.
4. Once a pink glow is established, let the glass slide be cleaned for 2 minutes.
5. Remove the glass slide from the plasma cleaner using a forceps.
   *Caution:* Be careful to touch only the edges of the slide.
   *Caution:* Once plasma-cleaned, the glass slide must be used as soon as possible (ideally within 1 hour).
6. In a chemical hood, place slides into a solution of 30% $H_2O_2$ and $H_2SO_4$ (3:1 ratio) or piranha wash for 5 minutes, then wash in $DiH_2O$.
   *Caution:* Be very careful when mixing $H_2O_2$ and $H_2SO_4$, slowly adding the acid while constantly stirring. Adding $H_2SO_4$ too quickly is dangerous and may cause the solution to boil and splash.
7. Place washed slides in 1 mM solution of 3-(trichlorosilyl)propyl methacrylate (TPM) for 5 minutes to add methacrylate groups to the glass surface. Wash with heptanes/carbon tetrachloride (80/20 v/v), then with another $DiH_2O$ rinse.

##### *10.2.3.2.2 Formation of PEG microwells*

1. Make PEGDA solution fresh for each day: 99.5% PEGDA and 0.5% photoinitiator, 2-hydroxy-2-methyl propiophenone.
   *Note:* Various molecular weights (MW) can be used, and in general, the smaller the MW, the stiffer the resulting hydrogel.
2. To make patterns with PEGDA as the bottom substrate of the well, pipette 300 $\mu$L of PEGDA solution onto a methacrylated glass slide.
3. To make patterns with the substrate as the bottom of the well, pipette 300 $\mu$L onto the PDMS stamp instead.
4. Place the PDMS stamp onto PEGDA solution while holding it on the sides. Lightly touch the PDMS stamp to ensure full contact.
5. Place under UV Light for 30 seconds at 365 nm, 300 mW/cm$^2$.
   *Note:* The UV light settings can vary based on total area or the amount of polymer used. The structures should form a solid PEGDA hydrogel. With too little UV light, the structures will remain liquid and give poor patterning.
6. Rinse with PBS and culture media.

##### *10.2.3.2.3 Formation of cell-hydrogel constructs within PEGDA microwells*

1. Place the PEGDA pieces within a 6-well plate and sterilize with UV light in a biosafety cabinet overnight.
2. Suspend ES cells in culture medium at $4 \times 10^6$ cells/mL.

*Note:* The general methods for preparation of embryonic stem cells, their expansion in monolayer culture (either on feeder layers or under chemically defined conditions), and dissociation with enzymes are not described here.

3. Allow the cells to aggregate in the microwells overnight.

   *Note:* It is important to leave the cultures undisturbed during this time.
4. Wash excess cells away using gentle flow from a 10 mL pipette.
5. To harvest EBs from the PEGDA wells, use a P200 pipette. Wash EBs with culture medium (200 $\mu$L, three times)

#### 10.2.3.3 Data acquisition, anticipated results, and interpretation

The use of PEGDA microwells resulted in a homogenous population of EBs, the size of which could be controlled by the size of the culture well [Figure 10.4(a, b)]. Notably, the initial EB size was maintained in culture for over 10 days, with excellent reproducibility from one well to another. EB diameters ranged from 40 to 150 $\mu$m. This added homogeneity of the size of EBs provides potential for reducing the heterogeneity of ES cell differentiation seen in suspension cultures of EBs [3].

The PEGDA material, being nonadherent, facilitates easy removal of EBs without affecting their size or structure. This ability to harvest the cells while maintaining their defined aggregate size promotes ease of analysis. As shown in Figure 10.3(a), each of the wells at day 10 contained harvestable EBs.

Notably, the expression of cell-differentiation markers was significantly different in suspension cultures of EBs and controlled-size EBs, and the expression levels were further modulated by the size of EBs. When the cells were stained for markers that are representative of endoderm (AFP), mesoderm (Brachyury), and ectoderm (Nestin), the well cultures had a tighter range of expression, showing their capacity to reproducibly influence ES cell differentiation [3].

#### 10.2.3.4 Discussion and commentary

Suspension cultures of EBs contain a wide array of cell types, but, concordantly, the yield for a specific cell type is relatively low. Therefore, scale up to generate larger amounts of cells for in vitro studies can prove costly and labor intensive. It would be beneficial to further direct and specify the range of lineages the cells may differentiate into. We tried to modulate the cellular microenvironment by restricting the size of the EBs and thereby modulating the cell differentiation. This technique is attractive since the microengineered culture system maintains the main features of standard EB culture with the ability to tightly control the size of EBs as well as the option to readily retrieve the EBs any time during cultivation. Notably, the uniformity in EB sizes was associated with the more uniform expression of cell renewal and differentiation markers. The expression of self-renewal markers could be further mediated by the EB size. The effects of size on cell-differentiation markers were less pronounced, except for much better uniformity of marker expression. This remains an area for future work with the potential for studying other environmental control factors such as hypoxia, cytokines, or physical stimuli.

Therefore, the use of microengineered wells is a simple, robust technique that can aid in controlling ES cell differentiation. The cells cultured with size restriction generated homogenous aggregates with added specificity of their self-renewal and differentia-

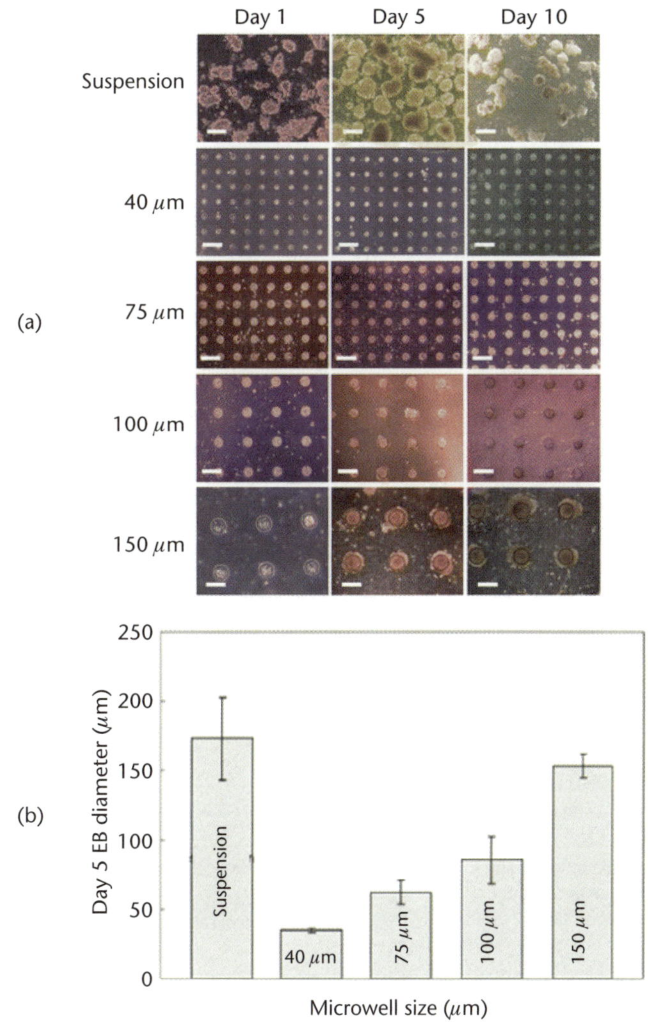


**Figure 10.4** 3-D patterning of ES cells. (a) Control of EB size by cultivation of ES cells in polyethylene glycol microwells. Rows show the EBs cultured in suspension (control group) and the microwells that were 40, 75, 100, and 150 $\mu$m in size. Columns show the representative images taken after 1, 5, and 10 days of culture. (b) The size of EBs was controlled by the size of the microwell; suspension cultures of EBs are used as controls. Data represent average ± SD of $n = 50$ samples per group. (Reproduced with permission from [3], adapted from Figure 3).

tion profiles. The technique can be scaled up to larger cell numbers and provides a versatile platform that retains most of the features found in suspension culture to work in concordance with future ES cell-differentiation studies.

## 10.3 Interactive Use of Substrate Topography and Electrical Stimulation for the Control of Cell Alignment

In contractile tissues such as myocardium, functional properties are directly related to cellular orientation and elongation. Thus, tissue engineering of a functional cardiac patch critically depends on our understanding of the interaction between multiple guidance cues such as topographical or electrical cues. In order to study the interactive effects of contact guidance and electrical-field stimulation on cell elongation and orientation, we prepared microstructured surfaces on which cells are cultured between stimulating electrodes. The cells were cultivated on 2-D abraded surfaces and field-stimulated using regimes of relevance to heart tissue in vivo as well as for cardiac tissue engineering (square suprathreshold pulses, 1 ms duration, 1 Hz). In our experimental design, the abrasions are oriented either in parallel to the electrodes (perpendicular to the field) or perpendicular to the electrodes (parallel to the electric field); nonabraded surfaces served as controls (Figure 10.5) [4].

Our main hypothesis was that substrate topology and electrical signals will interactively determine cellular orientation and elongation and that the same molecular pathways may be involved in cellular response to both cues. We provide here a method to make abraded surfaces, perform cell culture in the presence of electrical-field stimulation, and determine cellular orientation and elongation. In our previous work, methods are available for pharmacologic studies to assess the relevance of actin cytoskeleton and phosphatidyl-inositol 3 kinase (PI3K) pathway in orientation and elongation response to the topography and electrical field [5]. Contact guidance on abraded surfaces more strongly determined cellular orientation than electrical-field stimulation, while on nonabraded surfaces, cell elongation could effectively be modulated by electrical stimulation [4].

### 10.3.1 Materials

| Equipment | Reagents | Disposables |
|---|---|---|
| Electrical stimulator (providing a sufficient amount of current, e.g., Grass Technologies's dual output stimulator with 600 mA current rating) | Bovine fibronectin, sterile (Sigma, catalog no. F1141) | 100 mm glass petri dish, nonsterile (e.g., VWR, catalog no. 89000-318) |
| Ultrasound cleaner (e.g., Misonix 1510R-MT) | 95% ethanol (Pharmco, catalog no. 111000190) | 80 $\mu$m aluminum oxide lapping paper, nonsterile (McMaster-Carr, catalog no. LPA80) |
| Class II biosafety cabinets | Carboxyfluorescein diacetate, succinimidyl ester (CFDA) molecular probes, nonsterile (catalog no. C-1354) | Polycarbonate sheet of 3 mm thickness |
| $CO_2$ incubators (Napco, Thermoelectron) | Propidium iodide (PI), nonsterile (Invitrogen, catalog no. P1304MP) | Poly(vinyl) coverslips, nonsterile (VWR, catalog no. 48376-049) |
| Dremel, #80 drill bits, disc attachment, and 1.8" drill bit, nonsterile | High-glucose DMEM (Invitrogen, catalog no. 11965) | Laboratory labeling tapes, nonsterile (VWR, catalog no. 36428-065) |
| Digital multimeter (with capability to test for short circuits, e.g., Extech True RMS Digital Multimeter, RadioShack) | Heat-inactivated fetal bovine serum (FBS) (Gibco, catalog no. 10082) | Graphite electrodes, nonsterile (Ladd Research Industries, 1/8" diameter, catalog no. 30250) |
| | N-2-hydroxyethylpiperazine-N'-2-ethane-sulfonic acid (HEPES) (Gibco, catalog no. 15630) | Platinum wire, nonsterile (Ladd Research Industries, 8 mil, catalog no. 30572) |
| | Penicillin (Gibco) | |

### 10.3.2 Methods

#### 10.3.2.1 Preparation of abraded surfaces connected to stimulation electrodes

1. For one setup, cut two carbon electrode rods to 9.7 cm, cut two 7 cm pieces of platinum wire, and prepare one 10 cm petri dish.
2. Using the disc attachment on the Dremel tool, cut the polycarbonate sheet into $2 \times 2 \times 1$ cm pieces. Using the 1.8" drill bit, make in each polycarbonate piece two holes for

the carbon rods. The holes should be spaced 1 cm apart (edge to edge) and as close as possible (e.g., 1 mm) to the bottom of the polycarbonate holder, so that the electrodes can be placed close to the bottom of the petri dish.

3. Place #80 drill bit into the tip of the Dremel, and drill two holes 2 mm apart from each other through each electrode, as close to the end of the electrode as possible (~2 mm from the end to maximize the space available between the electrodes).
4. Poke the platinum wire through the hole furthest from the end, allowing no wire to poke through the other end. Wrap platinum wire twice around electrode, thread through the other hole, and pull tightly to create a good connection between the platinum wire and the electrode. The amount of wire hanging off the electrode should be enough to reach out of the bioreactor with just enough room for connectors you will be using to connect to it.

   *Note:* Free wire ends allow easy electrical connection between the electrical stimulator and the bioreactor, but because they are not insulated, the free wire ends should be just long enough to connect an alligator clip.
5. Insert the electrodes into the polycarbonate holders (one at each end) and place them into the petri dish. For use with monolayer inserts (which will be 8 × 10 mm), space the electrodes so that there is 1 cm distance from the edge of one rod to the other (for the purposes of calculating electric field with this configuration, divide input voltage by 1.1 cm).

   *Note:* It is important to make sure that the attached wires are on opposite sides of the opposing electrodes as shown in Figure 10.5(c). This placement of wires balances the effect of electrode resistance and ensures that a uniform amount of electrode resistance will be encountered for any given current path through the system.
6. Wipe off the probes of a digital multimeter with an alcohol pad and allow to dry. Using the "short circuit" option, check that there is a short circuit between the furthest tip of wire and electrode, which means that you have made a good electrical connection.
7. Prebend the platinum wire so that it comes out of the petri dish neatly. This will also facilitate placing the petri dish lid over the chamber.

   *Note:* Keep in mind that bending the delicate platinum wire too much may cause it to snap.
8. Place the chambers into the autoclave bags and autoclave for 20 minutes on dry cycle at 121°C/2bar, followed by a 20-minute drying cycle. Store the chambers at room temperature until use.
9. Cover one half of your polyvinyl carbonate coverslips (22 × 22 mm) with laboratory tape, and use lapping paper to abrade one half of the coverslip in one direction (~30 strokes should provide good abrasions). Peel off the tape, cover the part of the coverslip just abraded, and abrade the other half of the coverslip in the orthogonal direction in order to observe the cells' response to two different topographical cues on the same coverslip. Observe under the microscope that the abrasions are straight (i.e., are all running in the same direction in a particular area and not crisscrossing due to accidental scratches).
10. Using scissors, cut the abraded coverslips into 8 × 22 mm rectangles.

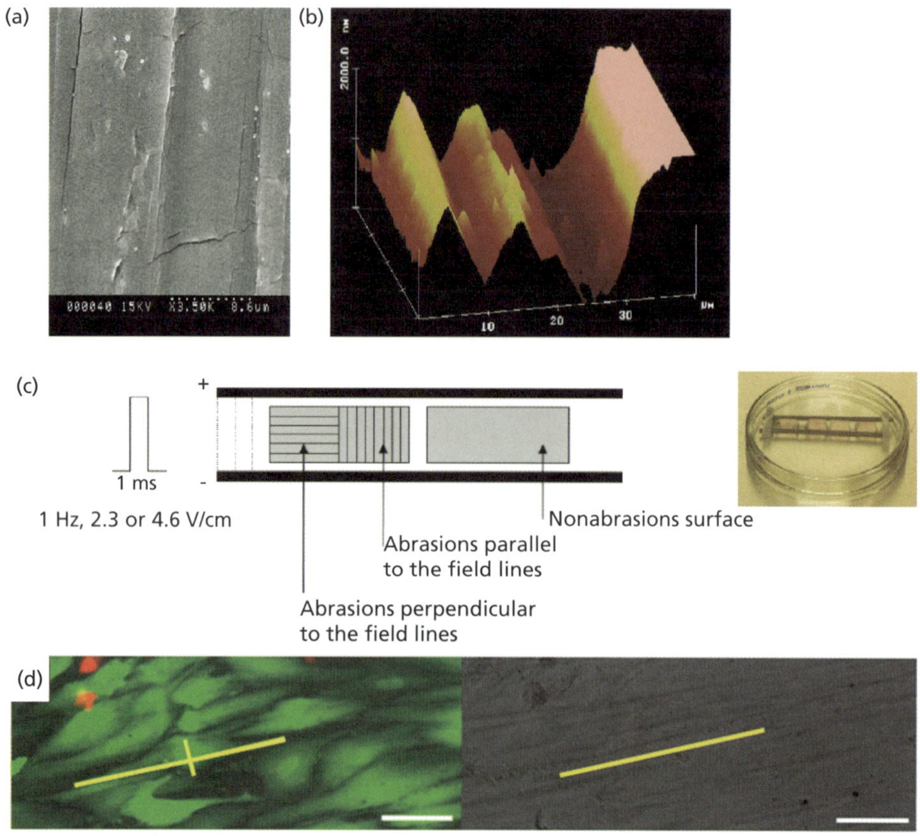


**Figure 10.5** Simultaneous application of topographical cues and electrical-field stimulation. Since surfaces abraded with lapping paper of 80 mm grain size (average abrasion width of 13 mm and depth of 700 nm) exhibited the strongest effect on cardiomyocyte elongation and orientation, as well as statistically significant effect on orientation of fibroblasts, they were used for electrical-field-stimulation experiments. (a) Scanning electron micrograph of a polyvinyl coverslip abraded with 80 mm grain size lapping paper. (b) Atomic force micrograph of a polyvinyl coverslip abraded with 80 mm grain size lapping paper. (c) Orientation of abrasions with respect to the electrodes in the stimulation chamber. (d) Live-dead staining (left) and bright-field image (right) of the same field for assessment of cell elongation and alignment. (Images a and b reproduced with permission from [4], adapted from Figure 1).

*Note:* Be careful to cut the strips so that *both* directions of abrasions are included in the strip, and not to cut a strip entirely out of a piece of the coverslip abraded in only a single direction.

11. Wearing tight-fitting gloves (for better control), fold along the two shorter edges by hand to form a "table" with a surface of $8 \times 10$ mm, and cut the "legs" to a height of 1.5 mm (to ensure that the cells cultured on the coverslips are placed at identical heights in the center between the carbon electrodes used for electrical stimulation).
12. Remove particulate debris by sonicating the surfaces using ultrasound cleaner in soap and water followed by rinsing in distilled water.
13. Sterilize the surfaces in 95% ethanol for 24 hours, followed by drying and UV irradiation for 30 minutes at room temperature (25°C)

#### 10.3.2.2 Cell culture

1. For cardiomyocyte culture, coat the surfaces at room temperature with 25 $\mu$g/mL of bovine fibronectin in PBS for 2 hours to enhance cell attachment (60 $\mu$L of solution per coverslip).
2. Remove excess fibronectin by aspiration.
3. Prepare cells (neonatal rat cardiomyocytes or fibroblasts) using the cell-isolation protocol described in Section 10.2.2.2.2.
4. Prepare cardiomyocyte culture medium consisting of a high-glucose DMEM containing 4.5 g/L glucose supplemented with 10% FBS, 10 mM HEPES, 2 mM L-Glutamine, and 100 U/mL penicillin. In order to make this, take a 500 mL bottle of DMEM, and remove 60 mL using a serological pipette. Add 50 mL of FBS, then 5 mL of HEPES, then 5 mL of penicillin.
5. Resuspend cells such that the desired number of cells for each coverslip is contained in a 60 $\mu$L volume.

   *Note:* We used 500,000 cardiomyocyte cells for the results shown here.
6. Pipette 60 $\mu$L of this suspension onto each coverslip, and incubate at 37°C, 5% $CO_2$, for 1 hour. After 1 hour, fill the chamber with 20 mL culture medium.
7. With sterile forceps, place coverslips between the electrodes in the electrical stimulation chamber [Figure 10.5(c)] with the "table legs" orthogonal to the electrodes (the way the "tables" were made should ensure that when they're placed between the electrodes, half of the abrasions will be parallel and half will be orthogonal to the electric field).

   *Note:* Up to four "tables" may fit in a 100 mm diameter petri dish.
8. Add 25 mL of cardiomyocyte culture medium to a 100 mm dish, making sure there is enough culture medium to cover the electrodes and ensure proper current flow. Cover and place chamber onto the lid of a 150-mm-diameter plastic petri dish to facilitate carrying and to provide an insulating layer between the dish and the metal incubator shelves.
9. Place the lid with the chamber into the incubator (37°C, 5% $CO_2$), and culture for 24 hours before applying electrical stimulation to monolayer culture inserts.

   *Note:* We usually cultivate the cells for 3 to 7 days in the presence of electrical-field stimulation.

   *Note:* During stimulation, pay particular attention not to allow constructs to float out from between the electrodes.
10. Make electrical connections by connecting electrodes to the electrical stimulator using alligator clips attached to the platinum wires.

    *Caution:* Take care to prevent any undesired electrical connections. If you connect more than one bioreactor to a single channel, make sure not to surpass the current limit of your electrical stimulator.
11. Use the following calculation to see how many bioreactors your system allows in parallel: $I_{max} = X \times V_{stim}/R_s$ where $I_{max}$ corresponds to the current limit of the electrical stimulator, $X$ is the number of bioreactors you intend to connect in parallel, $V_{stim}$ is the amplitude of the applied voltage, and $R_s$ is the resistance of the bulk solution in your bioreactor. For electrodes spaced at a 1 cm distance, with a length of 5.7 cm (out of which 4 cm is exposed to the culture medium), we have measured $R_s = \sim 20\Omega$; for

electrodes spaced at a 1 cm distance, with a length of 9.7 cm (out of which 8 cm is exposed to the culture medium), we have measured $R_s$ = ~10Ω).

12. Start electrical stimulation 24 hours after seeding the cells on tables. Stimulate with 1 ms duration square monophasic pulses delivered at 1 Hz.

    *Note:* Cells on surfaces are more sensitive to electric field, and too high a field can decrease cell viability. Biphasic pulses (1 ms duration per phase, 1 Hz) can also be applied.
13. Stop electrical stimulation 4 days after seeding (after 3 days of electrical stimulation).
14. For end point analysis of cell elongation and alignment, perform live/dead staining on cells cultivated on the "tables," and take both fluorescent and bright-field images of the same area. For live/dead staining, incubate the cells in a 0.7 mL volume (for one well of a six-well dish, or for two coverslips or "tables") of solution of CFDA (10 μM) and PI (75 mg/μL) in PBS for 60 minutes, followed by rinsing in PBS (two times) and imaging. Other fluorescent live/dead kits can also be used for this purpose.
15. Open a pair of images in ImageJ software. Enlarge the images until the cell outlines (bright green) are clearly visible. Draw lines corresponding to the cell's long and short axes in the fluorescent image (Figure 10.6).
16. For cells cultured on abraded surfaces, draw a line on the bright-field image corresponding to the direction of abrasions, and measure the angle between the abrasion and the cell's long axis.
17. For cells cultured on nonabraded surfaces, you can measure the angle between the cell's long axis and the axis that was either orthogonal or parallel to the electric field.

    *Note:* In our past studies, for example, we have compared the angle of the long axis of fibroblasts to the orthogonal direction and the angle for cardiomyocytes to the parallel direction. These directions were selected based on previous studies demonstrating that fibroblasts align orthogonally, while cardiomyocytes align parallel, to the electric field lines.
18. Determine elongation by measuring the long and short axes of the cell and calculating the aspect ratio (long-to-short-axis ratio). Exclude from analysis any cells of which the perimeters cannot be precisely determined (this number should not be more than 10%). A good number of cells to measure is 60 to 90 per group from at least $N = 3$ independent experiments (e.g., 20 to 30 cells per sample).

    *Note:* Immunostaining and electrical-excitability measurements can also be performed according to our published procedures [6].

### 10.3.3 Data acquisition, anticipated results, and interpretation

Abrasion of polyvinyl surfaces using lapping paper of 80 μm grain size, as described in this protocol, results in peak-to-peak distances of 13 μm between grooves and a depth of 700 nm [Figure 10.5(b)]. This abrasion type gave us the highest degree of alignment and elongation of cardiomyocytes (in comparison to abrasions created by using a finer lapping paper, consistent with previously reported studies that utilized grooves of precisely defined dimensions as well as rough surfaces).

After the application of electrical-field stimulation of amplitudes of 2.3 V/cm or 4.6 V/cm, either parallel or orthogonally to the direction of the abrasions (with nonabraded surfaces used as controls), we observed that cells (in all cases) elongated and aligned with their axis parallel to the surface abrasions (Figure 10.6). Because the abrasions were ori-

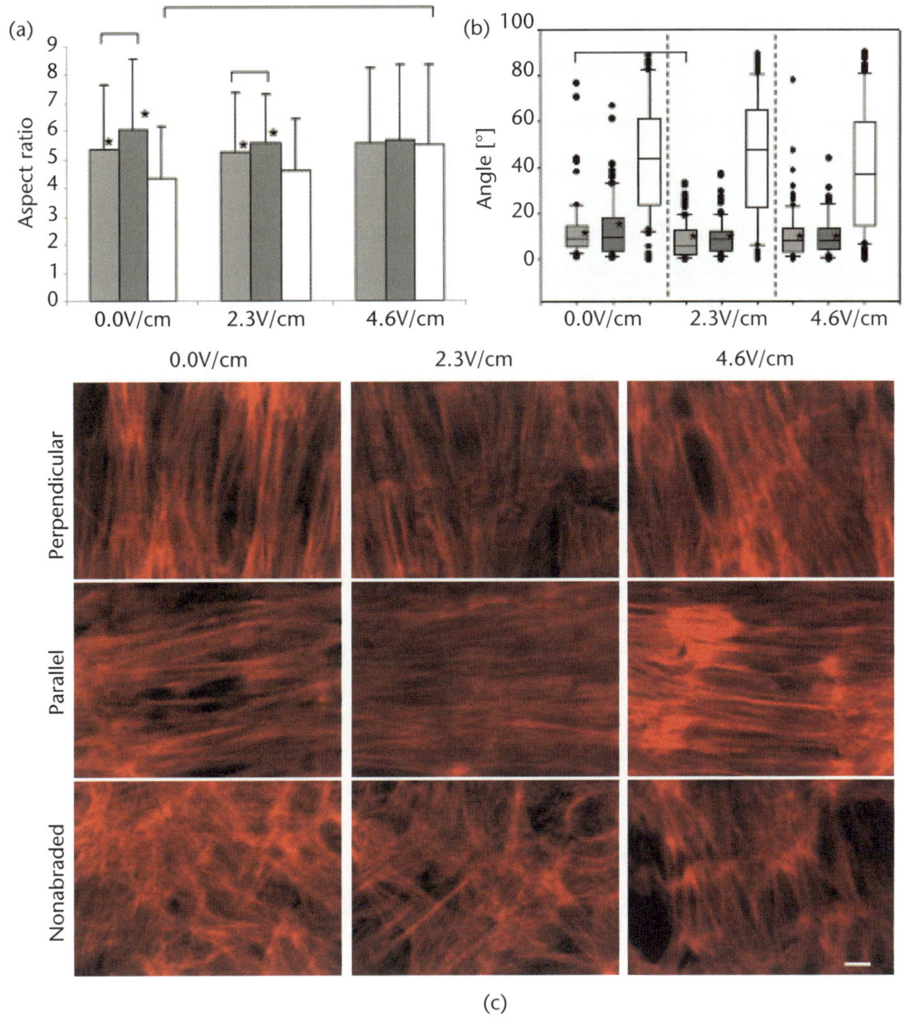


**Figure 10.6** Cardiomyocytes cultivated on abraded and nonabraded surfaces in the presence of electrical-field stimulation. The surfaces were obtained using lapping paper with grain size of 80 $\mu$m. Electrical-field stimulation using square pulses of 1 ms duration, 1 Hz, and 2.3 V/cm or 4.6 V/cm was initiated 24 hours after cell seeding and maintained for an additional 72 hours. (a) Cell elongation as defined by the aspect ratio. (b) Cell orientation as measured by the angle of deviation. (a and b) Light gray bars: abrasions oriented orthogonal to the electric field; dark gray bars: abrasions oriented parallel to the electric field; white bars: nonabraded surfaces. Data are represented as averages ±SD. Total $N = 2$ to 3 independent samples (coverslips) per group; 30 to 90 cells were analyzed per group. ($p < 0.05$ was considered significant.) *Significantly different than nonabraded surface at identical stimulation voltage. (c) Actin cytoskeleton as visualized by phalloidin-TRITC staining. (Scale bar = 10 $\mu$m; electric field lines are from left to right). (Reproduced with permission from [4], adapted from Figures 3 and 5).

ented either: (1) parallel to the electric field (i.e., orthogonally to the electrodes), or (2) orthogonally to the electric field (i.e., parallel to the electrodes), the two cues (topography and electric field) acted on the cells in either a parallel or an orthogonal direction [Figure 10.6(c)]. In case 1, the electrical field and topography were acting on the cell alignment in the same direction. In case 2, the field was acting on the cell in the orthogonal direction to the abrasions. On flat surfaces, pulsatile electrical field significantly enhanced elongation of cardiomyocytes [Figure 10.6(a)] to reach levels comparable to those achieved by surface abrasion.

The fact that field stimulation failed to promote elongation at higher levels than those obtained by topographical cues prompted us to hypothesize that the pathway for elongation is saturated by topographical cues and that the same signaling pathways could be involved in the cellular response (i.e., elongation) to topographical cues and electrical-field stimulation. In addition, pulsatile electrical-field stimulation significantly enhanced orientation of cardiomyocytes [Figure 10.6(b)] when they were cultivated on abrasions placed orthogonally to the field lines. Yet, within every voltage group, the nonabraded surfaces had approximately two times higher average orientation angle than the abraded surfaces [Figure 10.6(b)], indicating that topographical cues are overall a stronger regulator of cellular orientation than electrical-field stimulation.

Phalloidin-TRITC staining indicated that actin filaments generally followed the direction of surface abrasions [4]. For cells cultivated on nonabraded surfaces, actin cytoskeleton was disorganized. Higher-magnification images [Figure 10.6(c)] revealed remarkable differences in the orientation of actin filaments as a function of surface abrasion and electrical-field stimulation. On abraded surfaces, actin filaments were clearly aligned in the direction of surface abrasions (either orthogonally or parallel to the field lines). For cardiomyocytes cultivated on nonabraded surfaces at 0.0 V/cm and 2.3 V/cm [Figure 10.6(c)], the actin cytoskeleton was not aligned in any particular direction, and the overlapping filaments extended in multiple directions. Yet, at 4.6 V/cm on nonabraded surfaces [Figure 10.6(c)], there was an appreciable improvement in the organization of the actin filaments, which aligned in parallel along the long axis of the cell.

Morphometry following immunostaining for cardiac Troponin I indicated that 97% to 99% of cells on the surfaces were cardiomyocytes. Cross-striations in individual cardiomyocytes were generally oriented perpendicularly to the abrasion direction (Figure 10.7) with a remarkably well-developed contractile apparatus for cells cultivated at 4.6 V/cm on surfaces with abrasions perpendicular to the field lines [Figure 10.7(b)]. On nonabraded surfaces, there was no preferential directionality in cross-striation orientation due to the random cellular orientation, and cross-striations were present only in the short domains [Figure 10.7(b)].

Because we saw similar results with 3T3 fibroblasts and observed the abolition of the orientation and elongation response of cardiomyocytes to the abraded surfaces and electrical-field stimulation by inhibition of actin polymerization (and only partially by inhibition of PI3K pathway), we conclude that the surface topography more strongly determined the orientation of both fibroblasts and cardiomyocytes than pulsatile electrical field [4]. These findings have implications for the design of scaffolds for cardiac tissue engineering. A feasible strategy would be to design scaffolds of desired microarchitecture (e.g., by electrospinning) to drive cellular elongation and orientation in the desired direction, followed by the application of electrical-field stimulation for enhancement of these properties.

### 10.3.4 Discussion and commentary

In vivo, multiple guidance cues determine cell orientation and phenotype: topographical, adhesive, electrical, mechanical, and chemical. The interactive effects of topo-

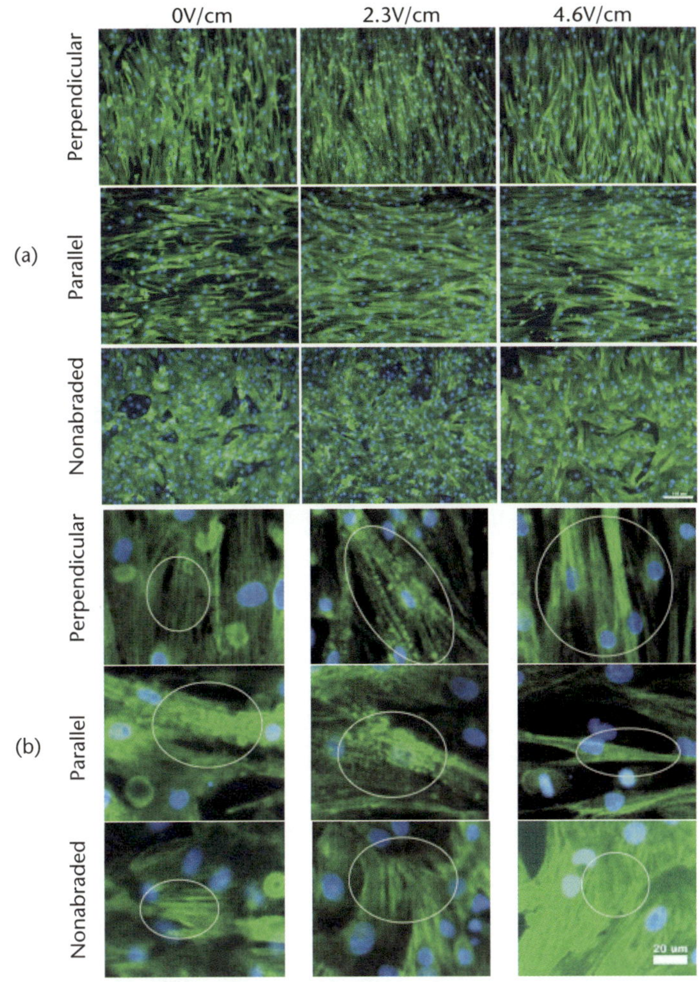


**Figure 10.7** Immunostaining for cardiac Troponin I of cardiomyocytes cultivated in the presence of electrical-field stimulation on abraded surfaces. (a) Orientation and morphology of cells expressing cardiac Troponin I. Scale bar = 100 $\mu$m. (b) Higher magnification images indicate the presence of contractile apparatus (cross striations). Scale bar = 20 $\mu$m. (Reproduced with permission from [4], Figure 4).

graphical and adhesive cues have been studied extensively. Lesser attention has been devoted to the interaction between adhesive and electrical cues, indicating that adhesive cues guide neurite outgrowth more strongly than electrical cues. Yet, the interaction between electrical cues and topography, especially in conditions relevant for tissue engineering, remain largely unstudied. In adult myocardium, cardiomyocytes are elongated and oriented in parallel and form a 3-D syncytium that enables propagation of electrical signals. Cardiac fibroblasts are scattered among the myofibers, secreting components of the extracellular matrix (ECM) and transmitting mechanical force via the receptor-mediated connections to the ECM. One of the challenges of cardiac tissue engineering is reproducing an in vivo–like orientation and elongation of the cells in an engineered cardiac patch.

The main objective of this study was to determine interactive effects of topographical cues and electrical-field stimulation on cellular elongation and orientation in conditions relevant for cardiac tissue engineering. We focused on neonatal rat cardiomyocytes, a contractile cell documented to be responsive to both contact guidance and electrical-field stimulation. We also studied the response of fibroblasts, a noncontractile cell type with documented ability to align in response to contact guidance and dc fields.

Mediation of cardiomyocyte elongation and alignment by topographical cues depend upon the orientation of actin cytoskeleton. Elongated cardiomyocyte phenotype and alignment can also be achieved by cultivating cardiomyocytes on microtextured silicone membranes that can additionally be microfluidically patterned with ECM molecules. The orientation and cardiomyocyte phenotype could also be improved by microcontact printing of ECM components (e.g., laminin) on thin films.

Approaches to applying electrical-field stimulation to cardiomyocytes during culture involved setups similar to the one described here (i.e., where pulses were delivered through a pair of electrodes). Suprathreshold electrical-field stimulation of cardiomyocytes in monolayers was shown to preserve cardiomyocyte contractility, maintain calcium transients, promote hypertrophy, increase protein synthesis, and maintain action potential duration and maximum capture rate. The beneficial effects of field stimulation were dependant on the occurrence of contraction following field-stimulated excitation, as evidenced by the lack of observed effects in the presence of excitation-contraction decouplers: verapamil or 2,3-butanedione monoxime. In 3-D studies, we demonstrated that electrical-field stimulation can be used to engineer functional cardiac tissue expressing hallmarks of cardiac differentiation. Stimulated constructs had thick elongated and aligned myofibers expressing cardiac markers, in contrast to nonstimulated constructs that contained round cells [5].

### Troubleshooting Table

| Problem | Explanation | Potential Solution |
|---|---|---|
| Microabrasions | | |
| My abrasions are not straight enough | You are not lapping in a consistent direction | Try lapping only in a single direction (not back and forth but lapping each stroke separately). |
| Contamination | The process of media exchange (transport and actual process is likely cause of contamination) | Be careful when changing medium not to allow any medium to splash onto the lid of the culture chamber. Placing the culture chamber on top of a 150 mm petri dish can facilitate carrying and placing the electrical setup within the incubator. In addition, wearing sterile gloves whenever handling the bioreactor during cell culture can help. |
| Electrical stimulation | | |
| During stimulation, culture medium becomes clear | The stimulation amplitude is too high for the amount of medium | Add more medium, change medium more often, or reduce the amplitude of stimulation. |
| When I look at the waveform applied to my cells, it is smaller in amplitude than I expected | Perhaps your stimulus generator has a current rating that is too low | Try reducing the amplitude of stimulation or shortening the electrodes. Or, consider buying a stimulator with a higher current rating or building an amplifier circuit to boost the power of your signal. Too many stimulation chambers may also be hooked up in parallel. |
| Cells are exhibiting signs of reduced viability, such as rounding up | The stimulation amplitude, stimulation frequency, or pulse duration is too high | Decrease the stimulation amplitude first, then reduce the pulse duration and frequency. Addition of 10 $\mu$M ascorbic acid as an antioxidant may also help. |

### 10.3.5 Summary points

- The methods described here enable studies of cell behavior, interaction, and functional assembly on a small scale and under conditions that are both biologically relevant and highly controllable.
- Surface patterning allows for two different cell types to be patterned using biocompatible materials, with any size or shape of the individual field occupied by one or the other cell type and with resolutions of cell patterning close to that of the size of an individual cell.
- Microfluidic patterning utilizes a native extracellular matrix material, hyaluronic acid, as a template to generate cardiac organoids composed of multiple cardiomyocytes functioning synchronously. This method represents a general technique to generate small, but functional, cardiac tissue organoids for potential application in biologic studies, in drug and toxicology screening, and as diagnostic models of normal and pathological cardiac function.
- Microarray systems for 3-D patterning of embryonic stem cells provide a simple, robust technique that can aid specificity to the embryonic stem cell self-renewal and differentiation profiles through geometric control of cell-hydrogel constructs.
- Surface topography strongly determines orientation of fibroblasts and cardiomyocytes. On abraded surfaces, the electrical-field stimulation enhanced orientation and elongation along the abrasion direction, but it could not reverse the effect of the cues provided by surface topography.
- Pulsatile electrical-field stimulation had appreciable effects on cellular elongation. On nonabraded surfaces, electrical-field stimulation significantly promoted elongation of cardiomyocytes and fibroblasts to the levels comparable to that obtained by the topographical cues.
- Recent advances in tissue engineering may change the way we conduct cell and tissue experiments by providing culture systems that combine the controllability of in vitro technologies with biological fidelity close to that seen in whole animals. We described here some of the culture platforms that were developed with this general goal in mind.

## Acknowledgments

The authors gratefully acknowledge research support by the National Institutes of Health (R01 HL076485 and P41-EB002520 to G. V.-N.), the National Science and Engineering Council (Discovery Grant to M. R.), and the Canada Foundation for Innovation (Leaders Opportunity Fund to M. R.).

### References

[1] Fukuda, J., "Micromolding of photocrosslinkable chitosan hydrogel for spheroid microarray and co-cultures," *Biomaterials*, Vol. 27, No 8, 2006, pp. 1479–1486.

[2] Khademhosseini, A, et al., "Microfluidic patterning for fabrication of contractile cardiac organoids," *Biomedical Microdevices*, Vol. 9, No. 2, 2007, pp. 149–157.

[3] Karp, J. M., et al., "Controlling size, shape and homogeneity of embryoid bodies using poly(ethylene glycol) microwells," *Lab Chip*, Vol., 7, No. 6, 2007, pp. 786–794.

[4] Au, H. T., et al., "Interactive effects of surface topography and pulsatile electrical field stimulation on orientation and elongation of fibroblasts and cardiomyocytes," *Biomaterials*, Vol. 28, No. 29, 2007, pp. 4277–4293.

[5] Radisic, M., et al., "Functional assembly of engineered myocardium by electrical stimulation of cardiac myocytes cultured on scaffolds," *Proc. Natl. Acad. Sci. USA*, Vol. 101, No. 52, 2004, pp. 18129–18134.

[6] Radisic, M., et al., "Cardiac tissue engineering using perfusion bioreactor system," *Nature Protocols*, Vol. 3, No. 4, 2008, pp. 719–738.

CHAPTER

# 11

# Subtractive Methods for Forming Microfluidic Gels of Extracellular Matrix Proteins

Gavrielle M. Price and Joe Tien

Department of Biomedical Engineering, Boston University, 44 Cummington Street, Boston, MA 02215

## Abstract

Microfluidic networks in extracellular matrix hydrogels hold great promise in tissue engineering. We have established two methods for forming microscale channels within collagen and fibrin gels. Both methods rely on molding gels around removable elements: stainless steel needles for creating single channels and gelatin meshes for forming networks of interconnected channels. These methods can form single channels 50 $\mu$m or more in diameter or networks 6 $\mu$m or more in width, at a rate of 5 to 10 samples per day. Potential applications include using microfluidic gels for engineering microvascular networks, perfusing cultured cells within native scaffolds, and controlling interstitial flows in cell culture models.

Key terms
microfluidic networks
gel
fibrin
collagen
gelatin
channels

## 11.1 Introduction

This chapter focuses on techniques recently developed in our group to construct single channels and complex networks within gels of the extracellular matrix (ECM) proteins type I collagen and fibrin [1, 2]. It details methods for establishing and maintaining perfusion in these gels. The methods described here are suitable for forming networks of different sizes and shapes, and we provide illustrative examples.

Microfluidic networks, primarily in polydimethylsiloxane (PDMS) and silicon, are widely used in many applications, including disposable chips for medical diagnostics [3], microreactors for performing small-scale chemical synthesis [4], and devices for controlling cell behavior [5]. Only recently have these structures been extended to hydrogels [6–8], largely because gels are difficult to manipulate: Unlike PDMS and silicon, gels need to be hydrated to retain their shape, are often highly sensitive to temperature or pH, and can collapse irreversibly under mechanical agitation. ECM hydrogels are particularly fragile, and special care must be taken to avoid deforming any channels formed within them.

These materials complement previously described microfluidic gels comprising alginate or poly(ethylene glycol) [6, 8]. Because ECM is the natural substrate upon which cells adhere and spread, microfluidic ECM gels are well suited for cell culture. For instance, we have used ECM gels to form microvessels in vitro by seeding vascular cells in or on the gels [1, 9]. Microfluidic networks also enable fast convective delivery of fluids within hydrogels and allow spatial and temporal control of solute transport [10]. We have used our method for fabricating microfluidic networks to study diffusive and convective transport in ECM gels [2].

## 11.2 Materials

In our design, the gel lies on top of a supporting dish and is held in place by a PDMS housing. Channels in the gel are made by molding the gel around removable elements. If desired, addition of a lid and tubing enables perfusion under high flow and/or high pressure.

### 11.2.1 Supporting dishes

1. 100 mm–diameter tissue-culture polystyrene dishes (Corning)
2. No. 1.5 glass coverslips, 22 $mm^2$ (Corning)
3. Liquid PDMS (Sylgard 184, Dow Corning), freshly mixed as 1 part catalyst to 10 parts prepolymer and degassed in a vacuum

### 11.2.2 PDMS housing

1. *For single channels*
   i. Silicon wafer (Montco Silicon) patterned with approximately 1-mm-thick negative photoresist (e.g., SU 8-50), using standard photolithographic techniques [see Figure 11.1(a) for pattern dimensions]

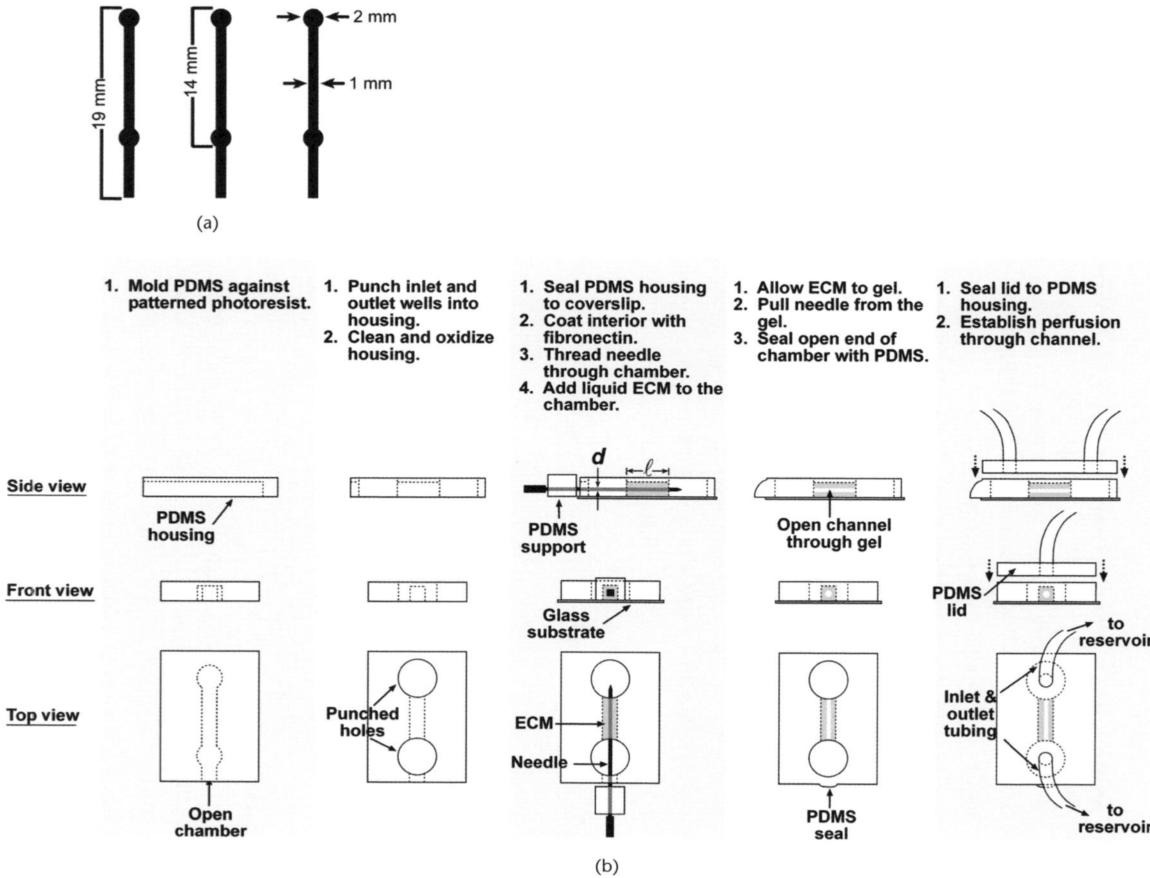


**Figure 11.1** Forming ECM gels that contain single channels. (a) Lithography pattern for making the PDMS housing. The rounded features serve as guides for the hole punch. (b) Schematic for forming single channels by using a small-diameter needle as a removable element. The parameters *d* and *l* refer to the diameter and length of the channel, respectively. The last panel illustrates how to establish high-flow perfusion in the microfluidic gel; for further details or for perfusion under low flow, see Figure 11.3.

   ii. Single hole punch that produces circular or square holes (available at arts and crafts stores)
   iii. Liquid PDMS (see item 3 in Section 11.2.1)
2. *For networks*
   i. Blocks of PDMS, approximately 1 mm and approximately 3 mm thick, cured at 60°C for several hours
   ii. Hole punch

### 11.2.3 Removable elements (needles and gelatin mesh)

1. *For single channels*
   i. Needles, 0.12 mm in diameter (Seirin J-type; Health Point Products)
   ii. Liquid PDMS (see item 3 in Section 11.2.1)
   iii. Sterile phosphate-buffered saline (PBS; Invitrogen)
   iv. 0.1% w/v bovine serum albumin (BSA; Calbiochem) in PBS
2. *For networks*
   i. Silicon wafer patterned with negative photoresist such that features have a rectangular cross section (50 to 100 $\mu$m thick) [see Figure 11.2(a) for pattern dimensions]
   ii. Liquid PDMS (see item 3 in Section 11.2.1)
   iii. 1% w/v Pluronic F127 (BASF) in PBS
   iv. 10% w/v type A gelatin from pig skin (Sigma) in PBS, filter-sterilized at 60°C and stored as a gel at 4°C
   v. 0.1% BSA and PBS

### 11.2.4 ECM proteins

1. *Collagen*
   i. 8 to 10 mg/mL type I collagen from rat tail (BD Biosciences), stored at 4°C.
   ii. 0.2M sodium hydroxide (NaOH; Sigma) in water, filter-sterilized.
   iii. 7.5% w/v sterile sodium bicarbonate solution ($NaHCO_3$; Invitrogen).
   iv. pH-indicator strips (Fisher Scientific), with a pH range of 5.0 to 10.0.
2. *Fibrin*
   i. Human fibrinogen (Sigma), freshly reconstituted at 50 mg/mL in water. The solution is clarified at 16,000g for 1 minute before use. We use fibrinogen that contains approximately 60% clottable protein.
   ii. Human thrombin (Sigma), reconstituted at 160 U/mL in water and stored at 4°C.
3. *Fibronectin*
   i. Human fibronectin (BD Biosciences), reconstituted at 1 mg/mL in water and stored at 4°C. Avoid excessive mixing of the solution that may shear-polymerize the fibronectin.

### 11.2.5 High-flow perfusion

1. Plain silicon wafer
2. PE-50 polyethylene tubing (BD Biosciences)

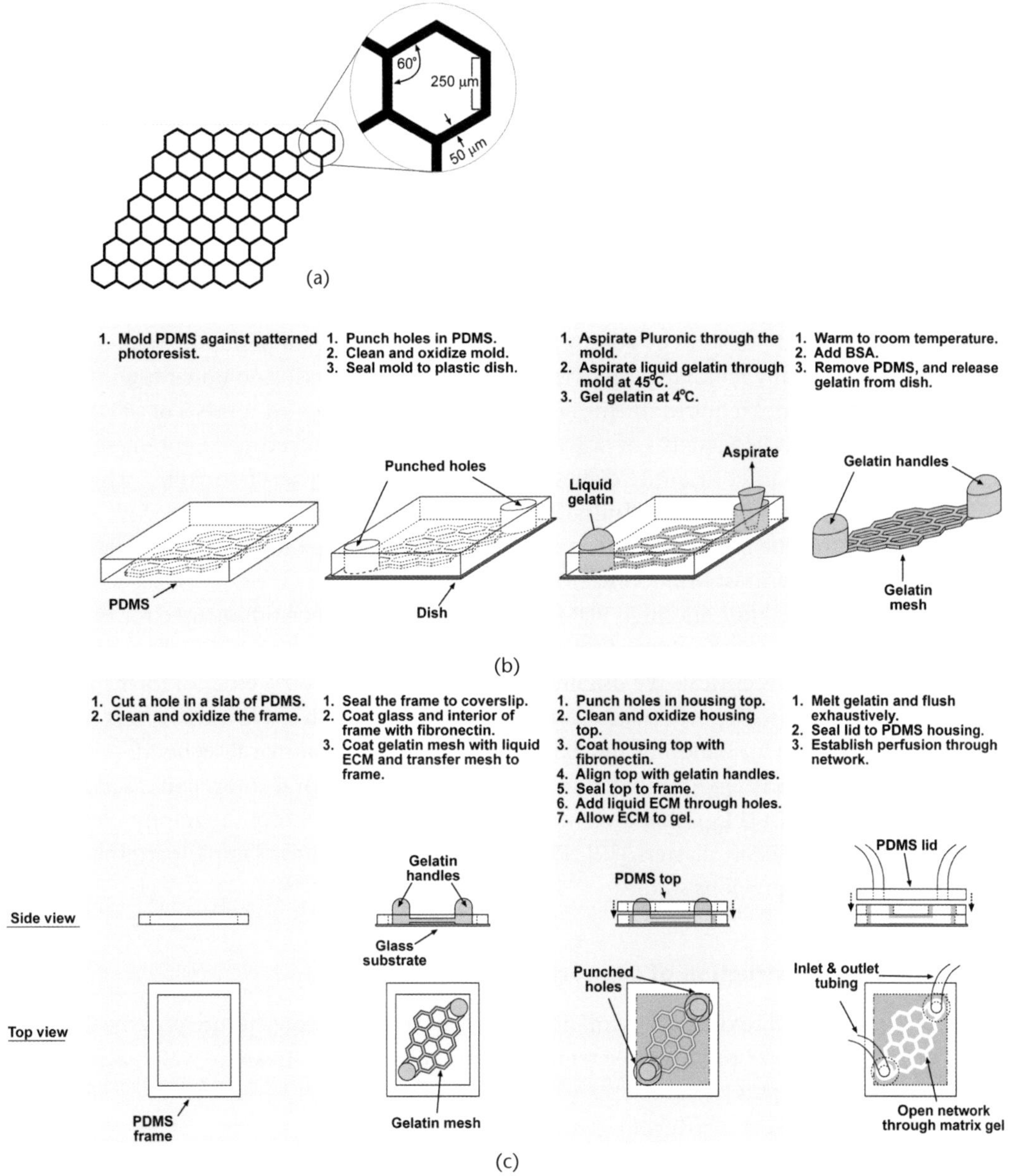


**Figure 11.2** Forming ECM gels that contain networks. (a) Lithography pattern for a hexagonal network. (b) Schematic for molding gelatin meshes. (c) Schematic for forming networks within ECM gels by using the gelatin mesh as a removable element. The last panel illustrates how to establish high-flow perfusion in the microfluidic gel; for further details or for perfusion under low flow, see Figure 11.3.

3. Liquid PDMS (see item 3 in Section 11.2.1)
4. 100-mm-diameter deep petri dishes (Fisher Scientific)
5. Glass slides, 1″ × 3″ (Fisher Scientific)
6. Small binder clips (Office Depot)

## 11.3 Methods

Both methods for creating microfluidic ECM gels are subtractive. In one method, partial encapsulation of a needle within liquid matrix, followed by manual removal of the needle after gelation, yields a single cylindrical channel. In the other method, partial encapsulation of a patterned gelatin mesh within liquid matrix, then removal of the gelatin after gelation by melting, yields an open network. In both cases, the channels are adjacent to inlet and outlet wells that enable perfusion of the system.

Surrounding each microfluidic gel is a PDMS housing, which serves as a mechanical support that is (nearly) impermeable to water. The system can be perfused under low flow rates and pressures by adding small volumes (~50 $\mu$L) of perfusate to an open inlet and periodically removing liquid from the open outlet. Perfusion under high flow rates and/or pressures requires attaching a lid to the housing; this lid consists of a PDMS block threaded with two lengths of polyethylene tubing. The polyethylene tubing is, in turn, connected to inlet or outlet dishes of perfusate that serve as "reservoirs." The relative heights of the dishes determine the pressure difference—and, indirectly, the flow rate—across the gel. Each gel lies on top of a glass coverslip for ease of visualization (e.g., with phase-contrast or fluorescence microscopy).

As a general rule, all cured PDMS, glass, and polyethylene tubing used for the system should be thoroughly cleaned and sterilized before use. Removal of dust that can cause leaks or clogs is critical. We usually sonicate each part in 70% ethanol for 1 minute or more, wash it with pure ethanol, and dry it with an aspirator. We use tweezers (as much as is practical) for handling and assemble the parts in a laminar flow hood.

The PDMS is replica-molded from a silicon master with features patterned in negative photoresist [11]. Other reviews provide details about photolithography used to fabricate these silicon masters [12]. Figures 11.1 and 11.2 show sample patterns that we have used in previous studies.

### 11.3.1 Construction of supporting dishes

1. Drill an approximately 1 cm$^2$ hole into the base of a tissue-culture polystyrene dish. All plastic debris should be removed manually under a stream of water. Sterilize the dish with pure ethanol, and allow it to dry.
2. Clean and dry a coverslip in a sterile hood.
3. Use a small amount of liquid PDMS to glue the coverslip over the hole in the dish, and cure. Additional coverslips may be used as spacers to tailor the height of the central coverslip.
4. Store the dish in a sterile hood.

### 11.3.2 Construction of PDMS housings

1. *For single channels*
   i. Pour the PDMS into a petri dish that contains the patterned silicon master. The PDMS should flood the top of the photoresist by approximately 0.5 mm. Use a dish that is sufficiently larger than the silicon master to prevent a meniscus from forming near the edge of the pattern (e.g., a dish 150 mm in diameter is large enough for casting on 4" silicon wafers).

ii. Cure, remove, and trim the PDMS into individual blocks, one feature per block.
iii. Use a hole punch to cut holes (~3 mm in diameter) at either end of the PDMS block, as shown in Figure 11.1(b). These holes will eventually form inlet and outlet wells next to a microfluidic gel.
iv. Sterilize the PDMS housing, and store it in a sterile hood.

2. *For networks*
   i. Cut an approximately 1 $cm^2$ rectangular hole from a 1-mm-thick PDMS block to create a "frame."
   ii. Use a hole punch to cut holes (~3 mm in diameter) near either corner of a 3-mm-thick PDMS block to create a "top," as shown in Figure 11.2(c). The frame and top, when combined, will form the PDMS housing.
   iii. Sterilize the frame and top, and store them in a sterile hood.

### 11.3.3 Preparation of removable elements

1. *For single channels*
   i. Deposit a thin stripe of liquid PDMS on a 100-mm-diameter petri dish, arrange 10 to 15 needles on the PDMS so that they lie parallel to the bottom of the dish, and cure the PDMS to fix the needles in place.
   ii. Add PDMS to cover the needles, and cure.
   iii. Cut out individual needles, and remove all PDMS from the needles, except for a rectangular block in front of the needle "handle," as illustrated in Figure 11.1(b). This PDMS support serves as a mount to stabilize and suspend the needle.
   iv. Sterilize the needle, and coat it with BSA for 40 minutes at room temperature (avoid exposing the PDMS support to BSA solution). Wash the needle exhaustively with water, and dry. The absorbed layer of BSA prevents ECM proteins from adhering to the needle [13].
2. *For networks*
   i. Cast PDMS on the patterned silicon wafer, and cure. The PDMS should flood the top of the photoresist by 3 to 5 mm.
   ii. Cut the PDMS into individual molds, and use a hole punch to cut holes at either end of a PDMS mold, as shown in Figure 11.2(b).
   iii. Clean the PDMS mold, oxidize it in a UV-ozone cleaner for 10 minutes, and place the patterned side down on a sterile petri dish.
   iv. Aspirate a solution of Pluronic F127 through the PDMS network, and allow the Pluronic to absorb for 1 hour at room temperature. Absorbed Pluronic prevents ECM proteins from adhering to the mold [14]. Flush the network exhaustively with water after adsorption.
   v. Warm the mold to 45°C, and aspirate liquid gelatin through the mold, as shown in Figure 11.2(b). Fill both holes in the PDMS mold with gelatin (after gelation, these gelatin cylinders will serve as handles for the mesh).
   vi. Gel the gelatin at 4°C for 15 minutes or longer. Warm the gelatin to room temperature for 0.5 to 2 hours.
   vii. Submerge the PDMS mold and gelatin in 0.1% BSA. Gently peel the PDMS mold from the dish, and release the gelatin from the mold and dish by repeatedly pipetting BSA solution onto the gelatin mesh.

viii. Store the gelatin mesh at 4°C in 0.1% BSA.

### 11.3.4 Formation of microfluidic gels

1. *Preparation of neutralized collagen solution*
   i. Keep all components—dishes, PDMS housings, removable elements, reagents, mixing tubes, and so forth—at 4°C or on an ice bath.
   ii. Prepare liquid collagen in this order: 12.5 parts 10× PBS, 1 part 7.5% $NaHCO_3$, approximately 12.5 parts 0.2M NaOH (the exact amount needed to neutralize collagen varies with the lot), and 100 parts collagen. Thoroughly mix the collagen without introducing bubbles into the mixture. Pipette collagen onto a pH-indicator strip—the pH should be 7.0 to 7.5.
2. *Preparation of liquid fibrinogen solution*
   i. Mix fibrinogen solution with 1 to 5 U/mL of thrombin (see step 2 of Section 11.5.1). It is necessary to work quickly after adding thrombin since fibrinogen rapidly polymerizes to fibrin (in less than 5 minutes at room temperature).
3. *For single channels*
   i. Oxidize the PDMS housing in a UV-ozone cleaner for 10 minutes. Place the PDMS housing patterned side down on the glass coverslip of the supporting dish.
   ii. Coat the housing chamber with fibronectin (~10 $\mu$g/mL) for 40 minutes. Aspirate the chamber to dry.
   iii. Carefully thread a needle into the PDMS housing, as in Figure 11.1(b). The PDMS support of the needle should firmly adhere to the dish, but the needle should be suspended above the dish.
   iv. Prepare the liquid ECM (see step 1 or 2 of Section 11.3.4), and pipette approximately 10 $\mu$L into one of the wells in the housing. Tap the dish to force liquid ECM between the PDMS housing and glass coverslip, and aspirate excess ECM from the wells.
   v. Add droplets of PBS to the tissue-culture dish (but *not* to the wells) to prevent dehydration.
   vi. Gel the ECM on a water bath to evenly distribute heat and to buffer the system from vibrations. We typically gel collagen and fibrin at 22°C to 24°C for 2 hours.
   vii. Add small drops of perfusate to the inlet and outlet wells, and make sure the liquid wicks along the side of the needle and contacts the ECM gel. Use two tweezers to remove the needle—one to hold the PDMS support and the other to withdraw the needle from the ECM. This process yields a single open channel within the ECM gel that is connected to two wells approximately 3 mm in diameter.
   viii. Seal the open end of the chamber with liquid PDMS, fill the wells with perfusate, and cure the PDMS at 37°C.
4. *For networks*
   i. Oxidize the PDMS frame and top in a UV-ozone cleaner for 10 minutes. Place the frame on the glass coverslip of the supporting dish.
   ii. Coat the interior of the frame and top with fibronectin for 40 minutes. Aspirate to dry.

iii. Prepare the liquid ECM (see step 1 or 2 of Section 11.3.4). Wash the gelatin mesh exhaustively with PBS, then one to two times with liquid ECM, and carefully transfer the mesh to the frame, as in Figure 11.2(c).

iv. Align the holes of the PDMS top to the handles of the gelatin mesh, and place the PDMS top on the frame. Tap the top to ensure good contact with the frame.

v. Fill the interior of the PDMS housing (the unit formed by the top and frame) with liquid ECM, leaving the upper portions of the gelatin mesh handles uncovered.

vi. Add droplets of PBS to the dish to prevent dehydration.

vii. Gel the ECM on a water bath to evenly distribute heat and to buffer the system from vibrations. We typically gel collagen and fibrin at 22°C to 24°C for 2 hours.

viii. Heat the dish to 37°C to melt the gelatin, and exhaustively flush the system with perfusate to remove the gelatin. This process yields an approximately 1 $cm^2$ network within the ECM gel connected to two wells approximately 3 mm in diameter.

### 11.3.5 Perfusion of microfluidic gels

1. *Dropwise perfusion*

   i. Add perfusate to the inlet of the gel to establish flow, as shown in Figure 11.3(a). To maintain perfusion, regularly add perfusate to the inlet, and aspirate it from the outlet. Keep a small volume of perfusate in the outlet to promote flow.

2. *High-flow perfusion*

   i. Cut approximately 1-cm-long posts of PE-50 tubing, dip their ends in liquid PDMS, and stand them upright on a flat silicon wafer. The spacing of the tubing should roughly equal the spacing between the inlet and outlet wells of the PDMS housing. Cure the PDMS to fix the tubing in place.

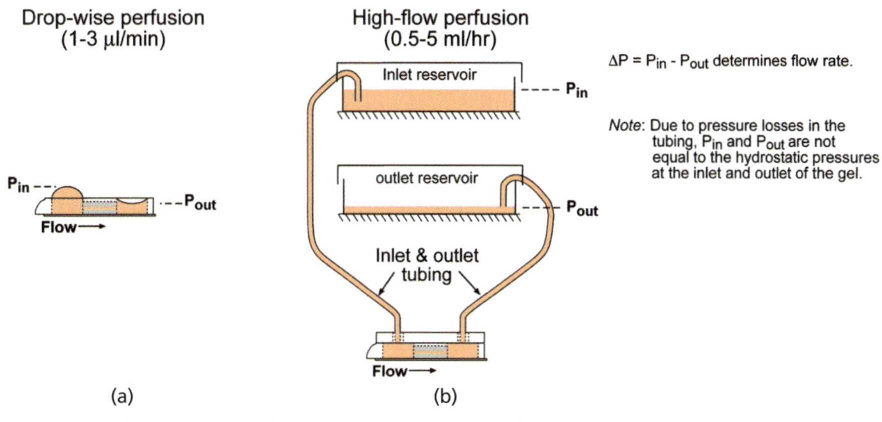


**Figure 11.3** Perfusing microfluidic gels. (a) In dropwise perfusion, small drops (~20 $\mu$L) of fluid are added to the inlet well of the system. Flow through the channel(s) occurs by the small pressure difference (<5 mm $H_2O$) between the inlet and outlet. To maintain continuous flow, additional media must be added at regular intervals to the inlet, while perfused media is removed from the outlet. (b) Under high-flow perfusion, a PDMS lid with tubing connects the microfluidic gel to inlet and outlet reservoirs of liquid held at hydrostatic pressures of $P_{in}$ and $P_{out}$, respectively. The pressure difference between the two reservoirs drives perfusion. The hydrostatic pressures $P_{in}$ and $P_{out}$ of the reservoirs do not equal the pressures at the inlet and outlet of the gel due to pressure losses in the tubing.

ii. Pour additional PDMS to cover the wafer (3 to 4 mm deep), but do not cover the top of the tubing. Cure and peel the PDMS from the silicon wafer, and remove the embedded tubing to yield PDMS with cylindrical holes. Trim the PDMS into blocks, two holes per block.

iii. Cut 40 to 50 cm lengths of PE-50 tubing, and thread the tubing into the holes in the PDMS blocks. The mouth of the tubing should be level with the flat underside of the PDMS.

iv. Apply a droplet of liquid PDMS to glue the tubing to the top side of the PDMS block, and cure. The liquid PDMS should wick completely around the tubing, but it should not wick down to the underside of the PDMS block.

v. If desired, calibrate the fluidic resistance of the tubing (see step 3 of Section 11.5.2).

vi. Sterilize the lid and attached tubing. Aspirate pure ethanol, then water, through the tubing. Aspirate to dry the inside of the tubing.

vii. Place sterile glass slides in each of two deep petri dishes. Place the two free ends of the tubing in separate dishes, with each end under a glass slide.

viii. Add perfusate to the inlet and outlet wells of the microfluidic gel, but do not fill the wells completely.

ix. Fill the inlet reservoir dish with perfusate, introduce perfusate through the inlet tubing by vacuum, and clamp the tubing with a binder clip to stop flow. Aspirate excess liquid from the underside of the PDMS lid. Align the two openings in the lid with the inlet and outlet wells of the PDMS housing, and place the lid on the PDMS housing. Tap the lid to make a good seal with the housing.

x. Unclamp the inlet tubing, and raise the inlet and outlet reservoirs, as illustrated in Figure 11.3(b). The perfusate should flow into the outlet tubing within a few minutes.

## 11.4 Anticipated Results

Figure 11.4 shows typical results from both methods. By light microscopy, the fibers of the gels are uniformly distributed and randomly oriented. Walls of the channels are clearly defined within the surrounding matrix. Single channels have circular cross sections, with diameters corresponding to the diameters of the needles [Figure 11.4(a), inset], while network channels have rectangular cross sections that reflect the shapes created by lithography [Figure 11.4(b), inset]. Perfusion with a suspension of labeled beads leads to flow with low velocities at the walls of the channels and high velocities along the center of the flowing stream (Figure 11.4).

## 11.5 Application Notes

### 11.5.1 Rate of gelation

1. For collagen gels, the rate of gelation largely dictates the resulting porosity. Increasing temperature (up to 37°C) or pH (up to approximately 8.0) increases the rate of gelation and yields gels with fine, faint fibrils [15, 16]. These gels are less

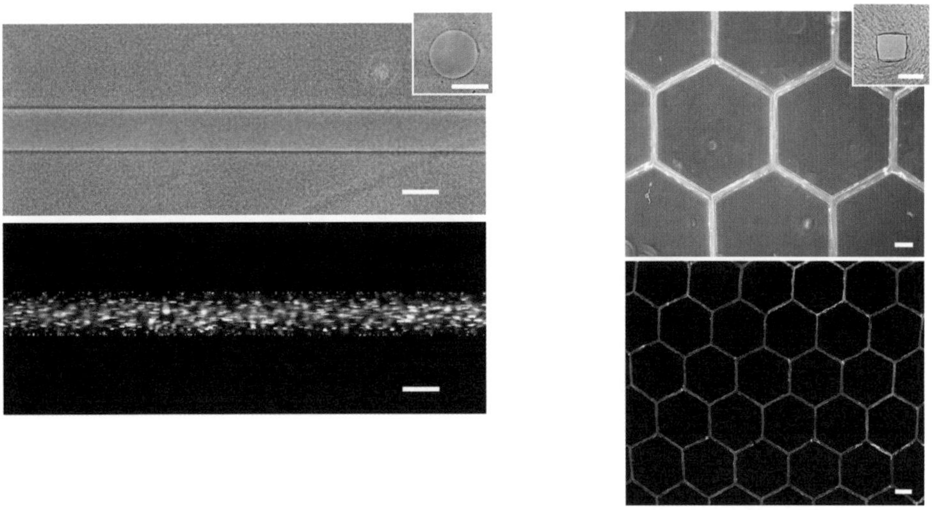

(a) (b)

**Figure 11.4** Phase and fluorescence images of microfluidic channels. (a) Phase-contrast image of a single 120-$\mu$m-diameter channel in collagen (top panel) and a fluorescence image of 1-$\mu$m-diameter beads perfusing the channel (bottom panel). Inset: cross section of a single channel in collagen. (Reprinted from [1] with permission from Elsevier.) (b) Phase-contrast image of a hexagonal network in fibrin (top panel) (reprinted from [2] with permission from the Royal Society of Chemistry) and a collage of fluorescence images of a single hexagonal network in fibrin (bottom panel). Channels are 60-$\mu$m-wide and perfused with fluorescently labeled 1-$\mu$m-diameter beads. Inset: cross section of a 100-$\mu$m-wide channel in collagen. Scale bars refer to 400 $\mu$m in the bottom panel of (b) and to 100 $\mu$m elsewhere.

porous and stiff than those prepared at lower temperatures or pH. In our studies, we primarily use gelation temperatures of 22°C to 24°C and pH of 7.0 to 7.5.

2. For fibrin gels, increasing the concentration of thrombin yields compact gels with smaller fibers and pores [17, 18].

### 11.5.2 Resistance of microfluidic gels and tubing

The flow rate through a microfluidic gel is the sum of the flow rates through the channels and through the gel (i.e., interstitial flow). Flow through channels is normally orders of magnitude larger than interstitial flow, but this rule may not hold for gels that contain channels on the order of 10 $\mu$m in width or for gels of large cross-sectional area. Flow through channels is inversely related to fluidic resistance, which depends on the lengths and cross-sectional areas of the channels.

1. Dropwise perfusion with PBS of a channel that is 6 mm in length and 120 $\mu$m in diameter yields flow rates of 1 to 3 $\mu$L/min. If the flow into the outlet well is not noticeable after a few minutes of perfusion, a constriction or clog likely exists at one end of the channel.
2. For the same channel dimensions and perfusate as above, perfusion under a pressure difference of 6 cm $H_2O$ between the inlet and outlet reservoirs yields flow rates of 1.2 to 1.5 mL/hr.

3. The tubing of the perfusion lid is a source of variation in perfusion rates between samples. Equal lengths of polyethylene tubing often do not have the same fluidic resistance. We always measure the resistance of each segment of tubing before use and alter it as needed by trimming the tubing. We normally adjust each tubing in a lid to have a resistance of 26.4 cm $H_2O$·min/mL at room temperature with PBS as perfusate.
4. In both perfusion systems depicted in Figure 11.3, the flow rate decreases over time as the difference in fluid height in the inlets and outlets decreases. Using reservoirs with a large surface area can lessen this effect.

## 11.6 Discussion and Commentary

The most common problems we have encountered are deformation of the ECM gels and development of leaks (see Troubleshooting Table). Contaminants can be avoided with standard sterile technique or by assembling parts in a clean room environment.

### 11.6.1 Enlarged and/or deformed channels

1. Single channels can easily be widened or deformed if the needle is haphazardly removed from the gel. The PDMS support should be held firmly in place with tweezers while the needle is removed, without applying excessive force that bends the needle. During gelation, floating the system on a water bath helps minimize any vibrations.
2. The gelatin mesh swells if it is released from the PDMS mold without first equilibrating to room temperature for at least 30 minutes. During gelation of ECM, swelling leads to enlarged and deformed channels in the microfluidic gel.
3. The gels deform if bubbles enter the channels or if the gelatin mesh has bubbles. Bubbles arise when insufficient liquid is present at the inlet or outlet or when the gel dries. We always make sure the gels are hydrated at their ends. Bubbles form during micromolding of gelatin if the PDMS mold becomes dehydrated.
4. Gelatin meshes, especially thin ones, are very flexible. We use minutien pins to maneuver the meshes and to stabilize the mesh during addition of ECM.
5. The basement membrane extract Matrigel (BD Biosciences) behaves strangely in our system. It swells significantly during perfusion, and channels often pinch off completely over several hours. We recommend using Matrigel only as a dilute mixture with collagen.

### 11.6.2 Leaks between the gel and PDMS or between the gel and coverslip

1. Leaks can form if the coverslip or PDMS is poorly coated with fibronectin. We always oxidize PDMS housings in a UV-ozone cleaner for at least 10 minutes to enhance the absorption of fibronectin onto the PDMS surface.
2. Mechanical stress can loosen the bond between the gel and surrounding surfaces. Excessive force when connecting the perfusion lid to the PDMS housing will cause the gel to detach from the housing.
3. Trace amounts of BSA from a gelatin mesh can inhibit the adhesion of the gel to surrounding surfaces. We typically wash meshes with PBS to remove BSA.

**Troubleshooting Table**

| Problem | Potential Cause | Solution |
|---|---|---|
| Deformation of single channel | Vibration of needle during gelation; vibration of needle during removal | Gel ECM on water bath. Firmly hold PDMS support during removal of needle. |
| Deformation of networks | Swelling of gelatin mesh | Equilibrate gelatin mesh at room temperature before releasing from mold. |
| Air bubbles in channels or networks | Insufficient liquid at inlet or outlet of gel | Ensure inlets and outlets are well-hydrated. |
| Leak between gel and housing | Weak adhesion between gel and housing | Oxidize PDMS housing and increase time for fibronectin absorption. Avoid applying mechanical stress to housing after ECM has gelled. |
| Leak between housing and perfusion lid | Liquid trapped between housing and lid | Dry surfaces of housing and lid before placing them together. |

4. Both the lid and housing must be dry to make a firm seal. Any liquid, especially culture media, trapped between the PDMS surfaces will weaken the seal. Catastrophic failure can then occur if the system is placed under high hydrostatic pressures (10 to 30 cm $H_2O$).

## 11.7 Summary Points

- Gelation around needles and their subsequent removal can yield single channels with round cross sections of 50 $\mu$m or more in diameter.
- Gelation around gelatin meshes and their subsequent melting can yield networks with rectangular cross sections of 6 $\mu$m or more in width.
- Perfusion of microfluidic gels takes place by introducing drops of media to the inlets of the gels (low flow) or by applying a perfusion lid and pressure difference to the gel housing (high flow).
- Maintenance of flow requires adequate adhesion between gel and surrounding surfaces and elimination of debris that can clog inlets.

## Acknowledgments

We thank Andrew Golden for providing images of fibrin networks. This work was supported by the National Institute of Biomedical Imaging and Bioengineering (award EB005792).

### References

[1] Chrobak, K. M., Potter, D. R., and Tien, J., "Formation of perfused, functional microvascular tubes in vitro," *Microvascular Research*, Vol. 71, No. 3, 2006, pp. 185–196.

[2] Golden, A. P., and Tien, J., "Fabrication of microfluidic hydrogels using molded gelatin as a sacrificial element," *Lab on a Chip*, Vol. 7, No. 6, 2007, pp. 720–725.

[3] Yager, P., et al., "Microfluidic diagnostic technologies for global public health," *Nature*, Vol. 442, No. 7101, 2006, pp. 412–418.

[4] Lee, C. C., et al., "Multistep synthesis of a radiolabeled imaging probe using integrated microfluidics," *Science*, Vol. 310, No. 5755, 2005, pp. 1793–1796.

[5] Whitesides, G. M., "The origins and the future of microfluidics," *Nature*, Vol. 442, No. 7101, 2006, pp. 368–373.

[6] Cabodi, M., et al., "A microfluidic biomaterial," *J. American Chemical Society*, Vol. 127, No. 40, 2005, pp. 13788–13789.

[7] Vernon, R. B., et al., "Native fibrillar collagen membranes of micron-scale and submicron thicknesses for cell support and perfusion," *Biomaterials*, Vol. 26, No. 10, 2005, pp. 1109–1117.

[8] Hahn, M. S., Miller, J. S., and West, J. L., "Three-dimensional biochemical and biomechanical patterning of hydrogels for guiding cell behavior," *Advanced Materials*, Vol. 18, No. 20, 2006, pp. 2679–2684.

[9] Price, G. M., Chrobak, K. M., and Tien, J., "Effect of cyclic AMP on barrier function of human lymphatic microvascular tubes," *Microvascular Research*, Vol. 76, No. 1, 2008, pp. 46–51.

[10] Choi, N. W., et al., "Microfluidic scaffolds for tissue engineering," *Nature Materials*, Vol. 6, No. 11, 2007, pp. 908–915.

[11] Whitesides, G. M., et al., "Soft lithography in biology and biochemistry," *Annual Review of Biomedical Engineering*, Vol. 3, 2001, pp. 335–373.

[12] Tien, J., and Chen, C. S., "Microarrays of cells," in A. Atala and R. Lanza, (eds.), *Methods of Tissue Engineering*, San Diego, CA: Academic Press, 2001, pp. 113–120.

[13] Tang, M. D., Golden, A. P., and Tien, J., "Molding of three-dimensional microstructures of gels," *J. American Chemical Society*, Vol. 125, No. 43, 2003, pp. 12988–12989.

[14] Tan, J. L., et al., "Simple approach to micropattern cells on common culture substrates by tuning substrate wettability," *Tissue Engineering*, Vol. 10, Nos. 5–6, 2004, pp. 865–872.

[15] Wood, G. C., and Keech, M. K., "The formation of fibrils from collagen solutions. 1. The effect of experimental conditions: Kinetic and electron-microscope studies," *Biochemical Journal*, Vol. 75, No. 3, 1960, pp. 588–598.

[16] Williams, B. R., et al., "Collagen fibril formation: Optimal in vitro conditions and preliminary kinetic results," *J. Biol. Chem.*, Vol. 253, No. 18, 1978, pp. 6578–6585.

[17] Blombäck, B., et al., "Native fibrin gel networks observed by 3D microscopy, permeation and turbidity," *Biochimica et Biophysica Acta*, Vol. 997, No. 1–2, 1989, pp. 96–110.

[18] Blombäck, B., et al., "Fibrin in human plasma: Gel architectures governed by rate and nature of fibrinogen activation," *Thrombosis Research*, Vol. 75, No. 5, 1994, pp. 521–538.

# About the Editors

**Yaakov Nahmias** is an instructor in surgery and bioengineering at the BioMEMs Resource Center at Massachusetts General Hospital (MGH), Harvard Medical School. He is a member of the scientific staff at Shriners Hospital for Children, a senior lecturer at the School of Computer Science and Engineering at the Hebrew University of Jerusalem, and an associate member of the Center for Bioengineering in the Service of Humanity. Work in Dr. Nahmias' laboratory is focused on the study of liver development, regeneration, and disease. His work on hepatitis C virus (HCV) infection was published in the leading journals of the field and was featured in the *Journal of the American Medical Association* (JAMA). Dr. Nahmias received a B.Sc., magna cum laude, from the Technion, Israel Institute of Technology, and a Ph.D. from the University of Minnesota. He has been awarded the National Institute of Health Mentored Research Scientist Career Development Award (K01) for the development of microfabrication technology for the study of liver development.

**Sangeeta N. Bhatia** is a Howard Hughes Medical Institute Investigator and a professor of health sciences and technology (HST) and electrical engineering and computer science (EECS) at the Massachusetts Institute of Technology. She is a member of the Koch Institute for Integrative Cancer Research and the Harvard Stem Cell Institute, an associate member of the Broad Institute and Brigham & Women's Hospital. The research in her laboratory is focused on the applications of micro- and nanotechnology for tissue repair and regeneration. Dr. Bhatia trained at Brown, MIT, Harvard, and MGH. She has been awarded the David and Lucile Packard Fellowship, given to the nation's most promising young professors in science and engineering, the MIT TR100 Young Innovators Award, and the Global Indus Technovator Award. She is a Fellow of the American Institute for Medical and Biological Engineering and of the American Society for Clinical Investigation. She is a frequent advisor to governmental organizations and is the cofounder of two start-up companies. She holds 15 issued or pending patents and has worked at Pfizer, Genetics Institute, ICI Pharmaceuticals, and Organogenesis.

## List of Contributors

**Rashid Bashir**
University of Illinois at Urbana-Champaign
Micro and Nanotechnology Laboratory, MC-249
208 North Wright Street
Urbana, IL 61801
e-mail: rbashir@uiuc.edu

**Sangeeta N. Bhatia**
Massachusetts Institute of Technology
77 Massachusetts Avenue, E19-502D
Cambridge, MA 02139
e-mail: sbhatia@mit.edu

**Lourdes M. Cabrera**
University of Michigan
Ann Arbor, MI 48109l
e-mail: lcabrera@umich.edu

**Christopher S. Chen**
University of Pennsylvania
510 Skirkanich Hall
210 South 33rd Street
Philadelphia, PA 19104
e-mail: chrischen@seas.upenn.edu

**Xuanhong Cheng**
Lehigh University
Whitaker Laboratory
5 East Packer Avenue
Bethlehem, PA 18015
e-mail: xuc207@lehigh.edu

**George Eng**
Columbia University
New York, NY 10032
e-mail: gme2103@columbia.edu

**Christopher V. Gabel**
Harvard University
Department of Physics
Cambridge, MA 02138
e-mail: gabel@fas.harvard.edu

**Yun Seok Heo**
University of Michigan
Department of Biomedical Engineering
1107 Gerstacker
Ann Arbor, MI 48109
e-mail: yunsheo@umich.edu

**Elliot E. Hui**
University of California, Irvine
Department of Biomedical Engineering
3120 Natural Sciences II
Irvine, CA 92697
e-mail: eehui@uci.edu

**S. Elizabeth Hulme**
Harvard University
Department of Chemistry and Chemical Biology
Mallinckrodt 227
12 Oxford Street
Cambridge, MA 02138

**Daniel Irimia**
Massachusetts General Hospital
BioMEMs Resource Center
114 16th Street, Room 1404
Charlestown, MA 02129
e-mail: dirimia@partners.org

**Rohit Jindal**
Massachusetts General Hospital
Center for Engineering in Medicine
51 Blossom Street
Boston, MA 02114
e-mail: rjindal@partners.org

**Andreja Jovic**
University of Michigan
Department of Biomedical Engineering
1107 Gerstacker
Ann Arbor, MI 48109
e-mail: andreja@umich.edu

**Salman R. Khetani**
Hepregen Corporation
200 Boston Avenue
Suite 1500
Medford, MA 02155
e-mail: skhetani@hepregen.com

**Kevin R. King**
Massachusetts Institute of Technology
Department of Health Science and Technology
Center for Engineering in Medicine
51 Blossom Street
Boston, MA 02114
e-mail: krking@mit.edu

**Kenneth T. Kotz**
Massachusetts General Hospital
BioMEMs Resource Center
114 16th Street, Room 1205C
Charlestown, MA 02129
e-mail: kkotz@partners.org

**Joseph R. Kovac**
Massachusetts Institute of Technology
Department of Electrical Engineering and Computer Science
Room 36-865
77 Massachusetts Avenue
Cambridge, MA 02139
e-mail: jrkovac@mit.edu

**Yi-Shao Liu**
University of Illinois at Urbana-Champaign
Micro and Nanotechnology Laboratory, MC-249
208 North Wright Street
Urbana, IL 61801
e-mail: yishao@uiuc.edu

**Yaakov Nahmias**
Massachusetts General Hospital
BioMEMs Resource Center
114 16th Street, Room 2450
Charlestown, MA 02129
e-mail: ynahmias@partners.org

**Gavrielle M. Price**
Boston University
Department of Biomedical Engineering
44 Cummington Street
Boston, MA 02215
e-mail: gprice@bu.edu

**Milica Radisic**
University of Toronto
Institute of Biomaterials and Biomedical Engineering
164 College Street
Room 407
Toronto, Ontario, M5S 3G9
Canada
e-mail: milica@chem-eng.utoronto.ca

**Sergey S. Shevkoplyas**
Tulane University
Department of Biomedical Engineering
624 Lindy Boggs Building
New Orleans, LA 70118
e-mail: shevkop@tulane.edu

**Michael L. Shuler**
Cornell University
Department of Biomedical Engineering
115 Weill Hall
Ithaca, NY 14853
e-mail: mls50@cornell.edu

**Gary D. Smith**
University of Michigan
Department of Obstetrics and Gynecology
6428 Medical Science I
Ann Arbor, MI 48109
e-mail: smithgd@umich.edu

**Jong Hwan Sung**
Cornell University
Department of Biomedical Engineering
Ithaca, NY 14853
e-mail: jhs45@cornell.edu

**Brian M. Taff**
Massachusetts Institute of Technology
77 Massachusetts Avenue
Cambridge, MA 02139
e-mail: taff@mit.edu

**Shuichi Takayama**
University of Michigan
Department of Biomedical Engineering
2115 Gerstacker
2200 Bonisteel Blvd.
Ann Arbor, MI 48109
e-mail: takayama@umich.edu

**Joe Tien**
Boston University
Department of Biomedical Engineering
44 Cummington Street
Room 715
Boston, MA 02215
e-mail: jtien@bu.edu

**Ronald G. Tompkins**
Massachusetts General Hospital
Center for Engineering in Medicine
Shriners Hospitals for Children – Boston
55 Fruit Street, GRB1302
Boston, MA 02114
e-mail: rtompkins@partners.org

**Mehmet Toner**
Massachusetts General Hospital
BioMEMs Resource Center
114 16th Street, Room 1402
Charlestown, MA 02129
e-mail: mtoner@hms.harvard.edu

**Joel Voldman**
Massachusetts Institute of Technology
Department of Electrical Engineering and Computer Science
Room 36-824
77 Massachusetts Avenue
Cambridge, MA 02139
e-mail: voldman@mit.edu

**Gordana Vunjak-Novakovic**
Columbia University
622 West 168th Street
Vanderbilt Clinic, 12th floor
Room VC12-234
New York, NY 10032
e-mail: gv2131@columbia.edu

**Liju Yang**
North Carolina Central University
BRITE
1801 Fayetteville Street
Durham, NC 27707
e-mail: lyang@nccu.edu

**Michael T. Yang**
University of Pennsylvania
510 Skirkanich Hall
210 South 33rd Street
Philadelphia, PA 19104
e-mail: myang2@seas.upenn.edu

**Martin L. Yarmush**
Massachusetts General Hospital
Center for Engineering in Medicine
Shriners Hospitals for Children – Boston
51 Blossom Street
Boston, MA 02114
e-mail: ireis@sbi.org

# Index

**D**